Understanding Sustainability Principles and ESG Policies

Colin Read

Understanding Sustainability Principles and ESG Policies

A Multidisciplinary Approach to Public and Corporate Responses to Climate Change

Colin Read
The State University of New York at Plattsburgh
Plattsburgh, NY, USA

ISBN 978-3-031-34482-4   ISBN 978-3-031-34483-1   (eBook)
https://doi.org/10.1007/978-3-031-34483-1

Cover illustration: Barry Winiker/GettyImages

This Palgrave Macmillan imprint is published by the registered company Springer Nature Switzerland AG
The registered company address is: Gewerbestrasse 11, 6330 Cham, Switzerland

This book is dedicated to the current generation of young people who, perhaps more than any previous generation, sense that humankind is not living sustainably, and are prepared to make the personal sacrifices to make the world a more sustainable place.

Preface

The world is increasingly concerned, and many young people gravely so, about our collective inability to live sustainably. The clarion call for sustainability began half a century ago with the publication of Limits to Growth[1] and has reached a more strident pitch of late with the accelerating rate of global warming and its consequences.

There are few subjects of greater multidisciplinary complexity than sustainability. To comprehend and converse intelligently about global warming requires more than a shallow understanding of science, energy, geology, the ecosystem, economics, meteorology, biology, politics, and public policy, among other disciplines. The vocabulary of sustainability and of global warming is precise, and the stakes have become so high that dialogues between the disciplines are critical.

This book offers a primer on the various issues and concepts of importance for a broad understanding of sustainability perspectives. It brings together a diverse set of disciplines across the sciences and social sciences to better understand and manage the existential issues of our day. For our understanding of the economic theories invoked in sustainability and of the philosophical discussions that come to bear, the treatment here brings the reader to the state of the social sciences. In other areas, such as the increasingly rich knowledge of the science of global warming and its myriad subtleties and recent insights, the reader is presented a broad landscape from which they can delve more deeply as more is discovered every year.

The book emphasizes the necessary vocabulary and concepts so a reader can engage intelligently and at a moderately high level with a wide variety of scientists, social scientists, and policy-makers concerned about sustainability and global warming. Some concepts such as the difference between power and energy, or the importance of innovations in economics may appear pedantic or technical. But it is this understanding and ability to converse relatively precisely that will help broaden the dialog. In doing so, I hope to broaden the discussion of pressing issues of public policy that will only become increasingly critical over time.

Those interested in further study in sustainability, in enhancing a dialog on sustainability and the Environment, Social, and Governance (ESG) Paradigm within their institutions and corporations, and those who feel passionately about sustainability and need the tools to become credible participants in the discussion will hopefully find the depth of treatment from these multiple disciplinary perspectives helpful in your life studies.

Plattsburgh, USA   Colin Read

Introduction

In 1823, two centuries ago, Thomas Malthus published an article that summarized and made broadly accessible a conclusion he had been discussing in his learned circles. Our human population will perpetually gyrate between periods of feast and famine. His discouraging prophecy has caused economics to be labeled *the dismal science* ever since.

Perhaps the label of the dismal science is not as unfortunate as it sounds. Malthus outlined policies that can restore natural, societal, and economic imbalances to ensure our planet and economy can continue to function at a sustainable steady state. While the world paid scant attention to his sometimes-controversial solutions to such dismal prophecies until lately, his Malthusian Trap captivated our attention. It at once warns us of the risks of unsustainable practices and motivates us toward the path of sustainability. Malthus' prophecy has also spawned a great debate over the meaning of sustainability.

A year later, in 1824, Joseph Fourier proposed what we now call *The Greenhouse Effect*. This now-familiar physical phenomenon was verified by Eunice Newton Foote in 1856 and Svante Arrhenius in 1896 as a relationship between atmospheric carbon dioxide and global warming. Scientists have verified this phenomenon myriad times since, with each subsequent improvement on our understanding tending to worsen predicted long-term effects of anthropogenic emissions of greenhouse gases.

The blessing of Malthus' curse is that concerns about the future of our physical world have now transcended science and the environment, and mobilized action toward social justice and sustainable organizations. Sustainability has moved from the laboratory to the classroom, the board room, town halls, and the United Nations. In doing so, it has unleashed the power of people gravely concerned about our collective future.

This book explores from an interdisciplinary perspective the definitions and implications of sustainability. I also describe how the Environmental,

Social, and Governance (ESG) paradigm can be illuminated from multidisciplinary perspectives to unify the various definitions of sustainability, from the economic to the ethical, the scientific to the environmental, and from the social to the managerial. What is it about the human enterprise that pushes us to excess, only to find ourselves in crises? What can we learn from science and economics that will allow us to establish a steady state path that is not prone to wild shifts, and instead can be maintained for an indefinite future? How can our organizations and corporations prosper through an appreciation of sustainability, and what social responsibilities do they have to mitigate global warming? How can sustainability contribute to economic and social justice? These are the questions we explore in this book.

Discussions of sustainability inevitably begin with an acknowledgment of human population pressures. Ironically, it is these same pressures that have created great prosperity for a few and poverty for many. The basis of the Malthusian Trap can best be described in Malthus' own words[2]:

> If the only check to population is misery, the population will grow until it is miserable enough to check its growth.

Malthus went on to say[3]:

> The power of the Earth to produce subsistence is certainly not unlimited, but it is strictly speaking indefinite; that is, its limits are not defined, and the time will probably never arrive when we shall be able to say, that no further labour or ingenuity of man could make further additions to it. But the power of obtaining an additional quantity of food from the Earth by proper management, and in a certain time, has the most remote relation imaginable to the power of keeping pace with an unrestricted increase of population.

Since the onset of the Industrial Revolution and in the wake of Malthus' dismal prophecy, the *More Developed Countries (MDCs)* have been able to skirt the direst predictions of the Malthusian Trap through myriad technological innovations, while the *Less Developed Countries (LDCs)* have been left to bear the brunt of unsustainability. However, affluence and technology are not antidotes to the Malthusian Trap and do not repeal Malthus' dismal theorem. They have merely delayed the inevitable. I remain optimistic over the prospect that humankind can become sufficiently educated about sustainability.

My goal for this book is to serve as a primer for those wishing to be immersed in all aspects of sustainability and a text for an interdisciplinary course in sustainability. I began my university studies with a Bachelor of Science degree in Physics with a specialization in sustainable energy. I followed up my science studies with a Master's of Science and then a Ph.D. in Economics as I realized that, while the analytic tools of physics permit us to better comprehend the physical world around us, to understand the interplay between scientific methods, the dismal science, social behavior, and politics requires one to also delve into the humanities, social sciences, and business.

I committed to a course of study in sustainability that eventually led me to a Master's of Business Administration, a Master's of Accountancy in Taxation, and a Juris Doctor of Law. I pursued this course of studies so that I may better understand the various facets of sustainability from a variety of perspectives. I believe that scientists and social scientists must share a common foundation and vocabulary if we are to communicate coherently in an increasingly complex interdisciplinary study of sustainability.

It was the 2022 run-up in global commodity prices that induced me to explore the degree to which humankind has exceeded the limits of sustainable growth. I realized that the disciplinary silos interested in various aspects of sustainability were not effectively communicating or sharing information across the diversity of groups and the general public that also share a stake in our collective future. This is a time for a more interdisciplinary analysis as our society is increasingly challenged by sustainability issues. In this book, I use the various tools of the sciences, social sciences, and business to create a more integrated understanding of the problems of and solutions to unsustainable practices.

The book is also informed by the passion of our current generation of students and leaders who understand and embrace the Environment, Social, and Governance concepts we collectively call ESG. From classrooms to boardrooms to the United Nations, we have come to see the consequences of our lack of regard for the environment, from climate change to resource depletion and economic injustice, and have developed a growing appreciation of the value of diversity of thought and experience. Concepts from the ESG paradigm are woven throughout this text.

I recognize the incredible advantages of free markets and capitalism when they work well. I also understand the limitations and challenges in managing the conflicts and risks, and the shifts in power and prosperity that decentralized markets create. A failure to sustainably manage our various environments is the product of troubling limitations of the market economy. In particular, I explore our failure to universally accept a common definition of sustainability that recognizes we must develop a capacity to meet the needs of the present generation without limiting opportunity for future generations.

This book defines sustainability and then documents why global economies are increasingly challenged to delay the Malthusian day of reckoning. I also describe the contributions of various disciplines and the successes of practitioners and corporations who create value through sustainability. The approach is positive rather than normative in that we describe the various prescriptions for sustainability drawn from the disciplines that scientifically study our environment. As we proceed, we will differentiate between the broader global environment, the human environment, and the environment that is the domain of policy-makers and managers.

We will discover that sustainability is simple to assert but difficult to define. I begin in Part I with a description of the various approaches to defining sustainability. I describe the significant awakening to sustainability issues arising from environmental catastrophes in the 1960s followed by strident expressions of concern of scientists and social scientists and the publication of the Limits to Growth in the 1970s.

In the second part, I describe the science of sustainability. I begin with the inviolate laws of thermodynamics and entropy. I then discuss the science and engineering of technologies that have perpetuated the Industrial Revolution and delayed the Malthusian Trap. I show how engineering innovations may create unintended consequences when imposed on complex economic and environmental systems. I close the part with a discussion of an emerging ethic among scientists and innovations in science and engineering that may aid in our transition from an unsustainable to a sustainable world.

I then turn to the essential role of sustainable energy in our pursuit of a sustainable ecosystem. There are few subjects as essential to our collective future as sustainable energy in our popular discourse. The study of sustainability is varied and taxing in the tools of science, business, and social science. Those who study sustainability may be experts in one or two facets of the multiple disciplines invoked in such issues as alternative energy or global warming. But effective discourse requires a modicum of familiarity in the various disciplines that contribute to the discussion. A shared vocabulary is then necessary, including a familiarity with the basic science of energy and resources.

The third part describes a fundamental flaw in the decisions of mortal humans that affect an immortal planet. A period of prosperity in the world's MDCs has created a false sense of complacency and economics optimism. If economics is labeled the dismal science, then the *Schumpeterian Hypothesis* may be the optimistic fiction. Too often we are shielded from the harsh reality of unsustainability only through the temporary reprieve of technology. I challenge the wisdom of blind acceptance of the Schumpeterian Hypothesis and its more recent extensions that technology guarantees economic progress, as compelling and as comforting as they may be.

In the economics Part I treat three types of resources that have challenged resource management; depletable resources such as oil; renewable resources that can be sustained only if managed properly; and fully sustainable resources. I include a description of backstop technologies that govern our transitions between different resources.

I also outline some of the fundamental flaws of laissez-faire free-market economics in pursuing sustainability, the role of government and regulation in correcting these failures, and the Tragedy of the Commons that causes us to abuse resources which belong to all and thus are respected by none. I describe the state of the social sciences that allow concepts from environmental studies such as the limits to growth to be placed in a firm economic context. I include a discussion of accelerating environmental degradation arising from our economic decisions. Its greatest perniciousness spans the last couple of generations who find humankind at the precipice of changes that will affect the ecosystem and humankind for centuries to come. In this part, I develop the intuition that allows one not well-steeped in economics to understand the concepts; I include the most mathematically rigorous treatments in appendices.

In Part IV I describe social, ethical, and public policy aspects of current issues in sustainability. Our environment was a subject dear to our Great Philosophers since ancient times. Indeed, they used the word *oikos*, from the Greek language word for *home*, that now has broader implications for our larger environment. This root was adopted by economics to describe the management of our home and surroundings, and ecology, the scientific study of our broader environment. I discuss how the Great Philosophers contributed to our current understanding of sustainability and document the cultural context to our modern understanding of our environment.

I acknowledge the immense power of humankind to profoundly affect our environment. These propensities to dramatically alter our surroundings as the human population increases has ramifications in the way peoples interact within our shared ecosystem. This increasing ability to dominate our environment may not always be an evolution in the Darwinian sense. The power of commerce, science, and economics now overwhelms the implications of Darwinian adaptations. We must understand sustainability by acknowledging social interactions within and between peoples and cultures.

These interactions evolve with changes in population density, sometimes destructively. These evolutions are not always Darwinian in that but they do not consistently sow the seeds for a more resilient future. Indeed, I conclude this section by documenting the growing clash of national ideologies and the challenging global politics of sustainability as we collectively strive for more effective environmental stewardship.

I end in Part V with a discussion of how public sector leaders, corporate managers, and policy-makers can both manage sustainability risks and capitalize on our sustainability values. While many businesses are discovering the value of sustainable practices in their internal operations, we observe that such internal practices are sometimes at odds with the immediate demands of financial markets. I challenge the assertion that businesses merely respond to current social and political fads. Instead, I describe the motivation for corporate sustainability efforts in theory and practice. I also document how many businesses are adopting sustainability values to better manage the resources that fuel the supply chain.

I close with a discussion of flaws in the traditional economic model that exacerbates the growing tensions between the values and needs of the More and Less Developed Countries. Global warming is an existential human threat that has pitted these MDCs and LDCs in opposition and has shifted the balance of power between people. I leave the reader with insights embodied in *ecological economics*. It offers provocative inquiries into the ways in which our private sector corporations can adjust their strategies to better reflect the Environment, Social, and Governance values in the ESG paradigm that act as a blueprint for a more innovative and sustainable world.

These are exciting and challenging times. Never before have management, science, sociology, environmentalism, and economics of sustainability come together in their mutual recognition of our current unsustainable practices. Nor are these lessons lost on commerce or government. Few topics or academic disciplines beg for the multidimensional and interdisciplinary approach required of studies in sustainability. Ultimately, each of our disciplines can contribute to the discussion and appreciate the varied approaches to the problem.

To deal effectively with sustainability, we must educate ourselves in the various meanings and applications of sustainable practices, while we simultaneously draw upon the science, economics, ethics, and business of our environment that increasingly creeps into our public debate. We must also understand the limitations of the classical economic model and embrace the normative aspects of ecological economics. If we can do so, we ensure that these values also enter our political dialog. The stakes are high, but the value in recognizing our mutual interdependencies is greater still. Our only solution is an environmentally literate global citizen. It is you, the global citizen and student, for whom I write this book.

NOTES

1. Meadows, D. H. (1972). The Limits to growth: a report for the Club of Rome's project on the predicament of mankind. Universe Books.
2. Malthus, Thomas, "The Malthusian Model as a General System," Social and Economic Studies, September, 1955; Collected Works, Vol. I, p. 455.
3. Malthus, Thomas Robert (1826), An Essay on the Principle of Population: A View of its Past and Present Effects on Human Happiness; with an Inquiry into Our Prospects Respecting the Future Removal or Mitigation of the Evils which It Occasions, (Sixth ed.), London: John Murray, Appendix I.17.

CONTENTS

About the Author

Colin Read has been teaching economics and finance for forty years, most recently as Professor of Economics and Finance at the State University of New York at Plattsburgh. He obtained a B.Sc. in Physics from Simon Fraser University, pursued graduate studies in Economics at the London School of Economics and in electrical engineering, and completed a Ph.D. in urban economics and finance at Queen's University in Kingston, Ontario. He also completed an M.B.A. at the University of Alaska, a J.D. in law at the University of Connecticut, and a Master's of Accountancy in Taxation at the University of Tulsa. He is a prolific author, with more than a dozen books in finance and economics, fifty journal articles and chapters, and more than 500 newspaper and magazine columns and blogs. His teaching is primarily in Money and Banking, Environmental and Resource Economics, and Sustainability. He is the principal of ESG Analytics Group that offers services to corporations and institutions concerned about their environmental footprint. He resides in Plattsburgh, NY with his wife, a new dog, and three cats, and enjoys flying, alternative energy, and skiing as hobbies.

List of Figures

PART I

A Definition of Sustainability

Worldly philosophers throughout the ages have recognized our accelerating pace of human development and interdependence. Free thinkers of the Renaissance era and social commentators concerned about the ravages of the Industrial Revolution first fomented our sustainability dialogue. But it was not until the last half century that broad swaths of society have recognized the incredible power of humankind to live far beyond the means of our natural capital to support us. Global warming, the existential issue of our day, has further sharpened our attention to sustainability.

This first section describes the emerging consensus on the meaning and significance of sustainability and on accelerating threats to it. Groups as varied as universities, corporations, associations, boards, and the United Nations have begun to develop frameworks they hope will weave the actions of public and private sector participants alike toward a greater recognition of each other and the environment around us. The *3Ps*, the *Triple Bottom Line*, the *Global Compact*, and the *Environmental, Social, and Governance (ESG) Pillars* provide foundations to incorporate the private sector as humankind seeks solutions to the existential questions of our day. Meanwhile, some are recognizing that the tools of economics can be extended to assist in new ways to enhance our understanding of the extent of the problem and the challenges ahead for all humankind.

CHAPTER 1

Competing Ideals and an Emerging Consensus

There are few terms in the English language that have the depth of significance as sustainability. Nor have many words been prone to evolve and deepen significantly over time from a scientific, ethics, and economic justice perspective. Sustainability invokes love and caring, stewardship and harvesting, growth and decay, and other words that affect the human condition and the health of the planet. Concerns over sustainability have a universality that has spanned generations. The concept of sustainability has also influenced the thinking of worldly philosophers, tracing back to *Aristotle* and perhaps even the dawn of language and cognitive thinking.

ARISTOTLE'S NICOMACHEAN ETHIC[1]

The *Nicomachean Ethic* was one of Aristotle's early explorations into the interaction of *humankind* and its *environment*. In Aristotle's day, the Greek term *oikos* and its root *eco* literally meant our house but metaphorically referred to our environment. In Aristotle's philosophy, the nature of human happiness was explored as a foundation for our motivations but also as a description of relationships with our surroundings. Our scientific study of the environment is the *logic* that gives us **ecology** while the *nomics*, its management, gives us *economics*.

Aristotle's theory asserted that the pursuit of happiness should be an ethic that guides our interactions, our spending, our politics, and our individual and collective efforts. We now know these aspects as the basis of economic decisions, but Aristotle was also subtle in his distinctions. He formulated the notion of justice such that each of us ought to have the freedom to pursue

© The Author(s), under exclusive license to Springer Nature Switzerland AG 2023
C. Read, *Understanding Sustainability Principles and ESG Policies*, https://doi.org/10.1007/978-3-031-34483-1_1

happiness. But, while he recognized our pursuit of personal property can contribute to our happiness, it cannot do so in an unbridled way.

For instance, Aristotle noted that **personal property** provides for human happiness but also allows us to express our benevolence to others through gifts. This was an individualization of Aristotle's mentor Plato's notion of communal sharing. But, Aristotle recognized that not all property should be diverted to human enjoyment. The Earth has fixed supply of some critical factors such as land that sustain life and happiness.

An aspect of virtue and justice requires that such **natural capital** should not be hoarded by one group in society, nor, by extension, enjoyed by only one generation. Certain resources that the Earth cannot replace must be considered part of a public trust so they can be shared, in Aristotle's times perhaps within generations, and now more generally across all generations. Once we make the leap in our recognition of mutual interdependence on Earth's resources, we see a greater need to live within the means that the Earth can sustainably provide.

The term sustainability itself is from the Latin rather than the Greek. Its root *sustinere* means to endure or maintain. This term also has meaning in the economic, ethical, societal, and ecological senses. It invokes the notion of a **steady state**, which describes actions and processes over time that can be maintained without degrading capacities. When applied to the Earth's carrying capacity to sustain life, it connects and unifies our individual decisions over time and across generations in a more holistic way.

THE MALTHUSIAN TRAP

Thomas Malthus was acutely aware of the consequences of exceeding the Earth's **carrying capacity** two millennia after Aristotle's observations. In his 1798 publication, An Essay on the Principle of Population[2] Malthus observed that humans can well outstrip the capacity of the Earth to support humankind. He outlined the consequences of living beyond our environmental means. Rather than the maintenance of a steady state **symbiosis** with nature, Malthus predicted that the population would grow and shrink in fits and starts, or feast and famine, primarily because of humankind's unique power to procreate beyond the ability of the Earth's natural capital to support us (Fig. 1.1).

Malthus was maintaining a discussion began in an **era of human enlightenment** within the Western World that other social commentators such as John Locke had initiated in the previous century. Locke spoke of the pursuit of happiness, perhaps even with less completeness than Aristotle had broached millennia earlier. Adam Smith, who we typically attribute as the founder of economics for his An Inquiry into the Nature and the Causes of the Wealth of Nations[3] in 1776, had also explored the motivations of humans in The Theory of Moral Sentiments in 1759.[4]

This work, over which Smith toiled to extend to the end of his life, argued that we have a naturally endowed conscience which encourages humans to

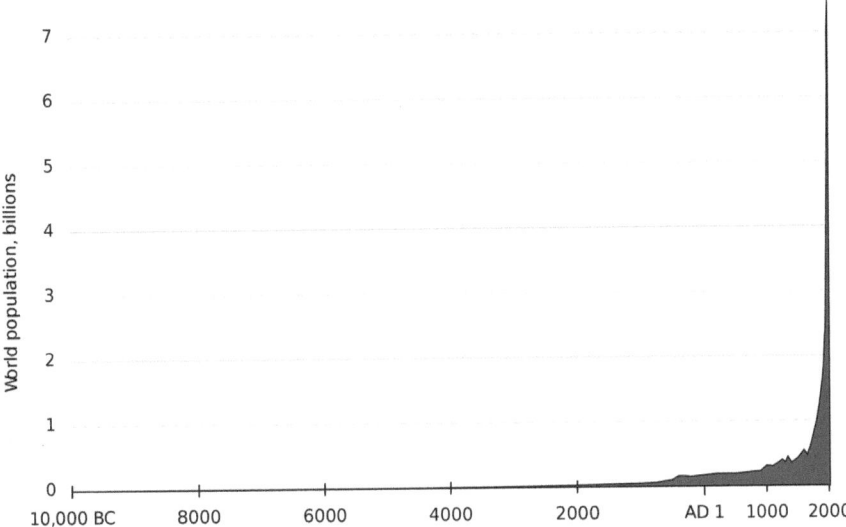

Fig. 1.1 World human population growth (Courtesy El T, Public domain, via Wikimedia Commons)

behave in the public interest. In doing so, humans create social rules and codes of conduct that help guide us along a moral and ethical path. In this *Enlightenment*, Locke, Smith, and Malthus, among many others, were beginning to explore the philosophical underpinnings of the relationships among humankind and within nature. Locke and Smith expressed optimism for the potential that we each can be well guided in our mutual interactions, but Malthus recognized a tendency unique among humans to abuse our relationship with nature.

We remember Smith's work on the Wealth of Nations that focused more narrowly on the means by which humans accumulate and distribute the fruits of its industry. However, he considered his earlier study into human motivations to be his greatest work. We also remember Malthus for his *dismal prophecy* that we constantly live in a *Malthusian Trap* of feast and famine, at least until humankind can evolve to develop methods and policies to live sustainably within our means, or, more precisely, the ability of the Earth to provide for humans and all creatures in a sustainable steady state. But, the greatest influence in our discussion about sustainability began a century after these early efforts established the discipline of economics.

The United States in the century of England's enlightenment and elucidation of the Malthusian Trap was a very different place. Pioneers of a new nation were not fretting over the frustrations of congestion and pollution, famine and disease that plagued *Industrial Revolution*-era England in Malthus' time. Instead, the profound power of nature in the New World struck some citizens and philosophers with awe. Henry David Thoreau, in his 1854 book

On Walden Pond,[5] marveled at the immenseness and beauty of the natural world from his lakeside vantage point. As he lamented the encroachment of a burgeoning Massachusetts population, Emerson fretted that some of his beloved species were disappearing. Emerson contemplated the distinction between our needs and wants within a fast-modernizing civilization. He was concerned about the ability of Mother Earth and its capacity to provide for us. Thoreau's writings were some of the first articulations of an environmental ethic.

A few years later, another American contemporary, the historian and diplomat George Perkins produced an even more starkly focused thesis. In 1864 his Man and Nature: Or, Physical Geography as Modified by Human Action,[6] vividly described his observations of the capacity of humans to destroy its environment. In a prophecy akin to Malthus' dismal predictions, Perkins observed that civilizations have come and gone because of their excesses and an inability to live in balance with nature. This notion combined the themes of Aristotle, Locke, Smith, and Malthus with the ethic of Thoreau to allow us to understand that it is a false economy to believe resources are solely for human enjoyment to enhance mortal life and comfort. If we fail to comprehend our relationship with the environment, it is at the peril of both humans and the planet.

By the end of the nineteenth century, we witnessed the emergence of a strong environmental ethic, even if humankind did not have the analytic and scientific tools we have today to describe our interaction more fully with nature. John Muir, who had explored many areas in the western part of the North Americas, from Alaska to California, founded the Sierra Club in 1892, and motivated the establishment of Yosemite National Park, implored us to see nature not primarily as a tool for the expansion of human wealth, but instead as a living organism in its own right.

A future U.S. President, Theodore Roosevelt, also marveled in nature and remained a lifelong advocate for its conservation. He helped create the U.S. National Parks system upon his election as president, established wildlife preserves to protect endangered species, and protected forest reserves so we can maintain nature, while we sustainably manage our forests. Roosevelt's notion of *conservation* carries with it not a nature lockbox, but rather a view that we can live in *symbiosis* with nature if we adopt scientific management practices.

THE SCIENTIFIC STUDY OF THE ECOSYSTEM AND THE ECOLOGY MOVEMENT

With an environmental ethic fast-developing, not solely for its own virtue, but also for humankind's need to conserve the ability of nature to sustain itself and hence humankind, the science of sustainability and the steady state was also developing. In the early part of the nineteenth century, James Fourier had described how the level of gases in the atmosphere can affect the ability of the Earth to reradiate heat. In 1856, Eunice Newton Foote, a self-trained scientist in New York State, performed experiments on the temperature of various gases contained in glass tubes and exposed to the Sun. She concluded that these gases indeed have an insulating effect, and noted that the presence of both carbon dioxide and water vapor could lead to higher atmospheric temperatures. The physicist John Tyndall affirmed this theory by showing how carbon dioxide and methane emissions can exacerbate global warming. By 1908, Svante Arrhenius demonstrated that the industrialized world's increasing reliance on coal to fuel industry and heat our homes was contributing to global warming.

But it was not until the 1960s that our scientific understanding of the effects of climate change really took root as people pondered the degree to which humankind was living in an increasingly precarious balance with nature. Throughout the nineteenth and early twentieth centuries, chemists and physicists were advancing our understanding of *equilibrium*. From the Latin *libra*, meaning balance or weight, it signifies when opposing forces are in balance. Chemists understood the concept to describe when a reaction is complete and the balance of compounds reaches a stable steady state, while physicists use the term to describe the balance of opposing forces. For instance, Isaac Newton described that an object at rest will stay at rest while an object in motion will remain in motion.

By the early twentieth century, these new concepts were applied to the balance of forces in nature. Such a natural extension views nature from a broader systems perspective of the Earth within the context of a complex and beautiful interaction of forces in essential balance. This *systems approach* gave rise to the term *ecosystem*, as described in 1935 in a paper by the botanist Sir Arthur George Tansley entitled The Use and Abuse of Vegetational Terms and Concepts.[7] In his book, he noted:

> Though the organisms may claim our prime interest, when we are trying to think fundamentally, we cannot separate them from their special environments, with which they form one physical system.

With Tansley's tool of the ecosystem and a new mindset it invokes within the sciences, economists and ecologists were given a much more expansive toolbox to explore the relationship between humankind and nature. Scholars became increasingly concerned that a simplistic view of economic attainment does not fully include hidden costs, especially in depletion of the Earth's carrying capacity, but also in the incomplete way in which external effects of our economic decisions are not fully internalized.

Economists were also exploring the interaction between human activity and the ecosystem. In the seminal 1920 book entitled The Economics of Welfare, British economist Arthur C. Pigou built upon an emerging and more holistic view of our environment and provided the economic concept of an *externality*. Pigou observed that some economic activities create consequences which are not properly priced in the free-market system and hence are not sufficiently considered in our economic decision-making. When we fail to incorporate the costs of our human decisions into our ecosystem, such externalities abound.

The classic example of a negative externality is *pollution*. If industries do not consider the cost of their pollution on the environment, they are inclined to overproduce, and in turn harm the environment and humanity itself. An extension of this concept is that if we live beyond the means of the Earth's carrying capacity to sustain the planet in a steady state, we are imposing upon future living beings a reduced capacity to create opportunities to thrive that we enjoy today. Our mortality hence induces us to live in an unsustainable way, sometimes out of ignorance, and sometimes through disregard.

By the 1960s, this recognition of our profound influence over the Earth, and the harm that arises when we ignore it, was penetrating our broader collective consciousness. The Earth witnessed environmental tragedies such as the perpetual combustion of the effluent-laden Cuyahoga River in Cleveland, Ohio. We also heard the ecological warnings in Rachel Carson's seminal Silent Spring[8] arising from dangerous pesticide use and environmental degradation. These economic concerns were beginning to seep into our collective awareness.

By the 1970s, increasing scarcity of natural resources, and the rising pressure of a burgeoning human population were also garnering society's attention. At a time when corporations such as AT&T, General Motors, and Exxon were growing to become corporate behemoths, a movement was afoot that challenged both the notions that bigger is better and that humans can successfully lord over *Mother Nature*. While cars lined up at gas pumps in 1973, the publication of Small is Beautiful: Economics as if People Mattered[9] in that same year demonstrated forcefully that humankind was veering away from environmental sustainability. In the same year, the dystopian movie Soylent Green described a planet a half century in the future that reprocessed human corpses into food for the living. The movie began with a two-minute montage of images that transitioned from humankind in harmony with nature to a world of overpopulation and urban decay.

Of course, this dangerous and fragile path had been occurring for some time, but the consequences of the Industrial Revolution that began in the eighteenth century, the development of coal and fossil fuels in the nineteenth century, and the rapid economic growth of the twentieth century were coming to roost by the 1970s. *Greenpeace, Earth Day*, and other social movements began to capture our collective imagination. A report in 1972 entitled The Limits to Growth,[10] authored by Meadows, Randers, and Behrens, used one of the first computer resource models, designed by Jay Forrester of MIT (described in his book World Dynamics[11]) to demonstrate the consequences of the path of humankind by simulating the interaction between human systems and the Earth.

The <u>Limits to Growth</u>, commissioned by the **Club of Rome**, was a culmination of work by seventeen scholars from a variety of disciplines. It demonstrated that, without dramatic changes in our pattern of consumption and resource extraction to better live within Earth's balance, the industrial capacity of humankind will be severely curtailed, in ways not dissimilar to Mathus's Trap. As one can imagine, the prophecies and policy predictions were troubling to many whose livelihood and wealth are related to resource overextraction. The methodology of the authors was challenged fiercely. Nonetheless, the book sold thirty million copies in thirty languages, and spawned updates <u>Beyond the Limits</u>[12] and <u>The 30-Year Update</u>[13] in 1992 and 2004.

The authors included population, food production, industrialization pollution, and the use of non-renewable resources as the primary variables in their model, and noted that the population's exponential growth demands translated into exponential growth of the extraction or production of these variables. Yet, our ability to leverage resources through better technology grows only linearly. Simulation demonstrated that our capacity to meet growing human needs could be exhausted within a century.

Instead, the authors assert that we must live within a sustainable world to avoid the overshoot and collapse predicted by Malthus' feast and famine prophecy. While not complete analyses of the ecosystem, the influence of these books and their approaches to scientific modeling of natural phenomena well into the future provided the framework for sustainability and climate change analyses ever since. They also challenged a generation of scientists to subscribe to a more complete ecosystem approach.

A Brief Political History
of Environmental Awareness

Politicians rarely lead social or scientific discourse, but they often follow once concerns captivate the broad public attention. In the United States, the cacophony of concerns over the environment in the 1960s led to the establishment of the *National Environmental Policy Act* in the United States in 1970, *the Clean Air Act* also in 1970, the *Water Pollution Control Act* in 1972, the *Endangered Species Act* in 1973, the *Safe Water Drinking Act* in 1974, and the *Clean Water Act* in 1977. The *Energy Policy and Conservation Act* was passed in 1975, followed by the establishment of the U.S. Department of Energy.

Meanwhile, the **United Nations** called to order the **Conference on the Human Environment** in 1972 in Stockholm, Sweden. European nations likewise passed acts to establish a better balance between humankind and nature. The post-World War II organization the **Organization for Economic Cooperation and Development** (OECD) established an environmental directorate and adopting a series of principles embodied in their **Polluter Pays Principle** that was subsequently adopted by the European Union in 1975. This principle influenced the deliberations of the United Nations' **Montreal Protocol** to create an international coalition of nations that culminated in the first global treaty to repair atmospheric ozone depletion.

These movements were captured in an image etched into our human consciousness. The memorable *Blue Marble* visualization of Earth's interrelatedness and fragility by Apollo 17 as it ventured to the moon in early 1972 fixed in our minds the need to collectively address global environmental challenges that respect no human-made borders. We became aware that that recent imposition of humankind in the life of the planet is of equivalent profoundness as the *Paleozoic Era* of the dinosaurs and their untimely end, and the *Mesozoic Era* that followed and created humankind. Our current *Anthropocene Epoch*, the period since the onset of the Industrial Revolution, marks the awakening of the potency of humankind over our environment and the consequences of unbridled industrial power as we wrestle with one of the most rapid changes in our environment of any era. The excesses of the Anthropocene Epoch created an awakening of the need to live within Earth's capacities (Fig. 1.2).

Fig. 1.2 The Blue Marble (Courtesy of NASA)

SUMMARY

The notion of sustainability did not recently dawn on humankind. Almost 2300 years ago Aristotle advocated for a greater sense of connectiveness and for higher human aspirations than mere consumption. As later social commentators observed the excesses of the Industrial Revolution, they too pondered the meaning and dangers of progress. It was not until the musings of scientists and social scientists finally created broad social awareness in the 1960s and 1970s that serious discussions of sustainability entered our collective consciousness. Since then, the United Nations and private institutions alike have called for greater dialog and, now especially as humankind is faced with the consequences of global warming, greater collective action.

An ESG Toolkit
The world has come to appreciate the nature of humankind's footprint on the planet. The book will later argue that corporations have values that reflect those of society.

To what degree does the corporate mission statement reflect these broader values for its stakeholders, including the planet and those who will follow in our footsteps?

NOTES

1. O'Neill, John. (2006). "Citizenship, Well-Being and Sustainability: Epicurus or Aristotle?" Analyse & Kritik. **28**: 158–172.
2. Malthus, Thomas. (1798). An Essay on the Principle of Population As It Affects the Future Improvement of Society, with Remarks on the Speculations of Mr. Goodwin, M. Condorcet, and Other Writers, J. Johnson in St Paul's Churchyard, London.
3. Smith, Adam. (1776). An Inquiry into the Nature and Causes of the Wealth of Nations, printed for W. Strahan; and T. Cadell, London.
4. Smith, A. (1759). The Theory of Moral Sentiments, Clarendon, 24.
5. Thoreau, Henry David. (1854). On Walden Pond, Sahara Publisher Books.
6. Perkins, George. (1864). Man and Nature: Or, Physical Geography as Modified by Human Action, Charles Schriener, Publisher.
7. Tansley, A. G. (1935). "The Use and Abuse of Vegetational Terms and Concepts", Ecology. **16** (3): 284–307. https://doi.org/10.2307/1930070. JSTOR 1930070.
8. Carson, R. (1962). Silent Spring, Penguin Books.
9. Schumacher, Ernst Friedrich. (1973). Small is Beautiful—A Study of Economics as if People Mattered, Blond and Briggs.
10. Meadows, D. H. (1972). The Limits to Growth: A Report for the Club of Rome's Project on the Predicament of Mankind, Universe Books.
11. Jay Wright Forrester. (1973). World Dynamics.

12. Meadows, D. H., Meadows, D. L., Jørgen Randers, & Club of Rome. (1992). Beyond the Limits: Confronting Global Collapse, Envisioning a Sustainable Future, Chelsea Green Pub. Co.
13. Meadows, D. L., Tapley, E., Jørgen Randers, & Meadows, D. H. (2004). Learning Environment: Limits to Growth: The 30-Year Update, Chelsea Green Publ.

CHAPTER 2

Population Dynamics and Systems—A Limits to Growth

In the first chapter, we introduced the innovative approach of <u>Limits to Growth</u> to help explain the Malthusian Trap. Their mathematical simulation provided a theoretical tool to explore how resource constraints impinge on human population growth. While humankind's access to some critical resources is fixed, or may expand only arithmetically, our population grows exponentially. These different rates increase resource pressure that challenges sustainability.

In this chapter we describe the mathematics of population growth. We shall see that the path of populations tends to grow exponentially while resources either decline, may remain constant, or grow only slowly. This exponential growth, at least to the point that populations become constrained by nature, can exhibit some surprising dynamics, just as Malthus predicted. This Malthusian Trap creates one of the fundamental themes of sustainability. Our effort to describe the basic mathematics underlying this tension between population growth and resource decline will provide us with a valuable tool for our understanding of sustainability.

THE ISOLATED STATE

Our definition of sustainability implies that we remain in equilibrium with our environment and its capacity to provide for us. We will see later the significant challenges when faced with resource constraints, even for resources that are not in fixed supply. But we begin with the first steady state economic model of human coexistence with its environment.

© The Author(s), under exclusive license to Springer Nature Switzerland AG 2023
C. Read, *Understanding Sustainability Principles and ESG Policies*,
https://doi.org/10.1007/978-3-031-34483-1_2

In 1826, an enigmatic Austrian named Johann Heinrich von Thünen was contemplating the optimal agricultural strategy for his land around Mecklenburg-Strelitz. While his vocation was farming, he was also a brilliant self-taught economist. The English-speaking world knew little about his work until The Isolated State was translated from German into English in 1966. With its translation, Paul Samuelson, the most eminent economist of our time, and the winner of the first Nobel Memorial Prize in Economic Sciences, proclaimed that Thünen's work places him in the pantheon of the world's greatest economists.

Thünen was interested in the ways in which our pattern of land usage evolves as the population grows. He did not explore the nature of all resources. Land and labor were the primary factors of production in what were primarily agricultural economies in Thünen's era.

In his isolated state, a sustainable and self-contained city does not rely on trade with any other region. Instead, as its population and hence demand grows, the isolated city satisfies its agricultural needs by expanding its hinterland outward in every direction in circles of increasing radii across a featureless plain. At harvest time, farmers take their crops to market in the central village, and maximize their profits by choosing the optimal crops, based on land and transportation costs, that can be delivered at harvest time to the central market.

From this very simple model, Thünen was able to make several important observations. Those products that are more fragile, perishable, and costly to transport will dominate the land closest to the central market. For instance, in an era before refrigeration, dairy, fruit and vegetable production, and meat packing must be closest to the center of the village. This is the first ring of intensive production.

Outside of this ring, timber and firewood, high-value consumables in civilizations that predated the age of coal, were heavy and needed constant replenishment for cooking and construction. These products were in the second ring around the city, outside of the intensive zone.

Still farther out are field crops that have a much lower yield and value per hectare of land, and are less costly to transport. These include grains and corn which may use more land than the other agricultural products per unit of value of production.

Finally, land of low cost due to its greater inaccessibility can be used for grazing and ranching. Ranchers employ large tracts of land of little value for animals that can be driven inexpensively into towns in herds. Hence, ranches locate in the fourth ring.

Beyond the fourth ring is land that remains unoccupied and hence is of little value for human commerce. In his model, it is these differences in market access and transportation costs that dictates land usage.

This model is rich with implications about how we manage an important natural resource. They presaged the contributions by the politician and self-trained economist David Ricardo in his theory of land rents described in On the Principles of Political Economy and Taxation,[1] and of Alfred Marshall's marginal analysis and demand curves in his Principles of Economics; an Introductory Volume[2] on the relevance of how differing costs determine economic choices.

We will delay until later in the chapter Malthus' reasoning of exponential population growth. Let us first explore the implications of an environment with sustainable production in which population grows exponentially but available land expands more slowly. In doing so we will gain Thünen's insight into Malthus' dismal prophecy, one which was available in Malthus' lifetime, albeit, in German rather than English.

To see Thünen result, consider the profit π to a farmer who is located a distance D from market and faces a freight cost f per unit of distance and unit of weight. Then a farmer able to grow a product of weight Y earns a profit of:

$$\pi = Y * (p - c) - Y * f * D,$$

where p is the value of the product per unit of weight in the central market, and c is its production cost.

Thünen noted that these profits earned from farming will be incorporated into the price of land of superior location. From this framework, Thünen's model shows that, if demand grows proportionally to population, and if population growth is exponential, the supply of land cannot keep pace with population and becomes increasingly expensive.

To visualize his results, observe Thünen's rings on an isolated state on a featureless plain (Fig. 2.1):

The central village is where the population resides and the markets exchange goods and services. Surrounding the village is a band of land that provides sustenance for the village. As the outer radius of this band expands linearly, the availability of the resource region then expands proportionally to the radius squared. To see this, recall the formula for an area within a circle:

$$A = \pi * R^2,$$

where A is the enclosed area and R is the radius of the circle.

We then see that the area of available land A increases at a rate:

$$dA = 2\pi dR$$

$$\frac{dA}{dr} = 2\pi.$$

Product costs increase linearly with transportation costs f in proportion to the radius of Thünen's rings, while resources availability increases linearly at a rate of 2π. However, we next show that human and biological populations rise exponentially rather than linearly, and hence quickly outstrip the availability and affordability of natural capital such as land.

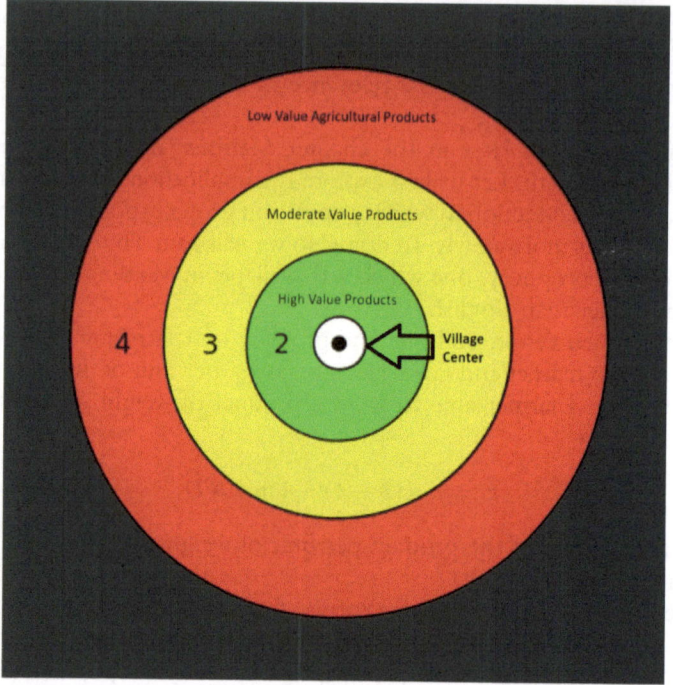

Fig. 2.1 Thünen's expanding rings of human activity around a village center

MALTHUSIAN POPULATION GROWTH[3]

To illustrate the exponential increase in population, consider the classic biological formula for reproduction. It is defined as a mathematical differential equation that is easy to derive, and describes a wide variety of other natural phenomena.

Biologists note that the population of a mortal species grows in proportion to its birth rate, net of deaths, in proportion to the size of its population. Demographers often quote these familiar rates as the number of births and deaths per 1000 population. For instance, Fig. 2.2 shows the birth rates per 1000 people worldwide, according to the United Nations Population Division:

Meanwhile, death rates per thousand people are shown in Fig. 2.3.

We can make an important distinction between these two demographic factors. Most countries experience a birth rate that exceeds the death rate. Let this birth rate, as a proportion of the population be denoted by k_{birth}, while the death rate be given by k_{death}. Then, in each period, we have:

$$\text{Births} = k_{birth} * P(t) \text{ and}$$
$$\text{Deaths} = k_{death} * P(t),$$

where $P(t)$ is the level of the population at any given time.

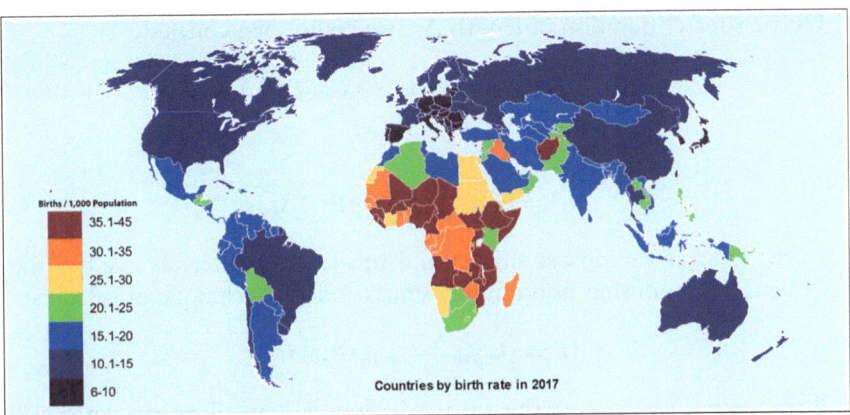

Fig. 2.2 Global birth rates per thousand people (*Source* Wikipedia Commons License)

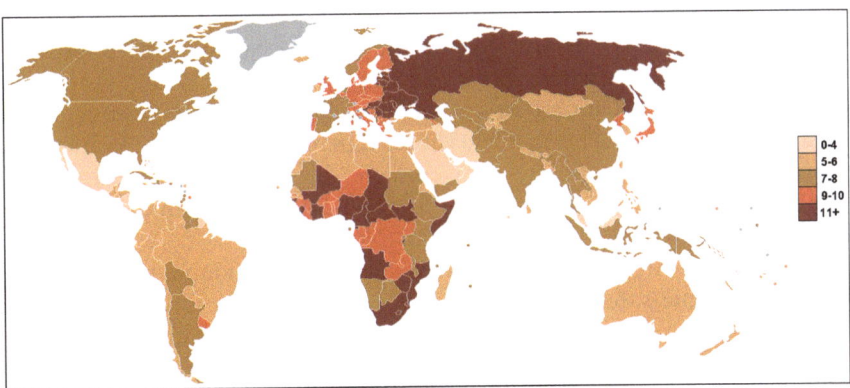

Fig. 2.3 Global death rates per thousand people (*Source* Wikipedia Commons License)

Note that this characterization is a simplification. Subtleties in demographics are certainly also important. Births and deaths are not only dependent on the size of the population but also on its composition. A population that is younger will tend to have a higher birth rate, all else equal, than an elderly population. Nonetheless, the relative size of these two coefficients of birth and death are the most important elements, even with the distinction of the importance of demographics. If so, we can determine the excess births, defined as the number of births less deaths in a given year according to:

Change in Population = Births − Deaths = $k_{birth} * P(t) - k_{death} * P(t)$.

Over a shorter duration of length Δt, we could then conclude:

$$\Delta P = k_{birth} * P(t)\Delta t - k_{death} P(t)\Delta t,$$

or,

$$\Delta P = (k_{birth} - k_{death})P(t)\Delta t.$$

In the limit as we look at smaller and smaller time intervals, we can then describe this relationship using infinitesimals and the technique of calculus:

$$dP = (k_{birth} - k_{death})P(t)dt.$$

Rearranging, we convert this into a differential equation that can quickly be solved using the tools of calculus:

$$\frac{dP}{dt} = \dot{P} = (k_{birth} - k_{death})P(t) = kP(t),$$

where the dot above P above signifies the rate of change of the population variable over time, and where $k = k_{birth} - k_{death}$.

This differential equation is of the family of equations $dy/dt = ky(t)$, where k is a constant. In fact, this family of linear differential equations is one of the easiest to solve, and is replete with examples in the biological and physical world. It applies to processes which grow at a constant rate in proportion to its size.

To solve, we often take an educated guess at a solution and then see if it works. Let us guess that the population equation takes the form $P(t) = Ce^{kt}$, where C is an arbitrary constant yet to be determined. Let us further specify our known population $P(0)$, often denoted P_0, at some starting time we can call $t = 0$. Then, differentiating our educated guess, we find:

$$\dot{P}(t) = Cke^{kt} = kP(t).$$

Note that, at time $t = 0$, this expression must yield a population $P(0)$, labeled P_0, typically called our initial condition. This differential equation then works if $Ck = P(0)$, or (Fig. 2.4):

$$P(t) = P_0 e^{kt}.$$

This equation is rich in implications.

First, it says that when the birth rate exceeds the death rate ($k_{birth} > k_{death}$), Malthus' prophecy of exponential population growth is verified. We also see that we reach a steady state population only if the birth rate converges to coincide with the death rate. Finally, populations crash if k is negative, as when the birth rate does not keep pace with the death rate.

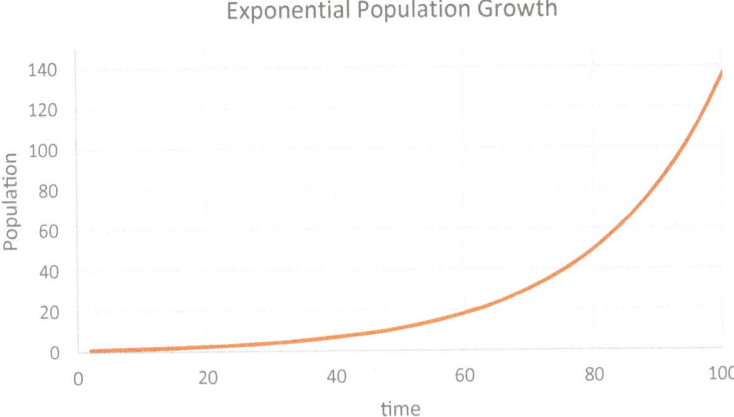

Fig. 2.4 Exponential population growth

Of course, there are many factors that impinge on such birth and death rates, often caused by humankind itself and our profound ability to dominate human and environmental variables. War, economic injustices, the level of affluence and technology, medical advances, illicit drug use, and myriad other human-derived forces affect the human condition and fertility and death rates. Meanwhile, famine, drought, pandemics and plagues, global warming, flooding, and population displacements are all natural causes for declines in populations. Technological advances can relax resource constraints, reduce the death rate, and extend longevity, which increases population pressures.

This simple model of unbridled population growth and possible decline for some species is part of what Malthus prophesized. Exponential growth beyond a certain point begins to accelerate rapidly if left unconstrained. Such unconstrained growth does not acknowledge any sort of natural constraint. Nor does it model the growth of decline of one element of an ecosystem as it interacts with another. In the appendix and the next chapter, we will explore these constraints to unbridled population growth.

For instance, Malthusian resource availability aside, species are interdependent, sometimes symbiotically and sometimes in an adversarial manner. As an example, the lynx population can grow or decline exponentially from its own procreation as the population model suggests, but will also grow as its prey, the hare, becomes more abundant. On the other hand, the hare population may also grow exponentially in its own right, but can decline as its predator, the lynx population, grows.

Likewise, human populations may grow exponentially through reproduction but can grow faster if a food source such as a fishery grows. On the other hand, while the fishery can also procreate and grow, human harvesting can cause the fishery to decline or potentially even crash, which may then hinder human population growth.

In a very simple linear extension of the simple population model above, we see in the appendix that the resulting interdependency between two interrelated populations creates a dynamic that can become quite complex once we acknowledge even the simplest interspecies interactions.

While we reserve of the appendix to this chapter a determination of population dynamics when species interdependencies exist, we find that, while the mathematics of population dynamics can be quite complex, once the subtleties of ecosystem interactions are fully explored, Malthus' prophecy is indeed plausible, and perhaps even likely. Populations can grow exponentially as they approach the limits of natural capital such as food supplies, but the growth is not necessarily smooth or simple. Indeed, it may gyrate just as Malthus prophesized in his stark warning of human feasts and famines, and can crash if these gyrations are sufficiently large.

The Club of Rome Analysis of the Limits to Growth

The Limits to Growth analysis of 1972 provided humankind with one of the first broadly distributed analyses that posed the question "How long can the Earth continue to sustain an increasing pace of economic activity?"

Some commentators at the time considered such a question pessimistic. However, others viewed it with some optimism. By showing how humankind is exceeding the ecosystem's capacity to sustain us, the enlightenment has helped spawn a movement toward more sustainable practices.

It might appear odd that this question had not been posed generations earlier. Recall Fig. 2.1 that showed the human population did not begin to grow explosively until well into the Industrial Revolution. While Malthus expressed concern about the Earth's capacity to support a rapidly growing human population, it took another century and a half for researchers to more fully explore the ability of natural capital to continue to support the increasingly rapid growth of humankind.

The Limits to Growth authors analyzed five forces that impinge upon the Earth's carrying capacity. These include population, food production, industrialization, pollution, and the consumption of natural resources. These variables that describe the state of the ecosystem at any point in time are obviously interrelated, but are too often treated in isolation and simplistically. Let us explore their results.

If there is a limited supply of a mineral essential for industrial processes, it was often assumed that the resource would still last many lifetimes. For instance, if the world consumed a million metric tonnes of an essential but finite element such as nickel or chromium each year, and the Earth held reserves of 500 million tonnes, then a simplistic estimate of the Earth's carrying capacity for such an element would be 500 million tonnes of reserves divided by use of one million tonnes extracted annually, to yield a resource lifetime of 500 years.

To most mortals, 500 years is an eternity, and certainly more than half a dozen lifetimes. However, this simplistic conclusion does not account for compounding growth.

Instead, consider a depletable resource with a fixed natural stock $\overline{S}$ (denoted by the bar above the variable to signify its fixed supply) that is to be consumed over time. We can compare that capacity to the flow of resource extraction and consumption that starts with a level X_0 for any arbitrary initial time $t = 0$.

Let the rate of extraction X of the resource grow at a positive rate G, perhaps proportional to the level of economic growth and activity. Then, using the same model we used for population growth, at any given time $t > 0$, the flow of resource extraction and consumption will rise to an amount $X_0 e^{gt}$ at any time t.

We can then quickly deduce the fundamental equation that caught the attention of the 1970s public. We do so by summing the amount of resource extraction and consumption to some time horizon T^* that is implicitly defined by the following integral that must sum to the resource constraint $\overline{S}$:

$$\overline{S} = \int_0^{T^*} X_0 e^{Gt} dt$$

We can solve this equation to yield an implicit value for the resource time horizon T^*:

$$\overline{S} = (X_0/G)\left(e^{GT^*} - 1\right).$$

Then, we can solve for the correct time period T^* when the resource is expected to sunset as:

$$T^* = \left(\frac{1}{G}\right)\ln\left(\frac{G\overline{S}}{X_0} + 1\right)$$

For instance, consider the previous example of a 500-year resource extraction horizon. Let us take the same assumptions of the extraction and consumption X_0 and a stock of fixed resources $\overline{S}$ and assume a modest rate of economic growth in consumption G of 5%. We find T^* is significantly lower than casually assumed:

$$T^* = \left(\frac{1}{0.05}\right)\ln\left(\frac{0.05 * 500,000,000}{1,000,000} + 1\right) = 65.2 \text{ years}$$

This dismal and dramatic proposition accelerates the sunset of an essential depletable resource from half a dozen lifetimes to within one lifetime. Such a dismal conclusion is disturbing to those unwilling to accept the consequences of our decisions today on future generations. Critics concerned about such a dismal prophecy often attack the assumptions. The easiest assumption to attack is to claim that the 5% annual consumption expansion rate is excessive.

To better allow critics the opportunity to make their own growth assumption, we can perform a sensitivity analysis by comparing the sunset year to various assumed growth rates (Fig. 2.5):

We see that the finiteness of resources conflicts with the rate of expansion of resource extraction and consumption. For instance, with consumption growth rates hovering around 8% for some major nations such as China, the world's second largest and populous country, the time horizon before resource depletion in this example would shorten from **500** years to about **46** years.

Given the profound sustainability implications of the simple <u>Limits to Growth</u> model, a couple of criticisms are often leveled that will be treated in subsequent chapters. First, known reserves of some resources may expand over time. Better technologies, in identifying and extracting fixed resources, or in the more efficient consumption of natural capital in industrial processes may be discovered. We will also treat the importance of some backstop technologies to which humankind may transition as a substitute for resources that dwindle over time as their prices rise. Some take as a matter of faith that new technologies shall be developed over time that provide us with remedies to these prophecies without describing the source of such technological salvation.

Note that these solutions are often stated as articles of faith and comfort upon which the fate of future generations will ultimately depend. Instead, if we seek solutions today, before it is too late, we can avoid the worst scenarios

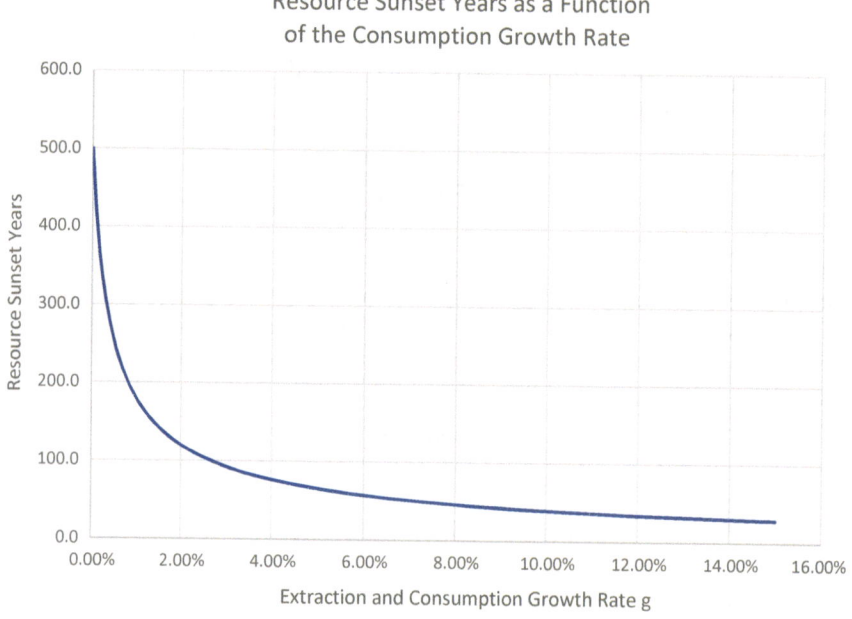

Fig. 2.5 The rapidly declining human planning horizon for even modest discount rates

of Jay Forester's World3 model incorporated in the <u>Limits to Growth</u> analysis. Indeed, the authors set such an optimistic tone. With better modeling and understanding about the profound effects humankind has on the planet and our future, our increased awareness today can guide us toward a pathway of sustainability. In doing so, we broaden the notion of *for all humankind* to include generations today and well into the future.

In the same year that <u>Limits to Growth</u> was published, William Nordhaus and James Tobin, two economists who would each go on to win a Nobel Memorial Prize in Economic Sciences over their subsequent careers, published one of the earliest papers on the social science of sustainability. Their 1972 paper <u>Is Growth Obsolete</u>?[4] questioned the premise of maintenance of the growth of **Gross Domestic Product** (GDP) as the basis for sound economic policy. They observed that GDP is an imperfect measure of the welfare of society and that it implicitly prescribes population growth as a means to realize economic growth. They note:

> Possible abuse of public natural resources is a much more serious problem. It is useful to distinguish between local and global ecological disturbances. The former includes transient air pollution, water pollution, noise pollution, visual disamenities. It is certainly true that we have not charged automobile users and electricity consumers for their pollution of the skies, or farmers and housewives for the pollution of lakes by the runoff of fertilizers and detergents. In that degree our national product series have overestimated the advance of welfare. Our urban disamenity estimates given above indicate a current overestimate of about 5 per cent of total consumption... There are other serious consequences of treating as free things which are not really free.

Their joint work provided a new moral imperative. They critiqued the traditional approach among economists to measure growth traditionally based on spending and production, rather than the growth of nature's assets that can contribute to production and sustenance over many generations. They also remind us of the need to include all costs when we weigh the benefits of production, not just the market costs of our actions.

By 1972, scientists and social scientists were increasingly lamenting the limits of traditional neoclassical economics, not based on inherent weaknesses of the tools, but how holistically they are applied. Tobin and Nordhaus, and their MIT contemporary, Robert Solow helped develop improved economic tools to allow us to better understand resource extraction and consumption over time. We shall describe these tools more fully in Section III.

SUMMARY

As resources grow arithmetically in the Thünen model, humankind tends to grow exponentially. Even in these simple examples, we see the resulting interplay between aspects of our environment and humankind are exceedingly rich and complex. We show in the appendix that subtleties and complexities among species even in a simple interrelated ecosystem can result in surprising gyra-tions in species populations. Such exponential growth and decline, and the gyrations that are induced through our interactions, can result in instabilities when nature's resilience is eroded.

Economists have advocated that the holy economic grail among nations should be growth in national income and output. In doing so, markets have aligned around the growth principle with little regard for far more important variables of long-term economic and natural assets. By focusing on the flow of income rather than the stock of natural capital that sustains life, we tend to manage natural capital for the present rather than for all generations to follow and for the entire ecosystem. Sustainability inherently views our economy and ecosystems from the longer intergenerational perspective.

ESG Toolkit

Population grows exponentially because every pair of parents contemplate the value for themselves of procreation, and may even explore their individual capac-ities to provide for a child. However, we do not collectively recognize the effects on each other and on the Earth when each of us makes our determination. Corporations too can explore the ethic of the footprint they impose on the Earth under the assumption that all other corporations behave as they do.

Corporations also often make predictions about future growth as part of their forward guidance to stakeholders. Do these estimates properly reflect the constraints to growth? If all competing corporations are also expected to grow in the same manner, are their sufficient resources and customers to sustain that growth?

What implications does that have on living individually and collectively within Earth's constraints?

Is your corporation's product market growing or declining exponentially?

What growth assumptions does your corporation use in its strategic planning, and is continuous growth realistic?

What mix of human-produced and natural capital do you employ?

Does your strategic plan include resource usage and does it differentiate between resource types based on their long-term supply?

Appendix: A Complex Ecosystem

Many species maintain their existence on the knife-edge of steady state between births and deaths. As a resource constraint is neared, deaths tend to rise and births fall until a population exhibits no net growth. This tapering toward a steady state equilibrium seems to deny Malthus' prophecy that popu-lations may oscillate in fits and starts. Yet, we see the population of some species such as hares and lynx gyrate over time. This is due to subtle inter-actions between components of an ecosystem. We next model the subtleties of even the simplest of linear differential equations that produce surprising complex gyrations.

Let us consider two species that interact. Let the variable x represents the size of a predator population, and let us assume that its population is self-correcting. In other words, deaths tend to exceed births if the species must compete for increasingly scarce subsistence resources.

Let us also allow this species to prey on a second related species population or resource, denoted by y such that the first species' population x tends to increase as y increases. We will use the shorthand *dot* over these population variables to denote its rate of change over time. Then, the path of the predator population is given by:

$$\frac{dx}{dt} = \dot{x} = -ax + by, \text{ where } a, b > 0.$$

The growth of the population x over time declines at a rate "$-a$" in proportion to its size but is enhanced at a rate "b" as its prey population y increases. Finally, we model the prey species y in this predator/prey model such that its population is self-correcting as its population grows, at a rate "$-c$," but it also suffers population decline if the predator population grows, at a rate "$-d$":

$$\frac{dy}{dt} = \dot{y} = -dx - cy, \text{ where } c, d > 0.$$

We have defined a mutually interrelated ecosystem in which each population depends on the other. Both populations are also partially self-correcting as they face other resource constraints. While intuition would suggest that this system of two linear differential equations should behave quite predictably, we find that simple systems can demonstrate very complex behavior not unlike that Malthus predicted.

To see this, let us determine a simultaneous solution to this system of two linear differential equations such that each parameter, a, b, c, and d are all positive numbers. It is easiest for us to solve this system using linear algebra that collapses the system into a simple matrix. Combining the two equations into the system of linear constant coefficient differential equations gives:

$$\dot{x} = -ax + by$$

$$\dot{y} = -dx - cy$$

In matrix form, this system reduces to:

$$\begin{bmatrix} \dot{x} \\ \dot{y} \end{bmatrix} = \begin{bmatrix} -a & b \\ -d & -c \end{bmatrix} \begin{bmatrix} x \\ y \end{bmatrix}$$

As before, with our population model, and as is standard practice in such *linear homogeneous differential equations*, we postulate a solution to the population of the following form that describes the evolution of the ecosystem over time. We then determine if this solution works:

$$\begin{bmatrix} x(t) \\ y(t) \end{bmatrix} = \begin{bmatrix} A_1 e^{\lambda t} \\ A_2 e^{\lambda t} \end{bmatrix}. \tag{2.1}$$

We will soon solve for our new parameters A_1, A_2, and λ. First though, substituting these potential solutions into the system of differential equations gives:

$$\begin{bmatrix} \dot{x} \\ \dot{y} \end{bmatrix} = \begin{bmatrix} A_1 \lambda e^{\lambda t} \\ A_2 \lambda e^{\lambda t} \end{bmatrix} = \begin{bmatrix} -a A_1 e^{\lambda t} + b A_2 e^{\lambda t} \\ -d A_1 e^{\lambda t} - c A_2 e^{\lambda t} \end{bmatrix}. \tag{2.2}$$

We can rearrange the two matrices on the right-hand side of (2.2) to determine:

$$\begin{bmatrix} -a - \lambda & b \\ -d & -c - \lambda \end{bmatrix} \begin{bmatrix} A_1 e^{\lambda t} \\ A_2 e^{\lambda t} \end{bmatrix} = \begin{bmatrix} 0 \\ 0 \end{bmatrix}. \tag{2.3}$$

The condition to a solution to (2.3) requires the determinant of the 2×2 matrix on the left-hand side to be equal to zero. The determinant equation is $(-a - \lambda)(-c - \lambda) - (-bd)$. Simplifying, we see that of characteristic roots λ can be solved according to:

$$\lambda^2 + (a + c)\lambda + ac + bd = 0.$$

Solving this quadratic equation for λ yields two possible solutions, an upper and a lower root:

$$\lambda = -(a + c)/2 \pm \sqrt{((a + c)/2)^2 - ac - bd}$$
$$= -(a + c)/2 \pm \sqrt{((a - c)/2)^2 - bd}.$$

Let us make a simplifying assumption that each population is self-correcting toward its mean at the same rate, i.e., $a = c$. Then, there are two possible values for λ:

$$\lambda = -a \pm i\sqrt{bd},$$

where $i = \sqrt{1}$, the imaginary number. Substituting the upper root into (2.3) gives:

$$\begin{bmatrix} -i\sqrt{bd} & b \\ -d & -i\sqrt{bd} \end{bmatrix} \begin{bmatrix} A_1 \\ A_2 \end{bmatrix} = \begin{bmatrix} 0 \\ 0 \end{bmatrix},$$

and the lower root gives:

$$\begin{bmatrix} i\sqrt{bd} & b \\ -d & i\sqrt{bd} \end{bmatrix} \begin{bmatrix} A_1 \\ A_2 \end{bmatrix} = \begin{bmatrix} 0 \\ 0 \end{bmatrix}.$$

The upper characteristic root requires $-i\sqrt{bd}\, A_1 + bA_2 = 0$. Let $A_1 = i\sqrt{bd}$ which implies $bd + bA_2 = 0$, or $A_2 = -d$. Similarly, at the lower root, if $A_1 = i\sqrt{bd}$, then $A_2 = d$. There are thus two possible solution vectors for x and y, the first corresponding to the upper, and the other the lower root for λ.

By taking a linear combination of these two solutions, we can determine the range of solutions for the populations x and y, subject to initial conditions that will allow us to calculate the arbitrary constants c_1 and c_2.

Our expression is simplified for the moment if we label $\omega = \sqrt{bd}$. From (2.1), this reduces to:

$$\begin{bmatrix} x \\ y \end{bmatrix} = c_1 \begin{bmatrix} i\omega e^{(-a+i\omega)t} \\ -de^{(-a+i\omega)t} \end{bmatrix} + c_2 \begin{bmatrix} i\omega e^{(-a-i\omega)t} \\ de^{(-a-i\omega)t} \end{bmatrix}. \tag{2.4}$$

On the first blush, the solution seems disturbing since it involves i, the imaginary number that represents the square root of -1. Fortunately, the great mathematical prodigy Leonhard Euler showed us how we can convert between the realm of imaginary numbers to the trigonometric functions.

Euler's formula, from his 1748 <u>Introductio in Analysin Infinitorum</u>,[5] is one of the most beautiful theorems and formulas ever created. It states:

$$f(\omega t) = e^{-i\omega t}(\cos(\omega t) + i\sin(\omega t)).$$

It is not difficult to derive Euler's formula, if one can for the moment simply consider the imaginary number i, the square root of -1, as a constant. Let us differentiate Euler's $f(\omega t)$ with respect to its independent variable ωt. Then, we have:

$$\begin{aligned} f'(\omega t) &= e^{-i\omega t}\left(i\cos(\omega t) - \sin(\omega t) - ie^{-i\omega t}(\cos(\omega t) + i\sin) \right) \\ &= e^{-i\omega t}\left(i\cos(\omega t) - i\cos(\omega t) + \sin(\omega t) + i^2\sin(\omega t) \right) \end{aligned} \tag{2.5}$$

The right-hand side of (2.5) cancels to zeros since $i^2 = -1$. We then see that $f(\omega t)$ must be constant since its derivative equals zero. We also see that $f(0) = 1$. It must then be the case that the exponential and trigonometric terms are offsetting, and:

$$e^{i\omega t} = \cos(\omega t) + i\sin(\omega t).$$

This simple deduction married the combination of imaginary and real numbers, called the complex number system, with our more familiar trigonometric functions.

In the form most useful to the problem above, it states:

$$e^{(-at \pm i\omega t)} = e^{-at}(\cos \omega t \pm i \sin \omega t).$$

Euler's Identity allows (2.4) to be expressed as:

$$\begin{bmatrix} x(t) \\ y(t) \end{bmatrix} = e^{-at} \begin{bmatrix} c_1 i\omega(\cos \omega t + i \sin t) + c_2 i\omega(\cos \omega t - i \sin \omega t) \\ -c_1 d(\cos \omega t + i \sin \omega t) + c_2 d(\cos \omega t - i \sin \omega t) \end{bmatrix} \quad (2.6)$$

With some simplification, we have found our general solution to ecosystem equilibrium:

$$\begin{bmatrix} x(t) \\ y(t) \end{bmatrix} = e^{-at} \begin{bmatrix} \omega i(c_1 + c_2)\cos \omega t - \omega(c_1 - c_2)\sin \omega t \\ -di(c_1 + c_2)\sin \omega t - d(c_1 - c_2)\cos \omega t \end{bmatrix}.$$

We next propose a simple but convenient substitution. Let $C_1 = i(c_1 + c_2)$ and $C_2 = (c_1 - c_2)$. Then,

$$\begin{bmatrix} x(t) \\ y(t) \end{bmatrix} = e^{-at} \begin{bmatrix} \omega C_1 \cos \omega t - \omega C_2 \sin \omega t \\ -d C_1 \sin \omega t - d C_2 \cos \omega t \end{bmatrix}.$$

Finally, we can impose the initial conditions for the respective initial starting populations $x(0) = x_0$ and $y(0) = y_0$ to require:

$$\begin{bmatrix} x(0) \\ y(0) \end{bmatrix} = \begin{bmatrix} \omega C_1 \\ -d C_2 \end{bmatrix}.$$

The complete solution for the adjustment path of the two markets is then given by:

$$\begin{bmatrix} x(t) \\ y(t) \end{bmatrix} = e^{-at} \begin{bmatrix} x_0 \cos\sqrt{bd}\,t + \sqrt{\frac{b}{d}}\,y_0 \sin\sqrt{bd}\,t \\ y_0 \cos\sqrt{bd}\,t - \sqrt{\frac{d}{b}}\,x_0 \sin\sqrt{bd}\,t \end{bmatrix}, \quad (2.7)$$

where we replaced ω with its constituent $\sqrt{bd}$. These solutions for populations $x(t)$ and $y(t)$ show that both the two populations tend to converge to a fixed size over time because of the exponentially declining term, but will gyrate toward equilibrium with a frequency f given by $\frac{\omega}{2\pi} = \frac{\sqrt{bd}}{2\pi}$, with the frequency of gyrations dependent on the interspecies interactions b and d per time period. A graph of these population dynamics appears as (Fig. 2.6).

While our mathematics is quite complex, the results are surprising and unnerving. Even simple linear interactions between species can result in quite significant ecosystem instability. If these population gyrations are sufficiently pronounced, they may even push species into mutual extinction in what can be described as a **population crash**. In fact, while we began by assuming that the own effects, denoted by the parameters a and c, tend to stabilize each

Fig. 2.6 Complex interactions in predator and prey populations

species, if the net own effects are positive, the system does not only oscillate, but may also continue to grow exponentially, as Malthus surmised. This growth amplifies the wild population gyrations.

We see that even simple interrelations between species in an ecosystem can result in very complex and gyrating population patterns over time. For instance, we can apply this model to a lynx population that grows exponentially from its own procreation but will also grow as its prey, the hare, becomes more abundant. On the other hand, the hare population grows exponentially in its own right, but can decline as its predator, the lynx population, grows.

Likewise, our human population that depends on *a biotic resource* such as a fishery that can be overharvested may also suffer the gyrations that Malthus predicted. We find that, even in a very simple model of linear interactions, the result can be population dynamics that are quite complex and often defy human intuition.

Once such interdependencies are modeled, Malthus' dismal prophecy is realized. Populations can grow exponentially as they approach the limits of natural capital such as food supplies, but the growth path is not necessarily smooth, and may even gyrate and crash within the simplest of biological circumstances.

Notes

1. Ricardo, D. (1821). On the Principles of Political Economy, and Taxation.
2. Marshall, Alfred. (1920). Principles of Economics; an Introductory Volume, Macmillan, London.
3. Malthus T. R. (1798). An Essay on the Principle of Population. Anonymously Published (reprinted by Cosimo Classics, 2007).

4. Nordhaus, William, & TobinJames. (1972). Is Growth Obsolete?, Yale University. https://www.nber.org/system/files/chapters/c7620/c7620.pdf, accessed December 13, 2022.
5. Euler, Leonhard, Introductio in Analysin Infinitorum, 1748.

CHAPTER 3

An Operational Definition of Sustainability

With some science and mathematics of population dynamics and the constraints of resources at hand, we can now understand the increasing awareness and concern beginning in the 1960s over nature's fragility in the wake of human population growth. Let us next document some approaches that emerged to allow us to better place the activities of humankind within a broader natural perspective. We then describe in Chapter 4 how natural capital must be differentiated from the *human-made capital* that we typically consider in our decisions to produce. In doing so, we see that our economic measures are tilted toward the purchase of human-made factors of production because their costs are easy to quantify, while we often take for granted the natural resources around us that are increasingly scarce and fragile.

Rachel Carson's <u>Silent Spring</u> published in 1962 and the Club of Rome's <u>Limits to Growth</u> in 1972 bookended a decade of increased awareness of humankind's influence on the environment and sustainability. This increased awareness motivated the United Nations to form the *Brundtland Commission on Sustainability* in 1983, which produced a report entitled <u>Our Common Future</u>,[1] four years later. The report has motivated individuals and institutions alike to become increasingly aware of the need to better manage our natural environment.

Recall the concerns of the Greek philosophers that humankind must acknowledge and function with an awareness of our role in the greater web of our ecosystem. Over the years, our understanding of the scope and strength of these interactions have broadened and deepened. The Greek metaphor of our environment, which they labeled our house, or *oikos*, remains relevant today. It is not only the root of ecology and economics, but it also invokes an increased

© The Author(s), under exclusive license to Springer Nature Switzerland AG 2023
C. Read, *Understanding Sustainability Principles and ESG Policies*, https://doi.org/10.1007/978-3-031-34483-1_3

immediacy regarding our interactions with the ecosystem. Economics, as a study of the management of our house, must now accommodate a broader meaning than the narrow neoclassical focus of the pursuit of financial profits and the realization of happiness through individual consumer sovereignty.

The business literature observes you cannot manage something you do not measure. We must then translate a broader set of economic principles into management goals and measurements. Indeed, we shall see in Section V that leaders in the corporate community have become champions in identifying the measures indicative of sustainability and guiding the private sector toward a more sustainable future.

A broader definition of economics, or, more accurately, a return to the *oikos* at its root, will require some vigilance, given the narrow lens through which traditional economics has been narrowed since Mathus' time. Social scientists have recently developed a much broader set of tools to better manage our house, not only for the good of humankind but also as an appeal to a broader set of ecosystem values.

Let us first explore some of these shared ecosystem and sustainability values. While there was initially some disagreement and lack of coherence in the definition of sustainability, we have gained more clarity by describing what we collectively wish to accomplish with its better management.

The strictest notion of sustainability could be labeled *preservationist*. We can attain sustainability if we are able to preserve natural capital. Better yet, we could return our natural state to a pre-industrialization level and then develop policies that prevent humankind from tampering with the natural balance. This repair of human damage to the Earth would come at such an overwhelming cost, in economic terms and in necessary interventions, that such a turning back of the clock would likely be impossible to coordinate globally, for politically pragmatic reasons.

We may instead be able to attain a weaker form of sustainability that emphasizes conservation rather than preservation by recognizing some human capacity can substitute for the degraded natural capital of the atmosphere we breathe and resources we extract. This is a theme we shall explore in subsequent chapters.

We are still left with the need to reach a consensus regarding how conservation should be defined. Since conservation invokes tradeoffs of gains in one area that may ameliorate abuses in other areas, there are as many pathways to *weak sustainability* as the values we wish to conserve.

The first approach in the accounting and measurement of sustainability documented the overlapping values of various stakeholders. The *Triple Bottom Line* (TBL) was developed in 1992 to demonstrate the interface between social and environmental factors and corporate values. While it had since been overshadowed by more recent approaches such as the **Three Pillars**, the *Environmental, Social, and Governance (ESG)* values, the TBL retains some advantage in its emphasis on measurability of values as an extension of accepted accounting practices.

Fig. 3.1 The ESG areas of the environment, social, and governance overlap for sustainability

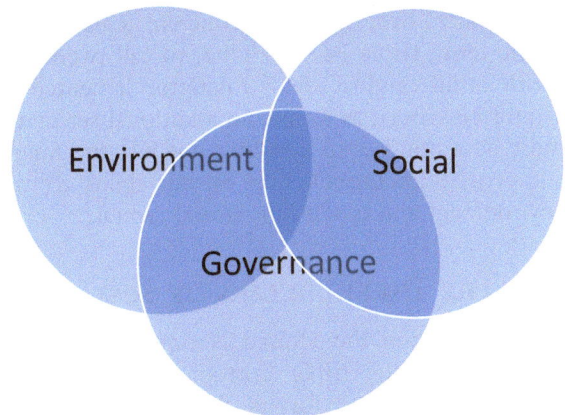

The triple bottom line accounts for our environmental, societal, and financial values. This framework is often alternatively described as *planet, people, and prosperity (3Ps)*. Let us frame these values based on what each dimension wishes to accomplish.

For the first element of the triple bottom line, we attempt to measure in myriad ways the benefits and costs of our business and human decisions on the environment and the planet. The second element includes the social benefits and costs of our decisions, while the third element includes the traditional financial profits arising from corporate decisions and the prosperity that benefits humankind.

The concept and application of a triple bottom line are a natural extension of the knowable and measurable, the traditional domain of financial accounting, to less familiar societal aspects that are more difficult, but not impossible, to measure. While measuring the cost of our decisions on the quality of the air and water, the sustainability of our resources, the fragility of our atmosphere to greenhouse gases, and the sensitivity of our ecology to pollution and pesticides may be challenging, we can nonetheless propose and improve on these various measures over time.

Similarly, we can acknowledge and attempt to quantify, or at least describe in meaningful qualitative ways, the effects of our decisions on our communities, our social structure, and our relationships, on the micro, the macro, and the global level. We can then use these measures and values to explore their effects on corporate financial, institutional, and organizational decisions on the economic system, rather than in isolation. This is also a goal of the modern ESG paradigm (Fig. 3.1).

While measuring allows us to manage, the simple act of measurement is insufficient to ensure good management. This triple bottom line approach also requires us to impose some set of values as our management goals. Without the articulation of concrete goals, we remain mired in normative debate rather than positive science.

A concern over the lack of concrete goals led John Elkington, the originator of the term Triple Bottom Line, to call twenty-five years later for a *product recall* of his original idea.[2] Elkington lamented that the Triple Bottom Line cannot be viewed as simply an additional set of accounting books. Instead, it must be a mindset to consider not just the financial but the true economic and ecosystem consequences on our ability to make decisions that harm the environment and erode our natural capital.

The Three Pillars of Sustainable Development

The commission the United Nations created defined *three pillars* in ways analogous to the triple bottom line. Formed in 1983, this Brundtland Commission, named for its chair, Gro Harlem Brundtland, former Prime Minister of Norway, considered the role of sustainability within the three pillars of environmental, economic, and societal values. Their work culminated in the Brundtland Report in 1987, at which time the Commission disbanded.

"Our Common Future"[3] advocated for the overarching principle of sustainable development. The Commission recognized a challenge. It could not freeze the natural human aspiration of growth in prosperity and well-being without also perpetuating the tremendous advantage MDCs have over LDCs because LDCs developed later and with fewer financial endowments.

In addition, the Brundtland Commission recognized the political reality of expecting a relatively small but wealthy segment of the global population to make immense wealth transfers to much larger and often poverty-stricken populations elsewhere in the world. This era was before the industrialization of China that remained mired in a significantly lower level of economic attainment. Populous countries such as India, Indonesia, and even the former Soviet Union were also languishing in the 1980s.

By speaking about the need for sustainable development for LDCs instead of global sustainability, the report could offer hope that the welfare of the LDCs could increase substantially each year, at perhaps 5% or 6% per annum in *Gross Domestic Product (GDP)* growth, while MDCs could maintain some growth, albeit at a much more modest pace.

The Environmental, Social, and Governance Approach

In this early era of sustainable development enlightenment, and despite the protests of economists Nordhaus and Tobin in 1972, mainstream economics continued to rely upon the manipulation of traditional financial variables to generate efficiency and profits. Maintenance of traditional economic tools had the advantage of the familiar, with variables that are simple to measure and easy to obtain (Fig. 3.2).

While environmental and ecological economics and other subdisciplines were increasingly identifying human well-being, educational levels, and social

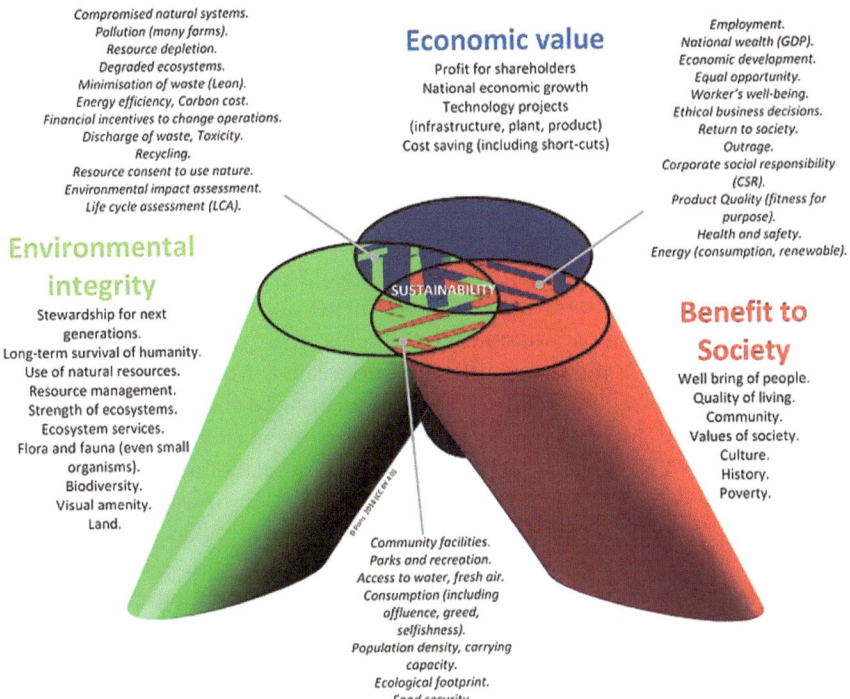

Compromised natural systems.
Pollution (many forms).
Resource depletion.
Degraded ecosystems.
Minimisation of waste (Lean).
Energy efficiency, Carbon cost.
Financial incentives to change operations.
Discharge of waste, Toxicity.
Recycling.
Resource consent to use nature.
Environmental impact assessment.
Life cycle assessment (LCA).

Economic value

Profit for shareholders
National economic growth
Technology projects
(infrastructure, plant, product)
Cost saving (including short-cuts)

Employment.
National wealth (GDP).
Economic development.
Equal opportunity.
Worker's well-being.
Ethical business decisions.
Return to society.
Outrage.
Corporate social responsibility
(CSR).
Product Quality (fitness for
purpose).
Health and safety.
Energy (consumption, renewable).

**Environmental
integrity**

Stewardship for next
generations.
Long-term survival of humanity.
Use of natural resources.
Resource management.
Strength of ecosystems.
Ecosystem services.
Flora and fauna (even small
organisms).
Biodiversity.
Visual amenity.
Land.

SUSTAINABILITY

**Benefit to
Society**

Well bring of people.
Quality of living.
Community.
Values of society.
Culture.
History.
Poverty.

Community facilities.
Parks and recreation.
Access to water, fresh air.
Consumption (including
offluence, greed,
selfishness).
Population density, carrying
capacity.
Ecological footprint.
Food security.

Fig. 3.2 The three pillars of the ESG paradigm (Courtesy of Dirk Pons, https://
commons.wikimedia.org/wiki/File:Three_Pillars_model_of_Sustainability.jpg)

investments as worthwhile of human pursuits, such variables were not typically included on financial statements and economic data and hence were
often omitted from mainstream economic analyses. By broadening corporate
and social values beyond traditional economic measures, the ESG movement
attempts to break out of the stranglehold that omits traditional measures and
more holistic indicators of social well-being.

The *Environment, Social, Governance (ESG)* approach categorizes our
collective values in three categories that are similar to the TBL, the Three
Pillars of Sustainable Development, and the 3Ps of people, planet, and
prosperity.

The Societal Pillar includes measures of quality of life, well-being, the erad-
ication of poverty, human dignity, sense of place and community, health, and
safety. This pillar places a premium on the attainment of basic human rights.
Some of these societal values intersect with the financial and economic values.
ESG proponents believe that education, health, economic development, and
corporate social responsibility often translate into stronger corporate balance
sheets and governmental ledgers.

The Environmental Pillar includes the use of natural resources, the strength and resilience of the ecosystem, effective ecosystem management, preservation of species of flora and fauna, biodiversity, and stewardship. This pillar overlaps with the Societal Pillar because the preservation and quality of our surroundings translate into human happiness. For instance, a sound ecosystem ensures that humans have safe drinking water, clean air, food security, and a sufficient natural carrying capacity.

In turn, this Environmental Pillar also overlaps with corporate governance in that degraded natural capacities reduce the availability and efficiency of resource usage, requires greater investments in the search for alternative resources, and imposes upon society higher costs to keep the workforce productive and the population sufficiently nourished. Meanwhile, the Governance Pillar provides insights into how the quality and diversity of leadership can lead to a more just and environmentally sensitive society.

The more recent ESG movement recognizes and elaborates on the various other three pillars approaches. While ESG explores how the pursuit of economic growth affects the Social and Environmental pillars, it is primarily concerned with how the degradation of elements within pillars hampers sustainability and the economy, finances, and the accounting bottom line. Using the triple bottom line paradigm, the ESG movement often attempts to quantify how environmental and social issues impinge on corporate profitability, the variance of profitability over time as a measure of risk, and the ability to sustain profitability for corporations, institutions, and organizations that are, or ought to be assumed to be infinitely lived.

While focusing on the effects of the pillars on finances and economics, the ESG paradigm also acknowledges the feedback between pillars. As an example, insufficiently or unpriced pollution by corporations degrades both the environmental and human rights protected by the social factors. This degradation in turn creates *market, operational, physical, and strategic risks* described in Chapter 26 that impose costs on corporations and ultimately erode profits over time. The ESG movement has a goal of modeling appropriate recognition of these forces and hence decisions of our institutions in ways that, if replicated broadly, will enhance sustainability over time.

ESG can also be expressed using the pillar paradigm (Fig. 3.3):

The goal of the ESG paradigm is to demonstrate to stakeholders of institutions how various non-market dimensions impinge upon their bottom line. The Environmental Pillar explores how an individual company's decisions may damage the environment, with the belief that if all institutions acknowledge and appropriately price this value not traditionally measured by the marketplace, the collective good and corporate sustainability will be enhanced.

Most typically, climate change and pollution, topics treated more thoroughly in subsequent chapters, receive the most attention. However, also included in the Environmental Pillar is our dependence on depletable resources such as fossil fuels, our use of other resources such as land and water,

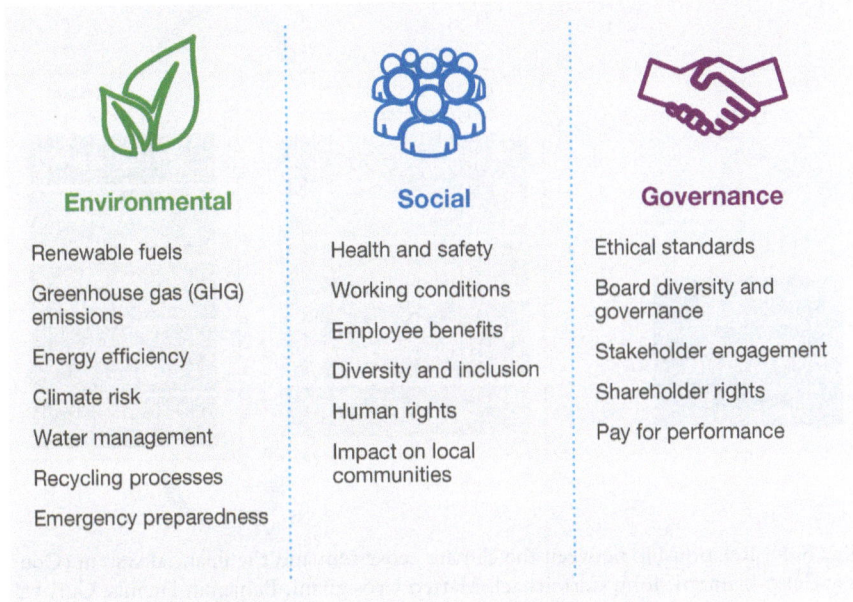

Environmental

Renewable fuels

Greenhouse gas (GHG) emissions

Energy efficiency

Climate risk

Water management

Recycling processes

Emergency preparedness

Social

Health and safety

Working conditions

Employee benefits

Diversity and inclusion

Human rights

Impact on local communities

Governance

Ethical standards

Board diversity and governance

Stakeholder engagement

Shareholder rights

Pay for performance

Fig. 3.3 Elements of the ESG paradigm (Courtesy of the World Economic Forum)

our disposal of hazardous materials, and the benefits of conservation measures and of the adoption of clean or renewable energy technologies.

The Social Pillar explores a corporation's behavior regarding social issues such as gender diversity, employment equality, worker safety, workplace discrimination, and investment in the education of their workforce. In addition, this ESG pillar recognizes that those who consume products or regulate institutions are increasingly concerned about the values inherent in a corporation's supply chain, human rights, and issues of privacy and equity. If markets increasingly value these social concerns, they are likely to purchase from corporations and affiliate with institutions that share their values.

The Governance Pillar explores how an institution or corporation operates internally in the development of strategies and implementation of policies. These include compensation equity, the ratio of compensation for employees of various ranks across the organization, board and corporate diversity, corporate ethics, internal corruption and the board's stance and vigilance against corruption and fraud, and their transparency internally and with regard to consequences on the rights of shareholders and stakeholder.

While the ESG pillars are more clearly focused on appropriate actions of corporations to enhance their reputation, ethics, resiliency, and the bottom line through the incorporation of various measures, the sustainability challenge that currently receives the most attention within the corporate community is the existential threat of climate change. The variables related to climate

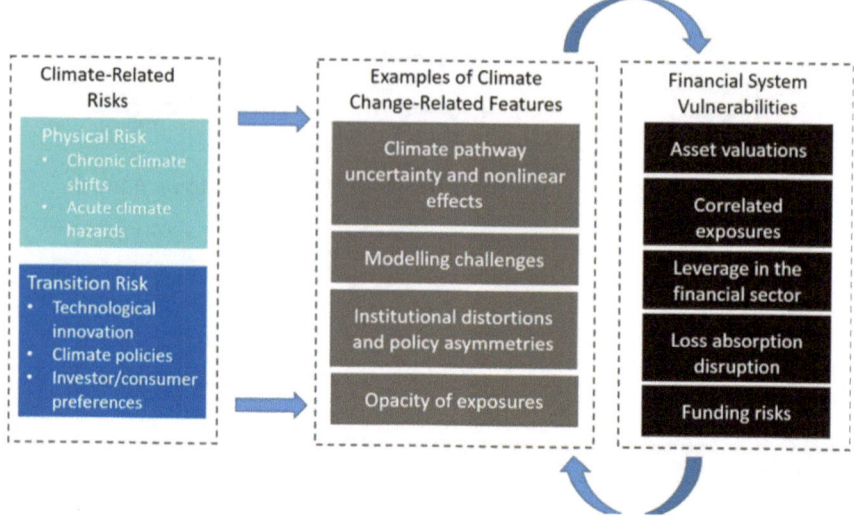

Fig. 3.4 Relationship between the climate ecosystem and the financial system (Courtesy Celso Brunetti, John Carmichael, Matteo Crosignani, Benjamin Dennis, Gurubala Kotta, Don Morgan, Chaehee Shin, and Ilknur Zer [2022], "Climate-related Financial Stability Risks for the United States: Methods and Applications," Federal Reserve Board Finance and Economics Discussion Series, Washington, DC[4])

change often transcend all three pillars: They are often straightforward to measure compared to many others included in the ESG paradigm; The costs and risks to corporations, institutions, and countries are most profound; And, the various dimensions and feedbacks arising from climate change and environmental sustainability are often treated more extensively in the ESG paradigm.

For instance, ESG feedback in climate change embrace aspects shown above (Fig. 3.4):

THE GLOBAL COMPACT

The United Nations developed an analogous paradigm. Their *Global Compact* was designed to apply to corporations and organizations so they may better align business goals with the greater goals of societies globally and locally (Fig. 3.5):

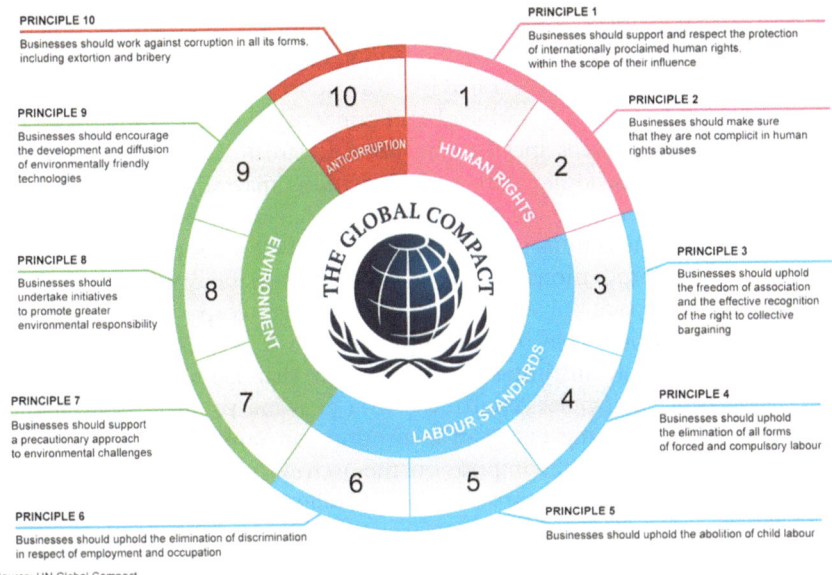

PRINCIPLE 10

Businesses should work against corruption in all its forms, including extortion and bribery

PRINCIPLE 9

Businesses should encourage the development and diffusion of environmentally friendly technologies

PRINCIPLE 8

Businesses should undertake initiatives to promote greater environmental responsibility

PRINCIPLE 7

Businesses should support a precautionary approach to environmental challenges

PRINCIPLE 6

Businesses should uphold the elimination of discrimination in respect of employment and occupation

PRINCIPLE 1

Businesses should support and respect the protection of internationally proclaimed human rights, within the scope of their influence

PRINCIPLE 2

Businesses should make sure that they are not complicit in human rights abuses

PRINCIPLE 3

Businesses should uphold the freedom of association and the effective recognition of the right to collective bargaining

PRINCIPLE 4

Businesses should uphold the elimination of all forms of forced and compulsory labour

PRINCIPLE 5

Businesses should uphold the abolition of child labour

Source: UN Global Compact

Fig. 3.5 The UN global compact

SIMILARITIES AND DIFFERENCES OF THE THREE APPROACHES

The three approaches share much in common. Each is designed to increase awareness about the carrying capacity of the planet, but each also speaks to a slightly different audience.

The Triple Bottom Line is an attempt to incorporate into corporate decision-making the notion of externalities. If a corporation or organization acts in ways that harms others, especially the Earth and future generations, there ought to be accounting for these costs. Likewise, under the triple bottom line, one can enhance assets and build goodwill that contributes to society.

The TBL approach depends on the ability to attach prices to our activities. It is hence rooted in the neoclassical economic notion of free markets and supply and demand, and may then, if not invoked carefully, produce a bias toward the measurable, and may even cause mispricing through omission of factors difficult to measure.

Likewise, the Brundtland Commission's Three Pillars enshrine the notion of the protection of the Earth's carrying capacity, and explicitly states that due consideration must be granted to diverse peoples today and to future generations. It is designed to promote *economic justice* by permitting and even advocating for economic growth. This approach enshrines a notion that continuous improvement in technologies and efficiency can induce greater demand for Earth's natural capital. As such, the Commission positions itself closer to the definition of weak sustainability, and offers little assurance that some of the worst artifacts of global warming shall be reversed.

Finally, ESG is primary a collective conscience for corporations and the stakeholders they serve. The principles of concern over environmental, social, and governance practices include the need to create strategies that reflect sustainability and economic justice. Corporations may adopt such principles from two parallel perspectives. It acknowledges that challenges to sustainability often begin with the actions of corporations and the functions of markets, and explicitly enlists corporations to be part of the sustainability solution.

The ESG approach incorporates the value that corporations which fail to acknowledge their role, or any role in such issues as climate change, that do not promote institutional diversity, or that fail to embrace demo-cratic governance principles, contribute to a growing problem for which we must all reckon. In some cases, an institution can remedy their actions and hence reduce the risks they impose on themselves and society by instituting protective practices that minimize risk and liability.

Institutions can then collectively recognize that, should they all move in a more sustainable direction, their institution, their industry, and the societies they occupy will become more efficient, sustainable, and perhaps even prof-itable. However, this broader sense of the benefits of collective sustainability suffers from classic free-rider problems described in Chapter 14. When such market failures exist, decision-makers may well depart from the socially optimal solution.

Institutions that adopt ESG principles may also do so to minimize oper-ational, legal, and reputation risk on behalf of a variety of stakeholders. A modern institution rarely views itself as serving only one monolithic stake-holder, their shareholders. In practice, stakeholders are often an economically diverse group. Some stakeholders, suppliers, staff, shareholders, and shoppers subscribe to environmental and societal concerns, while others may not, and still others yet may even harbor hostility to these paradigms.

Likewise, some boards of directors may not wish to lead with a heavy hand given such complexities, and may instead choose not to be particu-larly active in ESG matters. However, the ESG paradigm asserts that a board serves its shareholders and its stakeholders, the suppliers, staff, and its shop-pers and regulators. If these other constituents share concerns about ESG issues, they may implicitly or explicitly exert significant influence on boards of directors and management to incorporate ESG values or take their business elsewhere. In balance, many institutions may believe it is simply good business and reputation-enhancing to adopt ESG practices.

SUMMARY

1972 saw an awakening of interest in sustainability. Early in the year, the Blue Marble picture of the Earth from Apollo 17 demonstrated the isolation and fragility of our planet. Later that year, the Club of Rome published their Limits to Growth to show that our natural capital we had taken for granted cannot last forever, or perhaps even for very long.

Over the ensuing half century, a greater sense of stewardship has evolved, with collective leadership offered by the United Nations' Brundtland Commission and its successors, and with growing recognition that challenges to sustainability must necessarily involve the private sector. The 3Ps, the Triple Bottom Line, and the ESG Pillars were all developed to explore how private institutions and public governance alike must view their decisions from a perspective that acknowledges both the mutual interdependence of humankind and our environment. The goals these various approaches develop overlap significantly. What is less clear is the ultimate and unifying objective. The next chapter defines natural capital as the ultimate objective of these various approaches to sustainability.

ESG Toolkit
Corporations find themselves facing competing goals. They publish policies as part of strategic plans that are typically three to five years in duration but are also steeped in corporate history. Meanwhile, markets, stakeholders, and the Earth are evolving rapidly. How can a strategic plan articulate and incorporate the dynamic environment within which they operate?

NOTES

1. http://www.un-documents.net/wced-ocf.htm, accessed February 19, 2023.
2. Elkington, John. (June 25, 2018). "25 Years Ago I Coined the Phrase 'Triple Bottom Line.' Here's Why It's Time to Rethink It", Harvard Business Review. Retrieved October 11, 2019.
3. Gro Harlem Brundtland, Australia. (1990). Commission For The Future, & World Commission On Environment And Development, Our Common Future, Oxford University Press.
4. https://www.federalreserve.gov/econres/feds/files/2022043pap.pdf, accessed January 4, 2023.

CHAPTER 4

The Planet's Balance Sheet and Nature's Bank

The Triple Bottom Line, the 3Ps of People, Planet, and Prosperity, and the ESG paradigm each set a number of goals for organizations and corporations to create greater harmony between humankind and our ecosystem. However, these paradigms do not articulate a single overarching and unifying principle to which they all strive to enhance. The next challenge is the integration of these various values into criteria within the vocabulary of economics and finance.

We can frame the Triple Bottom Line, the Global Compact, the Three Pillars, and ESG with regard to capacities, or stocks of natural capital, optimized to satisfy the needs of the broadest definition of ecosystems. We may then manage humankind's metaphorical house using an inclusive approach, with the environment and society enhanced by our interactions and organizations. Humankind optimizes the benefits that arise from a broad definition of natural capital that encompasses the health of our ecosystem in perpetuity.

One could consider the environment, society and communities, and corporate finances as three somewhat intersecting sets of stakeholders. The missing stakeholders are the future generations not yet even born. We can ensure their viability through preservation of all resources for all humankind, now and in the future, but this strong form of sustainability precludes progress and economic development.

To enhance the triple bottom line framework, let us explore the term *capital* in more depth than most people contemplate. To economists, capital, either in the form of physical capital, as a means to produce, or the financial capital, as the ability to purchase, both provide the capacity for humans to satisfy our wants and needs. Nature is inherently utilitarian, with each component occupying a space, storing or transforming energy, and performing a

C. Read, *Understanding Sustainability Principles and ESG Policies*, https://doi.org/10.1007/978-3-031-34483-1_4

function. Our *natural capital* is an essential capacity that contributes to our collective ability to perform human functions within the ecosystem.

If, then, we include one additional element of this ecosystem, nature's ability to maintain its capacity over time, then we have a meaningful definition of sustainability. The preservation of natural capital is a measure of *strong sustainability*.

We can also define *weak sustainability* as the preservation of the combination of our various forms of capital in the aggregate, be they natural, financial, human, or human-made physical capital. Framing sustainability based on the requirement that each capacity contributes to the preservation of our collective capacities, rather than preserves the quantity of each type of capital, affords economists greater versatility by providing for the substitutability of capital. We may be able to preserve a function and satisfy a need without rigid preservation, but instead with the sensible exploration of different ways to accomplish the same goal.

Economists are well-versed in the contemplation of processes as measured by their outcomes and outputs, with the opportunity to substitute inputs as necessary. If one input becomes scarcer, while an alternative can still be employed to augment or substitute for its function, we can retain our aggregate capacity.

Economists speak of our collective capacity, enabled in this case by the combination of natural, social, and financial capital, as a *technology*. Let us label how human capital L, human-made physical capital K, the extraction X of non-renewable natural capital S, and the harvesting H of renewable resources R can combine to produce the goods and services we value.

This technology $\Upsilon = f(L, K, X, H)$ transforms various forms of human-made capital L and K and extraction X and harvest H of natural capital into a quantity Υ of something of value. Such production functions are the basis for much of economic analysis. Within this function, we allow for the substitution of the use of one form of capital for another as we maintain the overall level of the function. If we can be confident that the scarcity of one component of these various forms of capital does not hinder the entire ecosystem in the long run, our collective capacities are maintained and perhaps even enhanced. In addition, we can improve on the technology that combines these forms of capital.

This approach induces us to reconsider the preservation versus conservation debate over strong and weak sustainability. By allowing our various forms of capital to substitute for each other, we can maintain our collective capacity even if we diminish our natural capacity. Of course, this implicit assumption requires a rigorous and honest assessment of the employment of natural and human-made technologies and capacities. It also challenges us to manage the risk of uncertainties, errors, and randomness that arise when we do not know what we may not know. These subtleties are often glossed over through the adoption of traditional economics of an abstract production function $f(L, K, X, H)$ economists call a technology.

Under this rubric of environmental, social, and financial capacities, our goal becomes one of maintaining collective capacities. The triple bottom line is a caution not to consider just the collection of capacities in the aggregate, but instead challenges us to describe the ways in which one capacity interacts with and substitutes for another. With measurement comes a mission, the responsibility to address how measures help maintain the aggregate function within a systems approach. The three pillars then coexist in their shared goal of the expansion of our aggregate capacity.

Consider an example of the advantages of such an expansion of paradigms. The Three Pillar paradigm includes stewardship and the preservation of natural capital by incorporating into the Environmental Pillar the need to ensure that earlier generations do not consume for themselves a natural capital stock that future generations also have a right so they may attain similar economic and social aspirations. Knowing that to impose preservation of environmental and natural capacities would place a politically unpopular burden on the current generation, the Brundtland Commission instead asserted the need for stewardship of our collective capacities and better conservation rather than preservation.

Using this capital stock and flows approach, we see the fundamental challenges and potential inconsistencies that this paradigm imposes. Continuing with the analogy of capital and employing the metaphor of a **Natural Bank** to represent our stock of natural resources, More Developed Countries (MDCs) have been drawing down the bank of natural capital for generations. While a few MDCs now recognize they may be asked to replenish some of that missing stock of natural capital, Less Developed Countries (LDCs) look over the shoulders of their MDC counterparts and assert that they too should have the right to deplete natural capital stock for their own growth as the MDCs have done in their economic development past.

The late author and academic Herman Daly was most fervent in his assertion that economics must move beyond the notion of growth for growth's sake and instead consider as paramount the carrying capacities of the Earth. In his groundbreaking 1996 book Beyond Growth,[1] he was critical of aspects of the neoclassical model that is the foundation of traditional economics and is extended inappropriately to resource usage.

Daly was an economist well-versed in the neoclassical model. He had been taught the model of traditional economics, which dates to the nineteenth and early twentieth century, that states all forms of capital are inherently and consistently substitutable, even if each form of capital exhibits *diminishing marginal returns* if relied upon too intensively.

The concept of diminishing marginal returns was described by David Ricardo, a member of the British Parliament and a gentleman farmer from the early nineteenth century. His inquisitive mind developed economic tools to persuade his parliamentary colleagues about important economic and public policy issues of his day. Ricardo's intellectual curiosity in a rapidly industrializing England turned to the exploration of a newfound factor of production, called *physical capital*. These human-made machines revolutionized the Industrial Revolution by making labor more productive.

Ricardo understood that production could be maintained by combining an optimal amount of physical capital, labor, and land. Later in the nineteenth century, the eminent American economist John Bates Clark formalized this relationship by explicitly describing mathematical conditions for the optimal combination of two forms of capital, the physical capital of machines and the human capital of labor, to produce the items humans consume.

In Ricardo's day, land served a unique function in that it was an essential factor of agricultural production that represented the vast bulk of production in his day. Ricardo argued that the productivity of land could vary, but innate qualities of the Earth demonstrate one could not eliminate the need for essential natural capital by simply employing more of the other two reproducible factors of production.

In other words, while there is substitutability of labor and physical capital, production always required some land to produce agricultural goods. Economists speak of this unique natural capital as complementary to the other two factors rather than as a substitute for human-made capital. The need for land even increases as more physical and human capital is employed to produce agricultural output. The optimal combination of human-made capital is determined by the relative costs of factors of production and by the value of the goods produced. Meanwhile, the scarcity of natural capital such as land usurps any remaining profit once other facts are paid their worth.

Another essential difference with natural capital is its *renewability*. We can procreate more human capital through reproduction and humans can always build more machines. The payment we make to these factors are incentives to produce more of each of them. However, Earth capital is often fixed in supply. Some types of natural capital can forever remain productive on a sustainable basis without diminishment, some can be sustainable if not mismanaged, and some Earth capital, such as minerals or fossil fuels are in fixed supply that will eventually run out if extraction continues. Natural capital does not respond by increasing its supply with increased demand and price as do the human-made factors of production.

Daly argued that modern economic theory has forgotten the uniqueness of natural capital. This flaw in the neoclassical model arises in two ways.

First, the *production function* economists developed is in some sense a black box that takes human capital L, physical capital K, the extraction X of fixed natural capital S, and the harvest H of renewable capital R and combines them to produce output Y according to $Y = f(L, K, X, H)$. In describing capital so symmetrically and opaquely within a production function, neoclassical economics presents these forms of human-made and natural capital as essentially mutually fungible, with none exhibiting any unique and essential role that cannot be substituted with more of the other forms of capital.

In other words, all forms of capital are implicitly assumed to be infinitely substitutable.

Yet, even Ricardo knew that, while we can substitute away from human capital by employing more automation and machines, and we do not need machines if we have enough humans to toil away, we cannot produce much without a modicum of natural capital. However, by expressing the traditional neoclassical production function of economies so plainly as the production function suggests, we quickly lose touch of the unique aspect of natural capital, especially our inability to reproduce it like we can so easily with the human-made capital.

The second aspect in the sustainability of natural capital is the way it must be managed to preserve its capacity to produce, or at least conserve the stock necessary to ensure future generations are not deprived of its use. This is an essential element of environmental economics to be covered in later chapters.

Herman Daly challenged the notion of continuous growth for growth's sake and the assumption of natural capital substitutability. As the Brundtland Commission had noted, humankind has an expectation that each subsequent generation should be assured at least the prosperity of the last generation. This affluence is typically measured on the monetary value of the goods and services we produce, rather than the ability of consumption, and of all else we enjoy, to generate happiness.

Yet, while an adage says that money does not buy happiness, to many economists, money at least has the advantage of measurability. This measurability of the value of production and consumption, embodied in our Gross Domestic Product calculation, then creates a disproportional obsession with economic growth of income rather than our enhancement of capacities over time.

Such GDP growth by definition requires an increase in output Υ. But while the economy can reproduce human and physical capital to aid in the enhancement of GDP, given their complementarity of natural capital, this growth inevitably necessitates extracting even more natural capital to combine with increased employment of the human and physical capital to produce an ever-larger GDP. Our natural human tendency is to increase rather than reduce our demands on the ability of natural capital to provide for us and fuel growth, despite the limitations described in <u>Limits to Growth</u> discussed earlier. Sustained growth then translates into unsustainable natural capital depletion. In maintaining such an emphasis on growth for growth's sake, no longer does our species live in as steady state balance with its environment.

Over the course of the Industrial Revolution, humankind has learned to dictate our environment. While our species is relatively small, at a population that recently surpassed eight billion people, humankind already dominates a sustainable Earth by diverting to our use approaching 40% of the total carbohydrate production of plants and animals. Humans and our ecosystem also now represent more than 90% of the biomass of all mammals on the planet.[2] Humankind is unique in that no other species consumes the planet's natural resources in such an unsustainable way. Humankind has cleverly determined how to thrive in this imbalance by continually expanding our ability to extract natural resources and by developing through science and engineering increasingly clever ways to use our resources to expand output still further.

However, these efficiency improvements are ephemeral. We have seen humankind continues its march of progress in our affluence, but these great strides in production since the onset of the Industrial Revolution are no guarantee of increasingly significant strides in the future. Empirical observations should not translate into blind faith that this march of technology will continue forever unabated. And, by focusing on humankind's income statement without regard for diminishing natural capital on nature's balance sheet, we may be satisfying our insatiable appetite by eating the seed that should have sowed prosperity for generations to come.

Indeed, Ricardo described for us the concept of diminishing marginal returns. Movement from a low technology to a high technology world certainly improves the bounty for those fortunate enough to benefit from these innovations. But, diminishing returns state that we cannot reasonably expect such progress to grow continuously and undiminishingly. Eventually, even progress must taper off.

Humankind regularly experiences this phenomenon of diminishing marginal returns. A Less Developed Country can expect spectacular growth of 5–10% as it climbs toward a more industrialized state. A More Developed Country half a century ago had come to expect growth of 3% perpetually. Now, MDCs are finding that growth of around 2% is difficult to sustain. Even that modest growth requires the population to grow, and physical capital to increase in about that same proportion (Figs. 4.1 and 4.2).

This tapering of growth is simply a product of diminishing returns. If we fail to realize that both the Earth and our economic processes have limits on their capacity to produce, and certainly cannot increase their capacity indefinitely, then we run up against the wall that Daly and the Limits to Growth authors so eloquently described.

The challenges are significant. To now, we have regarded some LDC economies as labor-rich and physical capital-poor. While labor is difficult to move across domestic borders, physical capital is both very mobile and increasingly abundant, even as natural capital diminishes. In a world with abundant physical capital and often surplus labor as well, especially with increased automation that obsoletes human labor and expands output still further, natural capital is becoming even more scarce, both in absolute and relative terms. This creates a growing challenge and threat to sustainability.

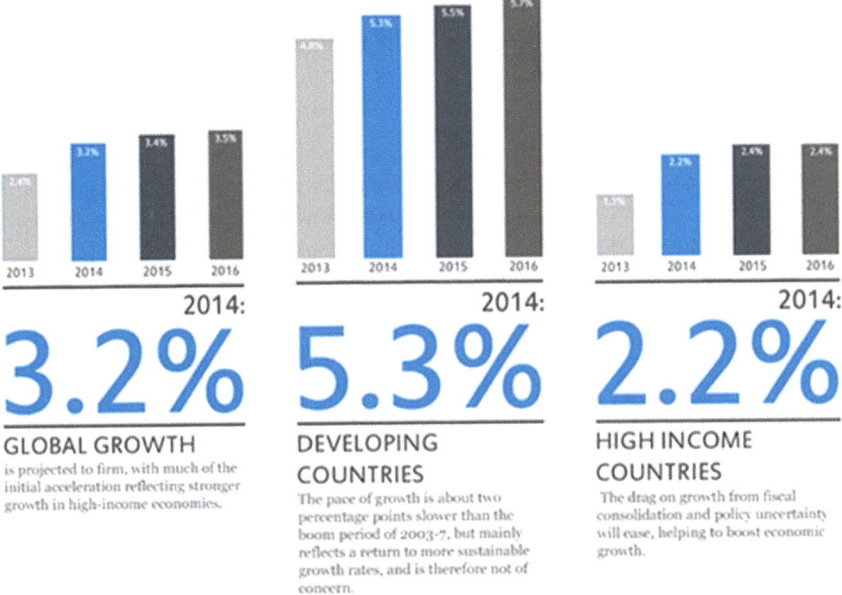

Fig. 4.1 Relative growth of less developed countries (LDCs) and more developed countries (MDCs) (Courtesy of the World Bank, Derek Chen [2014], www.blogs.worldbank.org, accessed January 3, 2023)

More Developed Countries (MDCs)	Less Developed Counties (LDCs)
• High per capita income and GDP	• Low per capita income and GDP
• High literacy and life expectancy	• Low literacy and life expectancy
• High level of urbanization	• High level of rural life
• Well-developed infrastructure	• Less developed infrastructure
• Low birth and death rates, low population growth	• High birth and death rates, rapid population growth
• High Human Development Index (HDI)	• Low Human Development Index
• Low rate of infant mortality	• High rate of infant mortality

Fig. 4.2 Qualities of MDCs and LDCs

Summary

Traditional economics has come to focus on the flow of income as we extract natural capital as the measure of humankind's well-being. But *ecological economists* are becoming increasingly aware of the folly of such a measure as we see diminishing our stock of natural capital. We will return to this theme in Section IIII when we construct economic models that are able to better explore the role of sustainability in the health of humankind and our ecosystem. First, though, we explore the scientific bounds of nature's bounty.

ESG Toolbox

Production and growth typically involve changes to the quantity and mix of factors used. These factors can include Earth resources R, human resources L, and human-made capital K.

To what degree does a corporation describe these resource mixes and how does the quantity and relative proportion of these resources evolve over time?

If corporate growth requires additional Earth resources, can this footprint be mitigated and can the Earth continue to provide these resources, especially as other corporations also expand their resource intensity?

Do corporate boards discuss this Earth resource ethic, especially when it involves unpriced or underpriced resources?

To what degree does your institution improve its income statement by depleting its balance sheet?

Notes

1. Daly, H. E. (1996). Beyond Growth: The Economics of Sustainable Development, Beacon Press.
2. https://wis-wander.weizmann.ac.il/environment/weight-responsibility-bio mass-livestock-dwarfs-wild-mammals, accessed March 22, 2023.

PART II

The Science of Sustainability

There are few topics in our present social discourse that produce so much misunderstanding, and such a mix of overoptimism, excessive pessimism, or ambivalence as does the role of nature and its sustainability in our collective future. While all of humankind has a stake in the future, our individual life experiences define our bounds of awareness, collective behavior, and sense of social responsibility. The debate over appropriately sustainable practices often then becomes a normative one that permits each of us to harbor our individual opinions, prejudices, and preferences. We then often resort to solutions or policy prescriptions that conform to our personal sensibilities.

Science does not function in such a normative realm. Instead, the scientific method relies on the development of a hypothesis, a theory to be tested, and then the creation of experiments that can either refute the hypothesis or permit the theory to remain viable, only to be again tested on another day and by other researchers. John Stuart Mill, in his On Liberty,[1] noted in 1859 that "The beliefs which we have most warrant for, have no safeguard to rest on, but a standing invitation to the whole world to prove them unfounded."

Science works best when subsequent attempts to disprove the theory are made by those who are most hostile to the theory's predictions. In this way, science focuses on verifiable fact rather than opinion. Science is hence, derived from data, rather than the *normative*, driven by opinion. Nonetheless, scientists must sometimes function in the realm of the normative, as we shall soon see.

The scientific method does not imply that theories developed are not biased by the prejudices of the theory's proposer. In fact, a diverse set of beliefs and biases will produce a greater diversity of proposed theories that can subsequently be tested. Such intellectual expansiveness is prized in the scientific method. Diversity of experience and unconventional thought are considered scientific assets.

The real work begins in the proof of theories. Some theories appear so robust as they have been verified time and time again to the point that their status is raised by some to a scientific law. For instance, school children are taught the law of gravity, often with the metaphor of an apple falling from a tree onto Isaac Newton's head. His subsequent development of a theory of gravity lasted for almost three centuries and was "proven" myriad times.

The scientific method does not hold a proof with the robustness laypeople believe. To a scientist, a theory's proof only means it has yet to be refuted by some experiment. If so, it is replaced by a new and improved theory. Theories that are proven and not refuted time and time again lend greater confidence in their accurate depiction of physical phenomena.

For instance, Newton's theory of gravity was certainly one of the most elegant and robust theories ever developed. Because our human, planet-bound realm never experiences speeds approaching the speed of light, our world never realized the limitations of Newton's "law of gravity" and his other laws of mechanics. They were approximations that were accurate so long as the velocities considered did not near the speed of light. Then, in one fell swoop, Albert Einstein, in a series of papers in 1905, and a subsequent General Theory of Relativity in 1915,[2] completely reoriented our view of the physics, not only of motion but of energy itself.

These distinctions are important in our understanding of sustainability, for at least two reasons. First, positive conclusions as a result of the scientific method, especially those that withstand the test of time and experimentation, are not comparable to normative opinions subject to public discourse, debate, and opinion. In addition, two alternative theories of a scientific process that may inform our notions of sustainability are not equivalent if one has substantially more instances of verification and proof than the other.

Certainly, alternative theories, such as Einstein's Special Theory of Relativity,[3] met with skeptics as first. This is an essential and natural part of the scientific process. However, over a period of months and years of understanding and experimental validation, his theory gained traction and replaced a theory as ingrained as Newton's laws of motion.

Ordinary public debate rarely works with such clarity. Many of our opinions are formed based on arguments we observe in the public forum. Moderators of public discourse who are not always well-steeped in the scientific method instead use the normative technique of reliance on dueling experts. Of course, this advocacy approach is the foundation of common law jurisprudence used in courts of law and typically in public debates. The technique is designed to influence opinion or reach consensus, but it cannot prove physical phenomena, even if public debate may be persuasive or even entertaining. Debates about sustainability fail to create public consensus despite the fervency of human opinions and do little to further the science of sustainability.

The second important element in our understanding of sustainability is that some of the underlying science is quite complex, and is often beyond the technical ability of many interested observers to understand its subtleties. For this reason, the public, sometimes even public policy experts, may not fully grasp that their opinions may not conform to prevailing scientific thought. For one to participate effectively in discussions of sustainability requires some knowledge of underlying physical phenomena, and investment in understanding the implications of these theories.

Many of the errors made in the analyses of issues relating to sustainability arise because of incomplete understanding of the science involved. Some errors are propagated intentionally to sway interest in a product, attract investment funding, deflect criticism of a particular industry or activity, or protect profitable status quos. There are few opportunities to correct these misconceptions without appeal to far greater scientific rigor than many members of the public are prepared to commit or comprehend. For these reasons, a basic understanding of the science of sustainability is essential for sound public discourse and policy.

The attainment of sustainability presents two potential challenges. The first is whether humankind can live sustainably within its environment from the present into the future. The second is whether we can also repair the damage and reverse the sustainability sacrificed during the Industrial Revolution.

In some critical ways, we cannot go back entirely. We can never remedy the extinction of species that has accelerated greatly since the onset of the Industrial Revolution and especially in the last half century. Once a species is lost, it cannot be reconstituted, at least without imposing ethical and ecosystem concerns.

We also know that the Earth has reached and even tripped a number of tipping points, which challenges the reversibility of natural capital depletion, as described in Section IV.

However, whether we remedy past environmental indiscretions or move forward in the best way humankind can muster, the primary scientific and technical challenge will be to convert current energy usage to sustainable sources. We must also generate sufficient new energy resources to reverse nature's feedback loops that give rise to tipping points that accelerate environmental degradation and challenge sustainability.

In other words, sustainability will necessarily require abundant sustainable energy. Yet, we live in a closed system. The fate of our Earth and the solar system is fixed. The resources we have available to us on Earth were essentially frozen within our solar system about 4.5 billion years ago. There will be almost no addition to the resources available on Earth within the remaining life of our solar system.

Fortunately, our closed system can provide humanity with the heat and light energy provided by the Sun on a relatively sustainable basis. That solar resource too will sunset once the Sun moves into its next stage of life and death, but not for another five billion years, and no doubt well beyond the time horizon of humankind.

This is because the Sun currently has a stock of about five billion years of remaining hydrogen to consume. As it nears the end of hydrogen consumption, the Sun's radiative force declines, which causes our star to collapse on its core and release its shell of outer plasma as a huge ball of fire. Our Sun will then form a red giant for another billion years. Meanwhile, helium in the dense core will fuse into carbon and oxygen. These fusion reactions do not radiate as much energy as does hydrogen fusion, so gravity will take over to eventually shrink the Sun into a white dwarf. Any remaining matter outside the core is shed off as our star becomes a nebula.

This entire process will take another eight billion years, with more than half that period maintaining a Sun about as we know it, radiating heat and light to the planets. Given the life cycle of all species that have come before us, if humankind follows the same fate, it is highly unlikely that our species will adapt and survive for as long as the Sun maintains its current form. From that perspective, sustainability can assume that the Earth will continue to have an influx of abundant but finite energy for the indefinite life of humankind, even if we may not live forever. The question then becomes one of how humanity can harness the power of the Sun to create a sustainable energy future without damaging the ecosystem for centuries to come.

Humans use an estimated 25,000 terawatt-hours (TWh) of electricity every year. About half of this amount is used by industry, and the other half for residential, commercial, and public services. The generational question is whether we can meet these energy needs in a sustainable manner. Should humankind make a wholesale conversion toward the employment of sustainable electricity, and use new technologies to heat homes rather than rely on natural gas, wood, fuel oil, or coal, and should transportation be provided primarily through electric vehicles, this annual sustainable electricity demand may well double to 50,000 TWh. The next section explores the science of sustainability and determines whether the mammoth energy needs of humanity can be met in a sustainable way.

NOTES

1. Mill, John Stuart (1859). On Liberty, London: John W. Parker & Son.
2. See for instance Einstein, A., & Lawson, R. W. (1916). Relativity: the special and the general theory. Ancient Wisdom Publication.
3. Einstein, Albert (1905) "Zur Elektrodynamik bewegter Körper", Annalen der Physik 17: 891.

CHAPTER 5

Cosmology, Entropy, and Sustainability

In our exploration of sustainability, we must first understand the nature of our natural capital; what created it; why is it in finite supply; and what we can do to create more of it as we consume what we have. These queries define the unique scarcity of the Earth's capital.

Scientists have developed a widely accepted model that shows the universe did not exist until about 14 billion years ago. Before the *Big Bang*, only energy existed. Then, an event occurred that converted some of this energy into mass, in the form of the simplest of all elements, the hydrogen atom, made up of one electron and one proton, held in attraction by their opposite charges but kept from collapsing because of the momentum of the electron as it spins around the proton. While the process is far more complex than this simple presentation, we have sufficient science to proceed. The most obvious question that arises in this magical transformation is whether energy can coalesce into matter and whether matter can be converted back to energy.

We feel the radiant energy of the Sun warming our face or we employ heat generated from electricity to warm a bagel in a toaster. We sense a jolt of energy if we receive an electrical shock. We know that energy from the heat generated in our internal combustion engines is converted into mechanical energy that propels our car, and its momentum and kinetic energy are dissipated as heat when we use the brakes to slow our car down.

We also experience mass. You feel the force of your body pushing back at you from the soles of your feet. You sense the centripetal force that makes you feel weightless at the apex of a rollercoaster ride and the centrifugal force that magnifies your weight at its nadir. This property of mass and force is a product of the attraction of masses first described by but not fully explained in Newton's laws.

C. Read, *Understanding Sustainability Principles and ESG Policies*, https://doi.org/10.1007/978-3-031-34483-1_5

Newton explained that two masses create an attraction. Our experience of weight is the sense of holding the Earth's gravitational attraction in equilibrium against our mass. In the case of our body weight, this attraction is proportional to the mass of the Earth and the mass of our body, and inversely proportional to the square of the distance between us and the center of the Earth. Our same body mass on another less massive planet will produce a smaller sense of weight. While weight is not mass, it is caused by mass, as Newton proffered in 1666 and Einstein eloquently corrected in 1905.

You sense a different aspect of mass if you try to quickly move horizontally a heavy weight suspended in front of you at arms-length. This massive object seems to want to resist sideways movement. You must exert energy to move it from side to side. Newton provided a description of momentum in his third law, which states a body that is at rest tends to stay at rest, while a body in motion tends to stay in motion unless a force is applied. This is what scientists label the conservation of momentum. It is these forces acting over distance and time and effecting change in a mass' position that is a form of energy.

Einstein took the Newtonian concept much farther in a way that helps us better understand the Big Bang. He spoke not of Newton's conservation of momentum, but of conservation of mass and energy. His theory of relativity showed that mass and energy are equivalent, with the conversion factor between the two physics entities described by the now-familiar $E = mc^2$ equation.

In fact, when scientists measure the mass of very small particles, for instance the mass of the electron or the elementary particles that make up an atom's nucleus, they find it more convenient to denote such minute masses as the equivalent amount of energy as Einstein's equation dictates. This energy is typically measured in electron-volts. Particle and energy physicists are accustomed to the mass and energy equivalency principle.

It is this equivalence of mass and energy that permitted abundant energy to precipitate into protons and electrons in the Big Bang that eventually created the plethora of elements we see in our universe today. Mass-energy equivalency also creates the potential to exploit the same phenomenon following the Big Bang to perhaps someday provide almost limitless power to fuel economic development and help sustain and repair our ecosystem.

In the Big Bang, energy was converted to create the simplest of all atoms, with two components, the elementary particle called an electron and the quarks that make up a proton, all contained in an atom of hydrogen. This sudden creation of a highly concentrated mass of hydrogen generated such a gravitational force that these hydrogen atoms collapsed upon themselves and begin to fuse into helium, the next atom on the periodic chart. Because one helium atom weighs less than two hydrogen atoms, the difference in mass following this fusion is given off as heat energy, which raised the primordial temperature as the mass of hydrogen and helium continued to fuse in a fiery ball.

This process of concentration of hydrogen through gravitational attraction, and the fusion of hydrogen into helium is precisely what still energizes our Sun today and provides the fuel for our solar system. We now know this process is spontaneous and self-sustaining, in the case of the Sun, if the temperature is 4 million degrees Celsius and with sufficient pressure as found in the core of the Sun or stars of sufficient mass.

The process does not stop there. In its infancy, hydrogen was abundant across the expanse of our universe. Even today, hydrogen still represents about 75% of our universe, with helium constituting almost all the rest of the atoms in the cosmos. Under the right conditions, helium can further fuse into carbon, nitrogen, oxygen, fluorine, neon, sodium, magnesium, aluminum, silicon, phosphate, and sulfur, to name the significant elements that we see around us. Cosmologists define various types of stars by their levels of maturity in their fusion cycle, as described by the stage of fusion that they primarily fuel. The remaining elements that are so familiar to us are in comparatively small proportions.

Generally, these heavier elements are rarer because of the long process to get from hydrogen through subsequently more massive fusions to fuse into all the elements around us. The early universe also created lithium, beryllium, and boron. Fusion can occur because of the energetic collision of all sorts of particles, not just the proton and electron of hydrogen. It turns out that lithium, for instance, can be created, but tends to blow itself apart at relatively low temperatures, compared to other elements in the early cosmic soup.

The early universe quickly consumed most of its lithium, beryllium, and boron, leaving but a trace of these elements compared to hydrogen and helium. Therefore, while sequences of increasingly massive elements are created over time, the lithium, beryllium, and boron on the Earth today formed in the earliest stage of the Big Bang.

Hydrogen fusion into helium can last billions of years. Gravity is concentrated in increasingly massive and more dense hydrogen–helium stars, balanced by the outward waves of energy and particle radiation. Eventually, once a star's hydrogen is almost consumed, gravity exceeds radiation energy and the star collapses into a tiny fraction of its original size. This compression induces another wave of extreme pressure and heat and can, under the right conditions, further fuse the helium into carbon.

For stars the size of our Sun, that is the end of the road, with only rare random fusions forming heavier elements. The outer layer of the star is stripped off into a gaseous nebula of relatively light elements, while the core shrinks to form a white dwarf star.

Such an element factory can continue for stars much more massive than our Sun, though. Their greater gravitational mass can fuse carbon into oxygen, oxygen into neon, etc., all through the common elements described below, and even into iron, cobalt, and nickel. When these processes are completed, massive stars explode into a super nova and expel their heavier elements outward into the universe, some of which find their way into the cloud of

Fig. 5.1 Elements of
the Earth

Atomic Number	Element	Share	Share Excluding Hydrogen and Helium
1	Hydrogen	73.9%	
2	Helium	24.0%	
8	Oxygen	1.0%	49.5%
6	Carbon	0.5%	21.9%
10	Neon	0.1%	6.4%
26	Iron	0.1%	5.2%
7	Nitrogen	0.1%	4.6%
14	Silicon	0.1%	3.1%
12	Magnesium	0.1%	2.8%
16	Sulfur	0.0%	2.1%
	All the Rest	0.1%	4.5%
	Total	100.0%	100.0%

dust and matter that eventually coalesced into our solar system's Sun, planets, asteroids, comets, and meteors.

Even rarer are violent collisions of white dwarfs and extremely dense burntout neutron stars that produce heavier elements yet, but in low amounts. These too are eventually expelled into the universe (Fig. 5.1).

This process of manufacturing the elements in the periodic table can continue until all hydrogen is consumed. But, since the universe is still 74% hydrogen, we will continue to see swirling gases condense into stars for tens and hundreds of billions of years to come. This constant birth and rebirth of stars can continue almost indefinitely. Our universe is still quite young and evolving.

The significant message is that the elements all around us have come from everywhere. They have been stripped off of exploding stars, carried to new solar systems by asteroids and cosmic clouds of dust, and condensed from cosmic gases orbiting a star such as ours. Once they coalesce into planets, these elements exist locally in fixed and an almost immutable supply.

While we can fuse minute amounts of hydrogen into helium through human-made fusion on Earth, and maybe into heavier elements someday, we can also go in the other direction. Atoms can be split from their heaviest and very rare radioactive form to produce lighter elements. This process of fission does not significantly modify the stock of the elements on Earth that the ecosystem and humans require. Because of the process that brought to the Earth our natural elemental resources, the elements on Earth are fixed in supply. They can be converted between elemental and molecular forms, from water to hydrogen and oxygen, or hydrogen and oxygen back to water, for instance, but they can only be transmuted from their elemental form through nuclear reactions, the dream of alchemists aside.

Entropy

Since the Big Bang, the universe moved from pure energy into a highly distributed and random state of matter, with clouds of hydrogen gases broadly dispersing over tens of billions of light-years (each representing 9.46 trillion kilometers), and also coalescing into more compact and organized states of matter. The process of reorganizing matter into more orderly states invariably requires energy to facilitate such cosmic reshuffling.

Nature and humans use energy to make more highly organized molecules in the world around us. Plants use photosynthesis to organize carbon dioxide into carbohydrates, with sunlight providing the energy for their molecular architecture. Other natural processes use chemical energy to convert simple amino acids into proteins and the deoxyribonucleic acid (DNA) and ribonucleic acid (RNA) that encode the next generation of life. The ecosystem and the environment around us are the product of nature's organization, mostly fueled by the energy impinging upon us from the Sun.

Humans too use energy to organize our environment. We take minerals, hydrocarbons, wood, water, and other natural resources to build our homes and factories and schools and roads. These conversions require human and physical capital, our natural capital, and energy to create the ordered world we enjoy around us.

But this world we build also requires energy to maintain our grand designs. If we fail to maintain our constructs and commit to the energy that such maintenance requires, our human-made artifacts disintegrate and then disappear. Consider the effort we make to build a sand castle. Even in the absence of wind and water, the sand castle will eventually succumb to the forces of gravity and to miniscule movements until it disappears into a pile of sand over time.

On the other hand, you will never see a pile of sand self-organize into a sand castle. In the absence of energy, this natural process of eventual movement from a higher degree of organization toward a state of increased randomness, or decreased organization, is the product of the unidirectional arrow of *entropy*.

Entropy is a concept intimately related to energy. As described by the **Second Law of Thermodynamics**, the natural world tends over time toward a thermodynamic equilibrium that corresponds to a maximum state of randomness and disorder. To imagine the concept, consider a gas contained in a bottle which, perhaps due to energy imposed on it, has organized all the molecules in clumps confined to one side of the vessel. From this initial state, the system will spontaneously spread out into the space available to end up in a more disordered and distributed state over time. This process will not spontaneously reorganize into clumped states of the gas without employing energy to recreate order.

Entropy then is a statistical concept that measures the microscopic tendency toward uniform disorder. The word *entropy* itself has as its root a Greek word meaning transformation. Our human efforts and the energy of the natural world lessen entropy and increase order only by using energy to build larger molecules from small ones, or ordered structures from random building materials. Meanwhile, the unidirectional arrow of entropy attempts to disassemble the complex world humans and nature construct.

Nature is in balance with entropy. Natural decay is a process that is reversed and recreated into new compounds and organisms of increased complexity only by tapping into the energy of the Sun or the geothermal energy emanating from the molten core of the Earth. Should the energy impinging on the Earth change, so too will the entropy balance and hence the equilibrium of the world around us. To simply maintain the equilibrium we enjoy which requires a constant inflow of energy.

The sustainability lesson from this basic science then is that we live in an equilibrium maintained by the bounty of the energy we receive from the Sun. Without it, we cannot create new elements that we rely upon to construct order in the world around us. We also need some energy to simply maintain our world in the balance we currently enjoy. The building blocks around us are ultimately in fixed supply, but we can, with sufficient sustainable energy, tailor them for our use, just as plants create order when they photosynthesize oxygen and carbohydrates from carbon dioxide, water, and the Sun's energy.

Chemical Reactions

The primary ability of nature and human ingenuity is in the conversion of simpler molecules into more complex and useful forms. Invariably, especially when we consider the unidirectionality of entropy, this process is intrinsically energy intensive.

There are some chemical reactions that are not reversible at the macroscopic level. For instance, one can make a souffle by combining eggs, butter, flour, salt, pepper, milk, and cream of tartar, but one cannot reconstruct eggs by deconstructing the souffle, at least without feeding the souffle to a hen.

At the microscopic level, chemical reactions constantly shuttle molecules to establish and maintain an equilibrium. The *principle of microscopic reversibility* views a chemical reaction as a steady state modification of its equilibrium. For instance, dissolved in water are some free hydrogen H_2 molecules and ions H^+, free oxygen O_2 molecules and ions O^-, and hydroxide ions OH^-, while the vast proportion is in the form of H_2O, our familiar water molecules. At room temperature, the water equilibrium has about one in one-hundred trillion molecules disassociated into ionic and elemental hydrogen and oxygen, and hydroxide. If the temperature is raised to those that may occur in a hot nuclear fission reaction, the equilibrium is shifted so that about half the superheated water vapor is made up of its elemental components.

This demonstrates the principle of microscopic reversibility. We observe water at room temperature as a homogeneous liquid because it is almost entirely made up of water molecules. More correct is that the equilibrium state of the hydrogen ions H^+, oxygen ions O^{--}, and water molecules H_2O reaction at standard room temperature and pressure strongly favors the molecular water side of its reaction equation.

At the microscopic level, water is constantly decomposed into its constituent elements and recomposed into its compound molecules. The rate of decomposition and recomposition are in balance in equilibrium, with the balance point dependent on temperature, and pressure, while the presence of other elements such as catalysts may help tip the equilibrium in one direction or the other.

Chemical reactions can also be described based on the second law of thermodynamics. Recall the law states that a closed system cannot evolve or react spontaneously in a way that reduces entropy without adding energy. However, adding energy to such a system can reverse a reaction. The concept of *Gibbs Free Energy* intimately relates entropy and energy in a chemical reaction. Specifically, the Gibbs free energy G is given by:

$$\Delta G = \Delta H - T\Delta S.$$

This equation states that, in a closed system, the change in Gibbs free energy ΔG occurs from a change in the amount of heat content, labeled as *enthalpy* ΔH, that can be extracted at a fixed temperature and pressure, net of the product of temperature T, and the change in entropy ΔS. This relationship guides chemical reactions. A system tends to a state of minimum Gibbs free energy once it is in chemical equilibrium. The Gibbs free energy equation gives us both a measure of the amount of energy we can extract as a chemical reaction moves toward increased entropy or disorder, or, likewise, the amount of energy we must provide to reverse the reaction. The enthalpy of a reaction H is then a measure of a reaction's capacity to release energy and depends on both the temperature and the change in order within a closed chemical system.

It is this heat of a reaction that creates most of the forms of energy familiar to us. Molecular constructs of nature and humankind rely on typically reversible reactions. For instance, one common reaction is the combination of two molecules of hydrogen and one molecule of oxygen to create water. We shall return to this important reaction later. It can be described as follows:

1 mole H_2 + 1/2 mole O_2 → 1 mole H_2O + 237 kilojoules of energy,

where a mole is simply a number of convenience, equal to 6.023×10^{23}. It measures the number of atoms or molecules of a given atomic weight that constitute an equal number of grams of molecular mass. It allows one to normalize the enthalpy of a reaction for a given mass of its components. In the case of water, a mole of water produced by oxidizing hydrogen with oxygen produces 237 kilojoules of energy. This reaction that creates water can be reversed to split water into its constituent components by providing 237 kJ of energy for each mole of hydrogen molecules and half mole of oxygen molecules that combine in an oxidizing reaction. These units of energy are described more fully below.

In theory, the energy given off by combining hydrogen and oxygen is not all fully convertible to other forms of energy, such as electricity. As we shall see, capturing even 60–80% of the heat energy to electricity conversion is challenging.

Nor can 237 kJ of energy be fully and efficiently employed to reverse the reaction by electrolyzing water into its constituent atoms. In electrolysis of water, some energy is lost to initiate the reaction; Some energy is lost to ensure the ions in water properly diffuse; Electricity is lost to wire resistance; Bubbles of hydrogen and oxygen at their respective anodes and cathodes hinder the reaction; And, the reaction must overcome the entropy change necessary. Hence, to perform the reverse reaction, one must provide overpotential, in the form of a higher than necessary voltage and energy, to ensure the reaction moves in the desired direction. This necessary overpotential results in less than 100% efficiency and prevents full reversibility because of the loss of energy, dissipation of heat, and creation of byproducts. While total energy is conserved, useful energy is not.

Knowing that usable system energy is lost through the arrow of entropy and that molecular conversions invariably have some efficiency losses, the myriad equilibria we see around us are not entirely spontaneous. Energy is constantly employed to overcome the principles of entropy and the efficiency losses that allow energy to escape from any system under consideration.

However, we are reassured that, should we enjoy sufficiently abundant and available energy, we can preserve the equilibria we value and the ecosystem we cherish. This is promising news. With sufficient energy available, we can lower entropy and increase order. We can also reverse the reactions that consume the resources humankind has extracted that we may otherwise regret from an intergenerational perspective.

This possibility of reversibility should not reassure us at the macroscopic level, though. Recall the mathematical appendix in Chapter 2 in which we demonstrated that even a simple interrelated ecosystem, of perhaps hares and lynx, the balance of nature over time can exhibit gyrations and either exponential growth or decline. Should system oscillations be so violent to cause one or the other component to crash completely into species collapse and extinction, the evolution of this interrelation ceases to exist. Extinction permanently prevents macroscopic reversibility.

Likewise, we shall see later that an otherwise stable system can pass a tipping point and begin to degenerate, sometimes dramatically, irreversibly, and rapidly. Just as a souffle cannot be reversed into its constituent eggs and flour, some assembled reactions cannot disassemble if the equilibrium is pushed too far.

Two lessons can be gleaned from scientific theory. The first is optimistic. Abundant energy can correct several unfortunate human interventions on the ecosystem and on future generations. The second is more pessimistic. Our actions can cause changes that irrevocably shifts the natural equilibrium by pushing species or reactions into extinction at best, or by pushing natural equilibria beyond a tipping point to the point they cannot recover from collapse.

Before we leave our discussion of the science underlying sustainability, let us understand some of the terminologies of energy. Since much of sustainability and remediation of past unsustainable practices comes down to energy and technology, it is important to have a rudimentary knowledge of the nature of energy and of some of its basic vocabulary.

ENERGY VOCABULARY

Energy is simply the ability to perform work. To a physicist, work is force exerted over some distance. Let us spend a few moments discerning between mass, force, work, energy, and power, the physical quantities that are required to understand energy sustainability.

We shall do so while staying within the most convenient and standard scientific measurement system, the metric **MKS units**, which stands for meter, kilogram, and seconds. A couple of generations ago, a British or American system may have been employed more commonly, which used feet, pounds force, the British Thermal Unit (BTU), and seconds or hours. Electrical analyses invariably use the metric system, and energy is increasingly denominated and transmitted in the form of electricity. Fortunately, the MKS measurement system is coherent and convenient, especially in comparison to the more arcane Imperial or American measurement systems.

Because so much of the study of sustainability requires a basic knowledge of the role of energy, let us take a moment to describe the measurement of energy more completely. You may recall Newton's Second Law from high school science classes:

$$Force = Mass * Acceleration.$$

On Earth, the most significant acceleration we feel is from gravity. It exerts a force on us which we experience as weight, defined by the product of our mass and by the accelerating gravitational attraction of the Earth's mass.

You may argue that you are not accelerating by merely standing still on the surface of the Earth. Actually, we are always accelerating, defined as the rate of change of velocity. Through our attachment to the Earth, in a broader frame of reference our speed is constant, at the rotational speed of the Earth, about 1675 kilometers per hour, or about 4700 meters per second. Gravity holds us to Earth, so we do not go flying off the Earth, given our incredible speed of rotation. Gravity is sufficient to keep us firmly footed by constantly accelerating us toward the center of the Earth in excess of the centripetal force as we spin around the center of the Earth.

Physicists use the MKS term newton to denote a force. For instance, if your mass is 50 kilograms (kg), your weight is the equivalent of your mass in kilograms multiplied by the Earth's gravitational acceleration of 9.81 meters/second^2 (m/s^2) to give your weight as the force you feel on your feet of:

$$\text{Weight} = \text{Mass} * \text{Gravitational Acceleration}$$
$$= 50\,\text{kg} * 9.81\,\text{m/s}^2$$
$$= 490.5\,\text{kg} - \text{m/s}^2.$$

This MKS unit of force is called the newton, or Nt, after Sir Isaac Newton.

While you are generating a force equivalent to your weight just by standing still, merely statically resisting gravity does not constitute work or the expending of energy. You are merely maintaining equilibrium between you and the ground upon which you stand.

To see the difference between force, in newtons or pounds force, and work or energy in joules, named in honor of English physicist James Prescott Joule, consider the effort you expend to climb a series of steps to the top of a 100-meter tower. As you rise, you are resisting the force of gravity exerted over the vertical distance you climb, which constitutes work. In climbing the steps, you exert the force of your weight of 490.5 kg-m/s^2 over the 100-meter climb to perform a total of 49,500 Joules of work.

We can also compare energy and power. If you climb to the top of the tower in ten minutes, the equivalent of 600 seconds, you have exerted 49,500 joules over 600 seconds to equate to a power level of 49,500/600 = 82.5 Joules per second. Power is the unit of energy per second. It is measured in the units of joules per second, called a watt, named for James Watt, the inventor of the steam engine.

You must sustain a power of 82.5 watts for each of the 600 seconds it might take to climb the stairs. The value of 82.5 watts is just over a tenth of a horsepower, in the more arcane, but often more familiar British scale. If you weigh 50 kilograms, you must exert more than a tenth of a horsepower for ten minutes to climb to the top of a 100-meter tower. In doing so, you expend 49,500 joules of work and energy, friction aside.

We can also convert this measure of energy dissipated to the units of kilocalories we may consume in our diet. This translates to a depressingly small extra 11.8 kilocalories in the tower-climbing activity, over and above your regular metabolic energy consumption and any necessary energy to cool your body because of my increased rate of exertion. If you are not familiar with the term kilocalorie, it is what dieticians typically call calories—they drop the kilo, but physicists do not.

These examples illustrate a couple of important points. First, force does not translate into work unless it is applied over a distance and time. In climbing the tower, you are converting the kinetic energy of movement into stored potential energy at the top of the tower that can be recovered on the way back down the tower.

To see why mere force does not equate to energy usage, let's say you have been charged to prop up a heavy weight on your shoulders. You may find that to resist the weight is exhausting. Instead, you could also prop the weight on a stand. The weight remains propped up, but obviously no energy must be consumed by the stand to hold it up.

We now have the two most important units for our purposes—the joule as a unit of energy, and the watt as capacity to expend energy each second. We often express the energy unit in kilojoules (kJ) and megajoules (MJ) for thousands and millions of joules, respectively, given the relatively small amount of energy contained in a joule. We also quote power as the energy moved per hour rather than per second, which is typically a more convenient measure for sustained power output or power requirements.

The corresponding unit of a kilowatt-hour (kWh) denotes total energy consumed to operate a device that draws a thousand watts of power for an hour. A small space heater may draw one kilowatt of power and hence use one kWh of energy each hour. Your electricity bill is typically denominated in this kilowatt-hour measure, which would be equivalent to 3600 kilojoules (3.6 megajoules MJ).

Large electrical generation facilities express their power output at each moment in megawatts or even gigawatts, which is a thousand megawatts or a million kilowatts, or terawatts, a million megawatts. The capacity of electric generation in the United States sums to about 11 terawatts. Figure 5.2 shows the energy capacity for various sources in the United States since 1990:

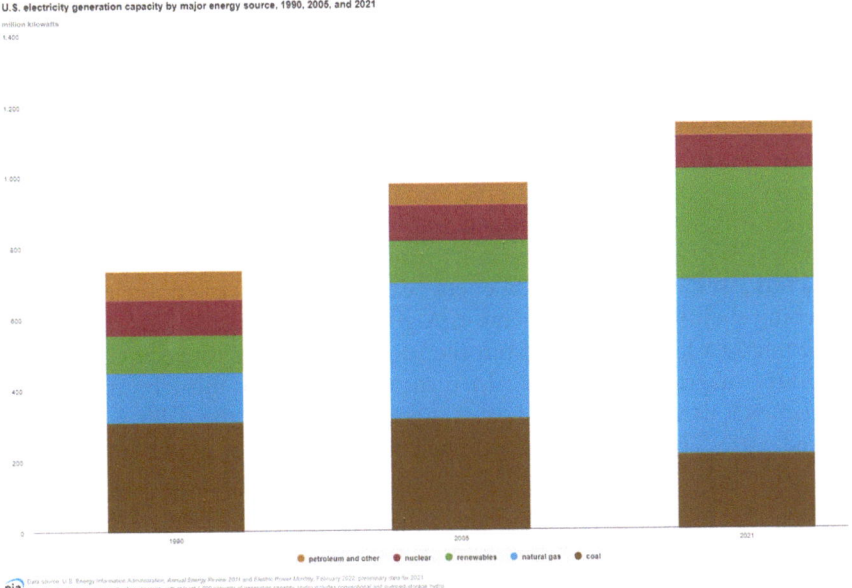

Fig. 5.2 United States energy generation capacity

Fig. 5.3 A steam turbine combined with a generator (Courtesy of the Nuclear Regulatory Commission)

These measures also allow us to convert between different forms of energy. Consider a traditional natural gas-fired electricity generation facility that provides power to a community. It converts the specific energy of combustion of methane into heat, which is then used to boil water to produce high-pressure steam that is directed to spin turbine blades that drive a generator. Figure 5.3 shows such a steam-driven generator:

The main energy component of natural gas, methane's **heat of combustion**, is about 50 MJ per kilogram of the fuel. However, not all that energy can effectively be converted into the mechanical energy of rotation of a turbine and, in turn, the generation of electricity. A good gas turbine design might recover 60% of this heat energy in its conversion to mechanical energy.

Let us assume the mechanical energy can then be converted to electricity by attaching the turbine to a generator. A good electricity generator can convert mechanical or kinetic energy from the spinning turbines to electricity with an efficiency of greater than 90%. In turn, this electricity must be transported to the end user to perhaps heat a home. Losses in these lines can easily exceed 10%. The overall heating efficiency is a product of the effectiveness of each stage of conversion and transportation, 60% for the heat engine, 90% for the mechanical to electrical conversion, and 90% for transmission. Overall efficiency is then the product of these conversions:

$$\text{Efficiency} = 60\% \times 90\% \times 90\% = 48.6\%.$$

We see from the example that a kilogram of natural gas can theoretically produce 50 MJ of energy but practically generates only 24.3 MJ of heat energy if converted first to electricity.

We now have the vocabulary of sustainable energy, with the basic units of work, energy, power, and efficiency, that allow us to proceed.

CHEMICAL ENERGY

Recall the principle of mass energy equivalence based on Einstein's Theory of Relativity that drives the processes of fusion and fission in which mass can be converted to energy. Chemical reactions can also give off energy, or absorb energy to reverse the reaction. The energy of chemical reactions is not a mass/energy conversion, though. Instead, it relies on the difference in the potential energy stored in chemical bonds. By changing the pattern of bonds through chemical reactions, this energy can be transformed, to give off heat in *exothermic* reactions or absorbs heat in *endothermic* reactions. Earlier we saw that the oxidation of a half mole of hydrogen molecules by a mole of oxygen molecules produces a mole of water and 237 kJ of heat.

The *heat of combustion* of methane also demonstrates the energy contained in chemical bonds. When methane combusts (is oxidized), it releases heat energy in an exothermic reaction. The internal bond energy of methane and oxygen molecules is higher than the internal chemical bonds of its combustion products, carbon dioxide and water. This difference in energy is the heat given off, all else equal.

One mole (6.023×10^{23}) of molecules of methane, oxidized by two moles of oxygen yields one mole of carbon dioxide, two moles of water, and 902.3 kJ of heat energy. More formally:

$$CH_4 + 2O_2 \rightarrow CO_2 + 2H_2O + 902.3\,kJ$$

Using the Gibbs equation, we know that this exothermic process also increases entropy as the reaction moves to a lower energy state. This process must first be initiated. Simply mixing methane and oxygen together in the necessary stoichiometric proportion does not spontaneously induce combustion. Methane and oxygen bonds are each very robust. This reaction must be sparked with additional energy to kick off the chain reaction. The spark is the necessary catalyst to the reaction that allows the reaction to overcome the necessary activation energy (Fig. 5.4).

Our knowledge of reactions, with the need to at times to provide sufficient activation energy or overpotential to allow a reaction to overcome energy hurdles, and the need to track efficiencies, gives us the necessary tools to understand the energy conversions upon which energy production and storage relies.

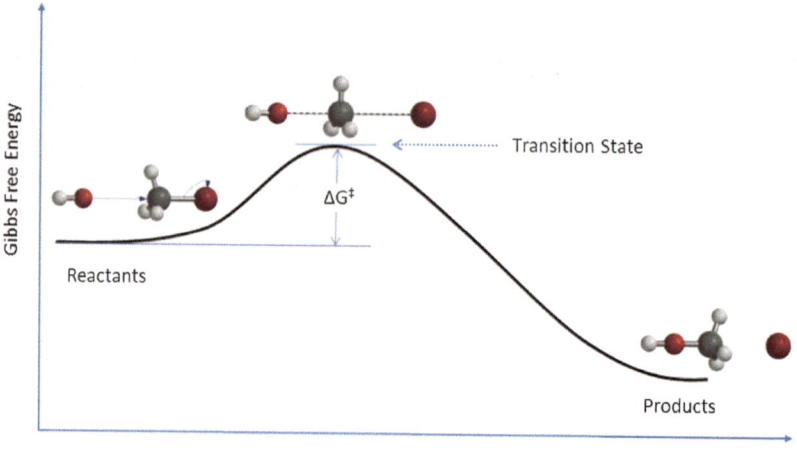

Fig. 5.4 Energy threshold for chemical reactions (Wikipedia Commons—Courtesy of Chem540grp1f08)

Summary

While the Big Bang converted energy to mass and our Sun regularly converts the mass of hydrogen to helium to create fusion energy we feel as heat, processes on Earth typically conserve mass and energy. However, the 2nd Law of Thermodynamics states that the entropy of a closed system will increase or stay constant. This physical law demonstrates that our world can enhance or maintain order only by consuming energy.

We need a constant source of energy to simply maintain our ecosystem and to disassemble and recreate the myriad chemical reactions we use as we transfer our natural capital into the consumption humankind enjoys. Only with abundant energy can we replenish natural capital and maintain our ecosystem. The creation of new and sustainable energy is the key to sustainability.

ESG Toolkit

A corporation uses various forms of physical energy a corporation consumes, both directly and through the factors of production it purchases.

Are the efficiencies and quantities of these energies tracked and considered?

Are there ways to decrease this energy budget?

Can energy consumption be reduced through a better accounting of the supply chain?

CHAPTER 6

Biology and Ecosystem Science

Growing concerns about the sustainability of our biosphere have increased interest in ecosystem science and management. Contributions from the life sciences, atmospheric science, Earth science, and various social sciences have afforded us a better understanding of the complex ecosystem relationships within which humankind thrives.

British ecologist A. G. Tansley is credited with coining the term *ecosystem* in 1935 as "a particular category of physical systems, consisting of organisms and inorganic compounds in a relatively stable equilibrium, open and of various sizes and kinds."[1] Awareness of such interrelatedness had been heightened in the previous century with the writings and explorations of Charles Darwin, but the fragility of our ecosystem is increasingly evident as species are disappearing at an alarming rate (Fig. 6.1):

The observations of Rachel Carson, described in her 1962 book Silent Spring,[2] raised our collective consciousness on the damage of pesticide use and extensive monocrop cultivation on the surrounding ecosystem. This awareness induced various scientific disciplines to weave together our knowledge of air, soil, water, and climate. Scientists began to model these aspects for the first time as systems rather than in isolation. A wealth of data from new remote sensing satellites has since added significantly to our understanding over the last half century.

We now know that a proper treatment of ecosystem complexity requires a broad and multidisciplinary scientific approach. There remain several dimensions necessary for the improvement of our understanding. These include the need to:

© The Author(s), under exclusive license to Springer Nature Switzerland AG 2023
C. Read, *Understanding Sustainability Principles and ESG Policies*,
https://doi.org/10.1007/978-3-031-34483-1_6

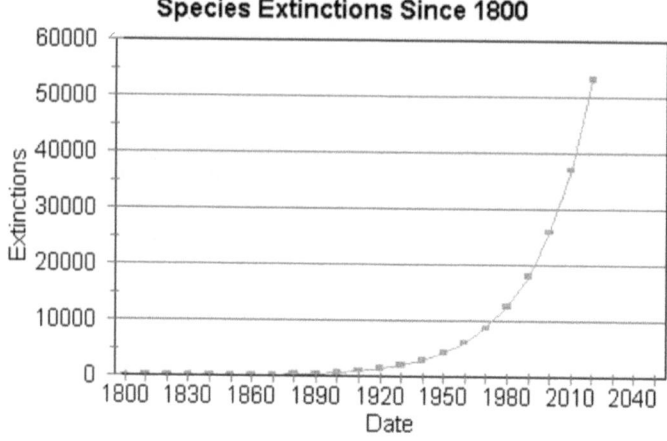

http://www.extinctionsymbol.info/graph.jpg

Fig. 6.1 The accelerating rate of species extinction

- Explore ecosystems and their stability and resilience
- Model and quantify the biological and chemical processes and their energy dependencies
- Incorporate stark observations regarding global warming and climate change.

These explorations have fostered entirely new subdisciplines, such as **ecoclimatology, ecological economics, ecohydrology**, and even business and social sciences into a new discipline of sustainability studies. The United Nations, in its *Sustainable Development Goals*, and their *Intergovernmental Panel on Climate Change* have elevated the profile and significance of such important cross-disciplinary work.

Not only are complex ecosystems explored, but now, more than at any time in our history, we have come to better understand the challenges to *ecosystem resiliency* to changes in the environment that are occurring at an unprecedented rate. The study of ecosystem resiliency depends upon new interfaces between scientific disciplines that have functioned until recently in isolation.

Basic Ecosystem Science

Ecosystem science is becoming increasingly sophisticated and better able to consider the nuanced degrees in which physical and biological systems interact. We shall explore some of these subtleties in our study of climate sustainability in later chapters. Consider for a moment the complexity of just the three components of our ecosystem that play prominent roles in the ***carbon cycle*** and affect or are most profoundly affected by global warming. These ecosystems are the oceans, the boreal forests, and the soil.

In the appendix to Chapter 2, we described how even a simple system of two related populations can interact in a way that produces surprising oscillations in populations over time. Earth systems are far more complex than that simple example, which makes the dynamics of ecosystem balance even more precarious. While the water cycle and differential warming of the Earth's atmosphere are the most significant physical phenomena, they also drive and trigger the balance of life in the ecosystem. The Earth's oceans represent the most significant ecosystem affected profoundly by global warming since the onset of the Industrial Revolution.

The Ocean Ecosystem Equilibrium

The Earth's oceans absorb more than 90% of the heat energy impinging on the Earth's surface from the sun's insolation, even though ocean water represents 71% of the Earth's surface. Its dark color and effective heat transfer results in a disproportionately high level of energy absorption. The oceans thus constitute the largest interface between the Earth's surface with our atmosphere. Ocean energy and temperature drive the water cycle and our climate, and oceans act as the most significant carbon sink that influences the Earth's temperature (Fig. 6.2).

In addition, given the temperatures, pressure, and atmospheric components found on Earth, about 30% of carbon dioxide emitted into the atmosphere

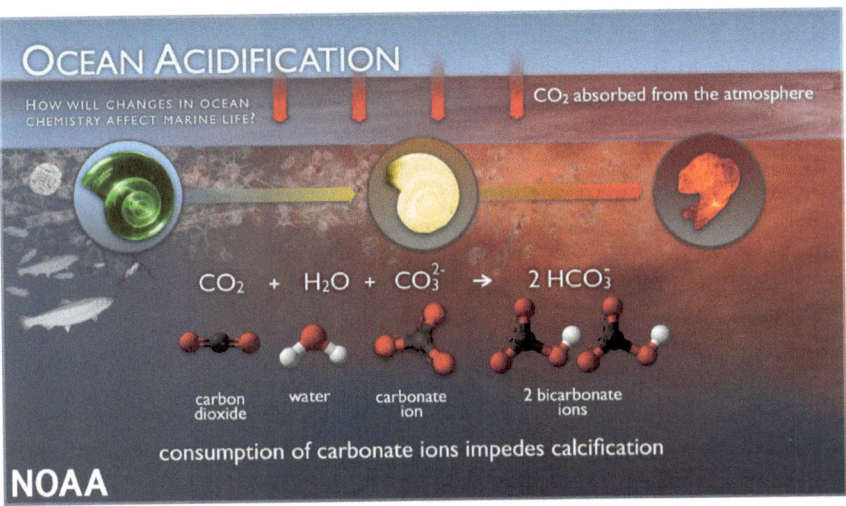

Fig. 6.2 Atmospheric carbon dioxide, ocean acidity, and calcification (Courtesy of NOAA)

from the combustion of fossil fuels has been absorbed by the Earth's oceans since the onset of the Industrial Revolution.

More specifically, ocean water absorbs and dissolves carbon dioxide to produce carbonic acid H_2CO_3 which then disassociates into bicarbonate ions HCO_3^- and hydrogen ions H^+. This higher concentration of hydrogen ions directly translates into the increasing ocean acidity we have observed since the beginning of the *Anthropocene Epoch* and the onset of global warming in 1850. The resulting shift in the balance of water-dissolved carbonate ions results in increased ocean acidity as indicated by a lower pH value, by about 0.1 pH unit on the logarithmic pH scale of acidity versus alkalinity.

As carbon dioxide levels increase, the abundance of dissolved hydrogen ions associates with more dissolved carbonate ions CO_2^{-3} to deprive organisms of this carbonate ion essential for their skeletal systems and shell protection mechanisms. The higher level of hydrogen ions also dissolves the carbonate in the shells and coral itself. The resulting greater acidity has profound effects on calcium carbonate components of coral and sea shells that constitute some of the most basic components of the ocean ecosystem.

Coral is one of the world's largest living organisms. Each coral structure is a colony of millions of individual living organisms sharing an ecosystem to which all kinds of ocean plants and animal life depend. Similarly, the pteropod population, a small sea snail that is a food source for a wide variety of marine animals, from small krill to whales, suffers as the lower pH dissolves their shells. Greater ocean acidity also affects the ability of many ocean species to procreate or to find suitable habitat for hatching of larvae.

Meanwhile, other ocean plant species that depend on photosynthesis may grow more vigorously with higher carbon dioxide levels. Seaweed expansion can occur, which further tips the delicate balance of species in the ocean ecosystem. These effects trickle farther up into the ocean ecosystem upon which people in coastal regions depend for sources of food (Fig. 6.3).

During the Miocene Epoch 23–5 million years ago (Ma) at the beginning of the Neogene Period, atmospheric warming raised ocean acidity and temperature to levels not seen for millions of years, until now. Ocean acidity that has worsened so dramatically in the past couple of centuries is now reaching a level not seen since the Miocene era about 15 million years ago. By the end of the twenty-first century, the Earth will be well on its way to demonstrating the most rapid change in ocean equilibrium since the earliest eras of Earth's geological formation. This shift in the ocean's equilibrium has caused the Earth to experience a dramatic chain reaction of species extinctions, from shell-based animals low in the food chain through fisheries, crab and other shellfish, and habitats upon which many humans depend, to the habitat and food that provides food energy to the world's largest mammals, the whales.

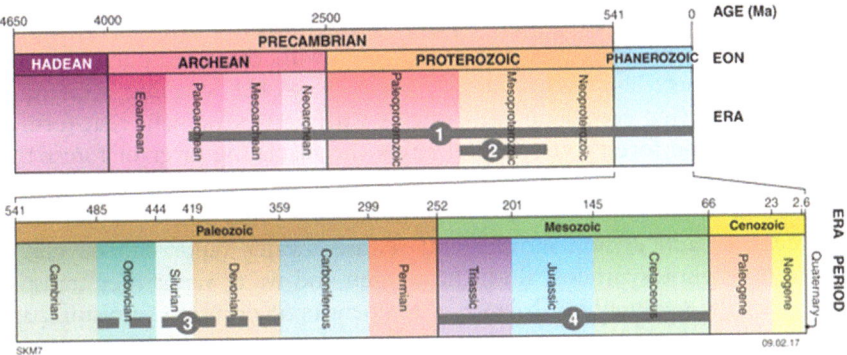

Fig. 6.3 Ages, eons, eras, and epochs since the earth's formation (Courtesy of the Department of Mines, Industry Regulation, and Safety, Government of Western Australia, https://www.dmp.wa.gov.au/Geological-timeline-1662.aspx, accessed March 23, 2023)

THE BOREAL FOREST ECOSYSTEM EQUILIBRIUM

The second largest natural atmospheric carbon dioxide removal influence arises from photosynthesis within the Earth's northern forests. Called the *boreal forest* in North America, and the Taiga elsewhere, like the oceans, these forests are changing dramatically, which in turn reduces their ability to act as a *carbon dioxide sink*.

Boreal forests constitute much of the northern forests of Canada, Russia, Alaska, and Scandinavian countries. They are often referred to as the Earth's lungs and are more significant in removing carbon dioxide from the atmosphere than the Amazon forest (Fig. 6.4).

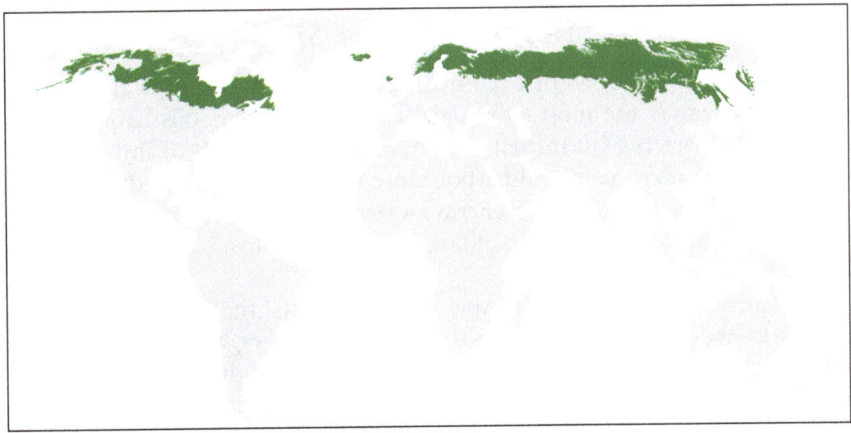

Fig. 6.4 The boreal forest (Wikipedia Commons—Mark Baldwin-Smith)

This immense northern forest depends not only on cool atmospheric conditions but also on its water cycle relationship with the atmosphere and the oceans and seas. The retreat of the forest, as global warming accelerates, replaces these vital ecosystem components with grass savannahs that form in the wake of the forest's retreat and strips the insulating layer of forest that keeps permafrost frozen.

The lost coniferous trees of spruce, pine, larch, and others had absorbed groundwater and carbon dioxide and released water vapor and oxygen as products of photosynthesis. Both the oxygen and water vapor they transpire also affect the geographical balance of temperatures, pressures, moisture, and molecules within the atmosphere.

This changing equilibrium, accelerated and augmented by global warming is far more profound in the Arctic regions than the tropics and temperate zones. Global warming results in changes in balance at the ice/land interface, the land/atmosphere interface, and even the jet stream that brings weather from the Arctic to temperate regions. By acting as the Earth's lung, the concentration of gases, and even the atmospheric pressure of the air above the massive forest is affected, with profound effects on the Earth's climate.

Such a massive but declining lung capacity arising from global warming also decreases the atmosphere's ability to cleanse itself of chemicals and volatile compounds. In essence, this massive portion of the ecosystem affects the atmosphere, the climate and weather, and the habitats that depend on it.

Methane that has been sequestered and insulated from release under the shallow roots of boreal forest trees for tens of thousands of years is released rapidly as shifts in the forest cause the underlying *permafrost* to melt. The boreal forest is retreating at an unprecedented rate because of global warming and because of the melting of the permafrost that underpins these forests, with far-reaching effects on the Earth's ecosystems and oceans.

The Soil *Carbon Sink*

A *carbon sink* is a process that removes carbon dioxide from the atmosphere. While the ocean is the most substantial carbon sink, our soils have captured about three quarters of terrestrial carbon, remove about 10% of anthropogenic carbon dioxide every year, and harbor more than half of the Earth's species.

Photosynthesis absorbs solar energy, water, and carbon dioxide to produce oxygen and carbohydrates, the building block of plant tissue that are eventually digested by animals.

As plants and animals die or expel waste products, the biomass that began with carbohydrates is deposited in the soil. This *soil organic matter (SOM)* then sequesters carbon until it is subsequently and slowly released over time. In some circumstances, this sequestration spans millennia.

These and other significant factors affecting atmospheric conditions have been profoundly altered over the *Anthropocene Epoch* and have, in turn, affected the ecosystem balance more dramatically than at any time since the end of the Mesozoic Era. The *Conference of Parties 15 (COP15)* of the *United Nations Biodiversity Conference* held in Montreal in December of 2022 acknowledged the rapid and irreversible loss of habitat and species arising from global warming. They affirmed the recent report from their science committee that voiced strong concerns about global ecosystem trends. For the first time, the attendees signed a *Montreal Pledge* that acknowledge the worsened ecosystem vulnerability and pledged to protect 30% of the Earth's land and ocean habitats by 2030.

Ecosystem Management

The *Intergovernmental Science-Policy Platform on Biodiversity and Ecosystem Services (IPBES)* reported in 2019 that[3]:

- One million species are threatened with extinction arising from human activity.
- Abundance of land-based native species has fallen by 20%, most of which since 1900, while 40% of amphibian species, about 33% of coral reefs, and a third of marine mammals are threatened.
- Two thirds of the marine environment and three quarters of the land environment have been altered significantly by human activity.
- Land degradation has reduced productivity of 23% of the world's land surface, with $577 billion US of annual crops at risk from pollination loss alone, while 100–300 million people face an increased risk of floods and hurricanes as a consequence of loss of coastal habitats.
- A third of fish stocks are being harvested at unsustainable levels.
- Urban populations have doubled since 1992.
- Current biodiversity and ecosystem trends will undermine 35 of 44 of the United Nations Sustainable Development Goals.
- Without transformative change, these negative trends will continue beyond 2050 because of increased land use, climate change, and exploitation of food stocks.

The IPBES notes that a more sustainable future requires an evolution of global economic and financial systems around sustainable economies and ecosystem. Current trends in human activities will fail to meet articulated ecosystem mitigation goals of 2030 and beyond. Instead, social, political, and technological transformative change is necessary to meet the Sustainable Development Goals formulated to slow the degradation of our ecosystem arising from global warming.

Summary

This chapter documented how even small changes in the molecular balance or temperature at the interface of the atmosphere, the ocean, and the landmasses can initiate processes that have profound effects on the Earth's ecosystems. Humans tend to think linearly and are often surprised when small incremental environmental shifts cause massive displacements of atmospheric natural capital. In some cases, as we will discuss later in the analysis of tipping points, positive feedback loops can be initiated that accelerate processes which normally occur over tens or hundreds of millennia into tens or hundreds of years. Such profound changes in the Earth's equilibrium and ecosystems have far-reaching implications on the sustainability of the planet and humankind.

ESG Toolkit

Corporations are inevitably part of the natural ecosystem through the resources it uses and the consumption patterns it establishes.

How do corporations track and acknowledge these interactions?
Do they track an ecosystem budget?
Are there ways to mitigate their ecosystem footprint?

Notes

1. Tansley, A. G. (1935). "The Use and Abuse of Vegetational Concepts and Terms", Ecology **16**(3), July 1, 1935.
2. Carson, Rachel. (2002). Silent Spring, Houghton Mifflin, Boston.
3. IPBES. (2019). Global assessment report on biodiversity and ecosystem services of the Intergovernmental Science-Policy Platform on Biodiversity and Ecosystem Services. E. S. Brondizio, J. Settele, S. Díaz, & H. T. Ngo (editors). IPBES Secretariat, Bonn, Germany 1148p. https://doi.org/10.5281/zenodo.3831673.

CHAPTER 7

Sustainable Energy from the Sun and Earth's Core

Our planet is a relatively closed system. The Earth has little interaction with the space beyond our atmosphere except the insolation impinging upon us from our Sun, net of the heat reradiated from its surface into space through our insulating atmospheric blanket. This net energy warms the Earth to a temperature that supports life and provides the energy on which our ecosystem relies.

While the Sun's radiated energy will cease entirely someday, we can expect about five billion more years of solar energy to power our otherwise closed Earth systems. The question is whether we can better use this almost endless energy to preserve and protect our atmosphere in the wake of significant damage that humankind has induced from its reliance on finite fossil fuels.

The Sun provides most all the energy needs of our ecosystem, and shall do so unabated for the conceivable future of humanity on Earth. It fuels photosynthesis in the most efficient solar energy conversion known to scientists and that sustains our ecosystem and humankind. With its ability to convert 60% of the Sun's radiative energy to chemical energy in the form of simple carbohydrates, photosynthesis is about twice as efficient as the best commercialized conversion of solar energy to electricity photovoltaic (PV) solar panels.

Scientific research may someday mimic nature's incredible efficiency through an artificial leaf to augment humankind's energy needs. In doing so, carbon dioxide in the air, and water in the air, oceans, lakes, and the ground, combined with abundant solar energy, are converted into carbohydrate molecules and oxygen. The carbohydrates are aggregated in sugars or in long chains called polysaccharides called starches that store energy in numerous carbon bonds.

C. Read, *Understanding Sustainability Principles and ESG Policies*, https://doi.org/10.1007/978-3-031-34483-1_7

These carbon bonds are the energy storage systems of the natural world. Animals store starches and metabolize them when needed to produce glycogens that can be readily used at the cellular level. This biomass can also be converted into sustainable liquid fuels that are able to concentrate significant energy at high densities.

At times in the Earth's history when carbon dioxide levels have been high, spurts of plant production and hence abundant feedstock for animals have created waves of immense and sometimes excessive biomass production. When these occasional waves of organic production die off, layers of biomass coat ocean floors and are covered with silt, only to decompose over time, heat, and pressure into fixed stocks of fossil fuels we rely upon today. This accelerating carbon sequestration a hundred million years ago is described more fully in Chapter 10.

Conversion of chemical energy from photosynthesis into other forms of energy, such as heat and electricity upon which humankind relies, is relatively inefficient. Overall efficiency as a share of solar incidence after necessary extraction, refinement and processing, distribution, and chemical-to-heat or electrical conversion results in energy yields of 50% or less. Some of that energy is renewable, as plants can be harvested and processed in a sustainable manner. However, the great concentration of biomass into hydrocarbons a hundred million years ago is an unusual and intense example of carbon-based energy sequestration. It induced our reliance on fossil fuels that is unsustainable since those fossil fuel resources took millions of years to produce and are being consumed by humankind over the course of a couple of centuries. Before then, the heat of the Sun and the reradiation from the surface of the Earth have interacted with the atmosphere to maintain our ecosystem in a steady state equilibrium that has changed only very slowly. Humans have burned wood, peat, and other organic matter to augment this atmospheric temperature to keep us warm day and night and across the seasons.

The Industrial Revolution has added to these energy needs. Currently, the most direct and efficient human-made way to tap the Sun's solar radiative power is through photovoltaic cells. Solar panels are able to capture the Sun's radiance and convert it directly to electricity with an efficiency now between 20 and 30% of the solar energy incident upon them.

The Sun's radiative incidence on Earth's surfaces, called *insolation*, represents a power level of more than one kilowatt per square meter on average. Insolation is calculated by measuring both the Direct Horizontal Irradiance (DHI) from light scattered in the atmosphere and the Direct Normal Irradiance (DNI), also called the beam radiation of energy directly from the Sun to the Earth. This latter DNI measure considers spreading of the Sun's rays the farther the angle of the surface of the Earth departs from orthogonal (90 degree) incidence to the Sun. The Sun's rays strike the Earth with the greatest incidence near the Earth's equator. Hence, total irradiance must take

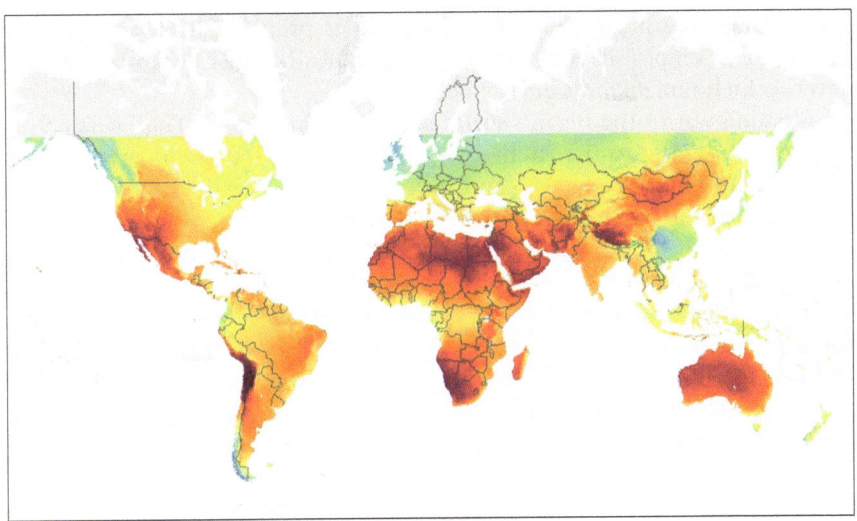

Fig. 7.1 Ideal locations for solar power generation (The map is obtained from the "Global Solar Atlas 2.0", a free, web-based application developed and operated by the company Solargis s.r.o. on behalf of the World Bank Group, utilizing Solargis data, with funding provided by the Energy Sector Management Assistance Program [ESMAP]. For additional information: https://globalsolaratlas.info)

into account the angle of the Sun to the Earth's surface, called the angle of zenith (z). Then, total Global Horizontal Irradiance GHI is given by:

$$GHI = DHI + DNI * \cos(z).$$

Total insolation is highest in locations where the Sun is highest in the sky and cloud cover is least significant. The map in Fig. 7.1 illustrates global solar energy potential.

Maximum direct and indirect insolation arriving at the Earth peaks at 1120 watts of power per square meter. This amount is reduced by atmospheric conditions such as clouds or water vapor in the air, departures from astronomical high noon when the Sun is not at its peak, and the difference in local latitude compared to the position of the Sun.

Given that the Sun's position changes throughout the day relative to a horizontal patch on the surface of the Earth, in the absence of clouds or atmospheric obstructions, this solar power impinging on the Earth begins near sunrise, increases until astronomical high noon, and then declines toward sunset. The total usable solar energy over the day is equivalent to approximately 6000 watt-hours per square meter (6 kwh/m^2) at the most ideal locations on the Earth. Figure 7.1 illustrates the geographical distribution of the Sun's ability to energize the Earth through solar panels.

While comparatively little solar capacity is captured by humans, as a share of the energy impinging upon the Earth, the energy needs of humankind have been substantially met in one form or another primarily from energy from the Sun. This empirical observation demonstrates the vast potential for solar power to fuel humankind's energy needs.

Insolation warms the Earth's surface and the atmosphere. This energy from the Sun, in addition to the latent energy of the Earth's molten core, the heat from natural radioactive decay, and even the friction of tidal forces, has for millennia kept the Earth at an average temperature of 15° Celsius. The Earth's temperature is in equilibrium between the Sun's insolation and the Earth's reradiation of heat back into space.

If one were to observe the Earth from outside its atmosphere, the Earth would appear to be radiating heat back into the solar system at a much lower rate than the temperature we experience, equivalent to a surface temperature of –21° Celsius. Greenhouse gases trap some of the energy the Earth attempts to reradiate solar energy back from the Earth's surface. Hence, the Earth's surface is about 36° Celsius warmer on average because of the insulating effect of our atmosphere. This differential is growing because of global warming as the Earth's atmosphere reflects more reradiated heat energy back toward the surface rather than into space.

Before we consider the full potential for solar power, note that the solar energy PV panels absorb does not mitigate global warming. About half of the electrical energy generated by solar panels is used for heating of one sort or another. Most all the remaining energy solar panels harvest is also reinjected into the atmosphere as heat generated by the electric devices we enjoy. Some of these devices convert electricity into mechanical or other forms of energy. The motors we operate, the friction in electric cars and in the wheel's rubber on the road, and even the electrons flowing in our smartphones or computers ultimately dissipate almost all the energy consumed in the form of heat. What is lost to heating the Earth's surface should humankind convert massively to solar power is still retained within the Earth's atmosphere. Hence, the energy balance is maintained even when solar panels absorb the Sun's rays.

This solar energy impinging across the entire globe at any instant totals about 1.73×10^{14} kilowatts of power, every hour, and sums to 1.52×10^{18} kilowatt-hours of solar energy each year. Recall that the total electricity demand could reach 5×10^{13} kilowatt-hours per year once humankind is weaned off fossil fuels. Even with PV panel efficiencies of 30%, humanity's electricity and heating needs can be met by harvesting less than 0.01% of the solar power incident on the Earth. This solar incidence alone, if effectively employed, can meet all of humankind's energy needs ten thousand times over.

A Brief History of Photovoltaics

While humans have understood the power of the Sun for millennia to warm buildings, heat food, and even as a weapon of war, the tie between light energy and electricity came only relatively recently. In 1839, Alexandre-Edmond Becquerel found that light impinging on a conductive liquid between two metal electrodes can generate electricity. Thirty-four years later, in 1873, Willoughby Smith found that the metal selenium has lower resistance when exposed to light.

Then, in 1905, as part of his miraculous year of research, Albert Einstein explained the photoelectric effect in which photons can impart energy that can cause electrons to flow. He went on to win the Nobel Prize in Physics in 1921 for his discovery of this interaction between energy and matter.

In 1954, two scientists at the legendary Bell Labs developed the first silicon photovoltaic cell to tap into Einstein's prediction that photons of light can cause electrons to flow. They were eventually able to produce solar panels that could convert 11% of the light energy of photons impinging on their surface to electrical energy. Over the next couple of decades, solar cells enabled satellites circling the Earth and Moon to remain powered for years.

It took a couple more decades to bring the cost of solar cells down from hundreds of dollars per watt of power capacity to $20 per watt, and another fifty years to bring the price down to thirty or forty cents per watt of power capacity (Fig. 7.2).

At these current prices, not only for the solar panels, but also for their mounts, and the electronics that convert their direct current (DC) output into alternating current (AC) output that can feed our AC electric grid, solar farms have become the lowest cost form of new electricity generation in some

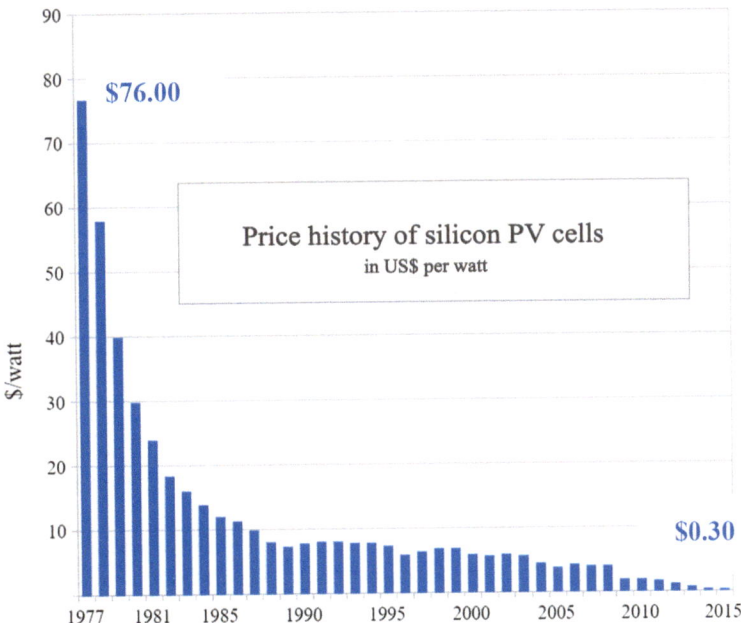

Fig. 7.2 Price drops in photovoltaic panels (https://commons.wikimedia.org/wiki/File:Price_history_of_silicon_PV_cells_since_1977.svg, accessed January 4, 2023)

of the world's most ideal locations. Obviously, locations that experience more overcast conditions benefit much less from the direct rays of the Sun, but still produce some electricity from indirect solar irradiation. In addition, extreme northern and southern locations are less productive because mounts that keep panels tilted directly toward the Sun take up much more horizontal space to prevent the shading of adjacent panels. In more extreme latitudes, the Sun's rays are deflected as they travel through a wider slice of atmosphere to reach the panels. Panels at such latitudes may need to move throughout the day to track the Sun and enhance output, which increases the cost of PV panel mounts in more northerly or southerly latitudes.

Ideal locations are in the regions of the Earth between approximately 30 degrees of latitude north and south. In these areas, given the current low price of panels, the tilt of the PV arrays can remain constant year-round and need not track the Sun throughout the day to still produce electricity cost-effectively. In these regions, a panel can reach an average of about six hours of power equivalent at the panel's rated output throughout the day. The shifting Sun across the sky causes a PV array to produce energy equivalent of its full power rating for only about 20–25% of the twenty-four-hour day at even the most ideal solar power locations. This output averaged over time relative to a panel's rated power is called the *capacity factor*.

Solar panel efficiency has risen to 25% conversion of light incidence into harvestable power. Innovations still in the laboratory may move this overall efficiency to 35% or 40% within the decade. The combination of insolation potential and cost efficiencies ensures that solar power is one of the most promising and least expensive forms of power, and is sustainable, with almost no ongoing variable costs.

While insolation could meet humankind's energy needs ten thousand times over, there exist several other technologies that can be even more cost-effective under certain circumstances.

Wind

Just as humans have harvested the heat of the Sun to augment residential heating and cooking for millennia, so too have humans harvested the power of the wind. In fact, wind power is also a form of solar power. The heating of the Sun is uneven, based on latitude, the tilt of the Earth throughout the year, and cloud cover. These differences drive the Earth's atmospheric temperatures, pressures, and water vapor content and hence produce our prevailing winds, our weather, and our climate.

The process is a familiar one. This differential heating causes the atmosphere in some parts of the planet to change density. Warmer air molecules vibrate more energetically and take up more space, and hence create a lower atmospheric density. This expansion of the atmosphere in warmer areas, and higher pressures in lower temperature regions elsewhere, causes differential pressures across the Earth. The atmosphere attempts to equalize such differential air pressures by the movement of air from high-pressure into low-pressure regions.

Adding to these movements in the atmosphere is the rotation of the Earth. The direction toward which winds move to equalize differential air pressures in the Northern or Southern Hemispheres appears to deflect as the Earth rotates. It is this Coriolis effect, named after the French mathematician Gaspard Gustave de Coriolis,[1] as the atmosphere circulates between the poles and the equator that deflects prevailing winds toward the east in the Northern Hemisphere, and to the west in the Southern Hemisphere, as shown in Fig. 7.3.

Explorers and traders have taken advantage of the constancy and predictability of these prevailing winds. It allowed explorers to travel to new continents and large sailing ships to move goods and people around the world until the invention of steam powered ships. Sails even propelled boats along the Nile millennia ago, while farmers have tapped windmill power to mill their grain and drain their fields for centuries.

Modern sails, wings, windmills, and wind turbines all work on a scientific principle that the Swiss mathematician Daniel Bernoulli (1700–1782) described almost three centuries ago. He reasoned that, if movement of air around both sides of an object such as a windmill blade must ultimately meet up again on the other side, air moving around the object along a longer path must travel a greater distance. If so, it must travel faster to catch up, which creates more space between these faster moving molecules in the air. The greater space translates into lower density of air, and hence a lower pressure.

Fig. 7.3 Reversal of Coriolis forces at the equator (Courtesy WikiCommons https://commons.wikimedia.org/wiki/File:Coriolis_eff ect14.png)

A blade or a wing creates lift as the resulting difference in air pressures sucks the wing toward its more curved side.

This differential force can spin a windmill or turbine as it extracts energy from the wind. To do so most efficiently, turbines in a wind farm are carefully positioned so adjacent mills do not obstruct the wind. Computer simulations and artificial intelligence (AI) can be employed to maximize the positioning, pivoting, and pitching of these turbines and blades for maximum efficiency.

Electricity generation from wind also relies on the smoother laminar flow of air above the ground, well away from the turbulence caused by waves, buildings, trees, and terrain undulation. Wind turbines that spin generators have been increasing in height and efficiency and hence are increasingly generating greater power. Such innovations have allowed the best wind turbine farms to now extract upward of 40% of their wind energy capacity.

In the early 1900s, Denmark had been employing numerous wind-powered electric generators of a capacity upward of 25 kW. By the 1930s, the USSR had designed and operated a 100 kW generator on a 30 meter tower, while by 1941, a wind-powered generator on top of Grandpas Knob in Castleton, Vermont, United States had a capacity of 1.25 MW.

By 2022, General Electric had designed a 14.7 MW GE Haliade-X offshore wind turbine that stood almost 300 meters high and had a rotor diameter of 220 meters, while the *China State Shipbuilding Corporation (CSSC)* has recently built an 18 MW wind turbine. A single such turbine operating at capacity could power almost 20,000 homes and avoid the emission of 60,000 tonnes of carbon dioxide annually. A quarter of electricity generated in the United Kingdom, and 10% of U.S. electricity needs were already met by wind by 2022 (Fig. 7.4).

The potential for wind power is immense. The National Renewable Energy Laboratory (NREL) of the U.S. Department of Energy estimates that the wind energy generation potential in the United States alone from regions capable of generating at least 30% of their wind potential over the day and night is 44.7 million GWh,[2] or 4.5×10^{16} watt-hours of electricity per year. This is more than ten times the total U.S. electricity consumption of 3.9×10^{15} watt-hours per year.[3] Aesthetic concerns aside, the height of turbine blades above the ground also does not preclude secondary uses of the land.

Much of the great potential for wind generation came about through improvements in technology over the last three decades. Wind stands with solar as the two sustainable energy investments that can dramatically decrease energy costs while they also allow economies to avoid the greenhouse gas emissions arising from the combustion of fossil fuels.

WATER POWER

Just as differences in air temperature and pressure create the wind, trade winds, and climate, so do temperature differentials create currents in the ocean. In addition, the effect of the gravitational pull of the Moon creates bulges in

Fig. 7.4 U.S. wind generation capacity growth

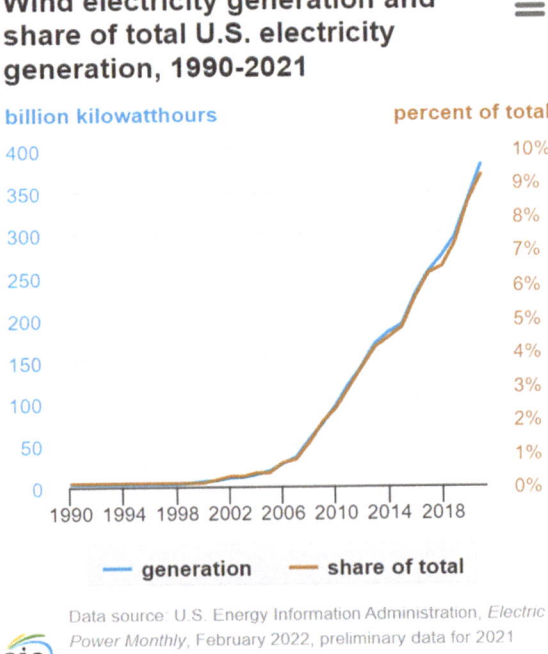

ocean water as the Earth rotates. This sloshing of seawater even modifies the gravitational forces between the Earth and the Moon to cause the Moon to synchronize its orientation around the Earth so that only one side of the Moon is exposed to the Earth at all times.

Finally, evaporation of water from the oceans, but also from lakes and dense forests, creates water-saturated air that is subsequently swept away by winds. As these winds pull moist lower density air into cooler regions unable to contain as much moisture, water vapor condenses into droplets which gravity then pulls back to the Earth to create rain.

This *water cycle process* is accelerated when winds force pockets of water vapor into cooler and less dense air as it is forced up hills and over mountains, and causes vapor to condense into droplets and rain on the windward side of mountains. These atmospheric rivers of moisture create our terrestrial rivers and lakes and replenish the ocean of water evaporated away as the Earth maintains its water cycle equilibrium.

This flow of water currents in the ocean, in the tides, and in our rivers can be tapped to spin turbines and produce electricity. Turbines driven by water flow through hydroelectric dams, or through the flow of tidal or river water or wave action generates a significant amount of the power we consume. Canada, Russia, and China, among other nations, have huge potential to tap into tidal power, from tidal streams, buoyant objects that rise and fall with the tide, and the flow into and out of lagoons and reservoirs.

One such tidal power station was built in Northern Ireland in 2007 between the Strangford Lough inlet and the Irish Sea. The fast-moving 4 meters per second water powered a device that could generate 1.2 MW of power. The Rance tidal river in Brittany, France was built in 1966 to harness tidal energy flowing in from the English Channel. Its 24 turbines can generate 240 MW of power, and are able to operate at about a 24% *capacity factor*.

Hydroelectric power from reservoirs and dams trumps these sources. They rely on water temporarily trapped in a higher elevation reservoir and permitted to flow out to a lower elevation outlet through pipes and water turbines. The potential energy of a metric tonne of water pulled by gravity downward a hundred meters can generate 1000 kg * 9.81 m/s^2 * 100 meters of work to provide almost 1 MJ of energy. If one tonne of water flows every second, and can be extracted with 100% efficiency, the flow can result in 0.981 MWh of energy each hour, enough to power about 1000 homes (Fig. 7.5).

Fig. 7.5 The Hoover Dam in Nevada, United States (Courtesy of Wikipedia and Mariordo [Mario Roberto Durán Ortiz], CC BY-SA 4.0, https://creativecommons. org/licenses/by-sa/4.0, via Wikimedia Commons)

Hydroelectric dams are about 90% efficient, which is comparatively high. In the United States, total hydroelectric power provides over 100 Gigawatts, with about 20% of that total from pumped storage, to be described later. Canada generates a similar capacity of about 81 GW of hydroelectric power, almost half of which is from one province, the Province of Quebec. Across the globe, total hydroelectric power constitutes about 1360 GW, equivalent to 1.36 terawatts.

GEOTHERMAL POWER

Geothermal power is derived from the heat contained below the surface of the Earth. This heat flow is derived from the molten core of the Earth that became encased in an insulating crust as the surface of the Earth cooled to insulate its molten primordial core. This high sustained temperature below the surface is also maintained by the release of heat in the natural splitting in the core of heavy atoms through nuclear fission.

These natural phenomena on occasion release heat, soot, gases, and lava when the molten core nears the surface erupts in volcanoes. In geothermically active areas, the higher temperatures held constant slightly below the insulating Earth's crust also contains energy that can be extracted more slowly and less dramatically.

The first geothermal power generator designed by Prince Piero Ginori Conti in Larderello, Italy in 1904 produced only a modest amount of energy. A later power plant at that location generated electricity from 1911 to 1958. Also in 1958, New Zealand's Wairakei flash steam geothermal plant was commissioned, while, in 1960, Pacific Gas and Electric began to operate power stations that tapped into the heat of The Geysers in California. The steam produced by water injected into the geyser heat source drove steam turbines that had a capacity of 11 MW.

A low 57° Celsius[4] water temperature is sufficient for a plant in Chena Hot Springs, Alaska near the Arctic Circle to generate power using a technique called the binary cycle. But, since the extractable temperature of geothermally heated water is not nearly so high as fossil-fueled steam generation plants, efficiency is comparatively low, at approximately 10%. This low efficiency is compensated by high capacity. Unlike wind and solar, the heat inside the Earth always flows, which results in an average capacity factor approaching 75%.

Because about 44.2 TW of the Earth's core energy is conducted to the surface at any moment, the ability to generate geothermal power is immense. However, few existing plants are tapping this efficient source of heat and electricity. A 2006 study by the Massachusetts Institute of Technology estimates that 100 GW of generating capacity could be produced in the United States by 2050 if a modest $1 billion were invested every year for 15 years. This represents almost 10% of the 1.1 terawatts of electric power capacity in the United States.[5]

While the capacity to produce electricity is not large, the costs are comparably low, which makes geothermal energy a valuable element in the sustainable energy mix. In some areas that can produce steam from an underground source, this higher temperature heat can be used to warm entire towns, such as Iceland's capital of Reykjavik. Geothermal energy is an excellent heating source for those who live adjacent to a site, but their efficiency in energy recovery drops by 40% to 60%, depending on the available steam outlet temperature, if used to generate electricity. Fortunately, an estimated 50% of all human energy needs are for heating. Nonetheless, the location dependent-nature of geothermal energy from high temperature sources such as geysers and magma flows means that electric production from geothermal energy is somewhat limited.

There is much more significant potential for low temperature geothermal energy to heat homes and structures. A heat pump can transfer heat from below the surface into our homes. Even if the subsurface is below the desired temperature for human comfort, it still contains some heat that can be extracted and concentrated. The temperature just below the Earth's core at typical well depths is relatively constant year-round, at approximately 10–15° Celsius depending on the region (Fig. 7.6).

We are familiar with air conditioners that can pump excess heat out of a warm house or building and into an even warmer exterior. Such air conditioners do not produce hot or cool air. Instead, they are heat pumps that can use a modest amount of electricity (typically) to pump heat from a lower temperature to a hotter temperature area through a combination of a pump,

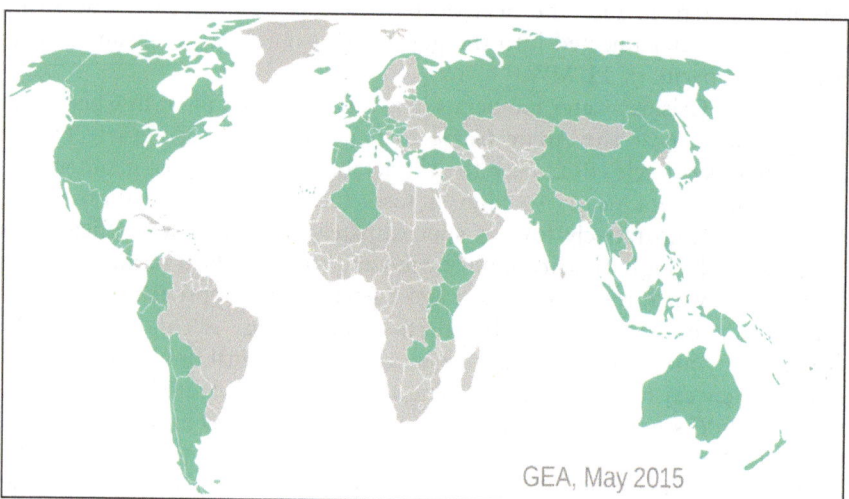

Fig. 7.6 Potential geothermal sites worldwide (Courtesy Wikipedia https://com mons.wikimedia.org/wiki/File:World_map_of_geothermal_power_countries_instal led_and_developing.svg#file, accessed January 22, 2022)

an evaporator, and special fluids that can evaporate and condense, and hence absorb and give off heat at the temperatures required for home comfort.

An air conditioner moves heat from a warm area inside into an even warmer area outside. The same principle can be reversed to warm a home to a comfortable 20° Celsius by reversing the heat pump to draw heat in from a colder area outside. Instead of heating by resistance, which returns 3.6 million Joules of heat for every kilowatt-hour of electricity consumed, a heat pump can move an equivalent of 3.6 million Joules of heat with about 0.9 to 1.2 million Joules of electricity.

Heat pumps typically take advantage of the latent heat exchanged when molecules experience a phase change. A fluid is pressurized by a pump to keep it in a liquid state on the high-pressure side of the circuit. The fluid is then passed through an evaporator into the low-pressure side that reduces the fluid's pressure and facilitates a phase change to a gaseous state. In the process, heat energy is absorbed in the same way as your hand or forehead feels cooler if liquid sweat evaporates off it. By placing the evaporator indoors, heat can be drawn from indoors and the fluid then releases the heat outdoors once it is recompressed into a liquid. In the process, heat is *pumped* from the indoors to the outdoors, which allows the indoor area's air to be conditioned.

The process can also occur in the reverse. In the heat pump, the evaporator is outside and hence any heat from outside can be drawn inside, even if the outdoor temperature is cooler than the indoor temperature. A heat pump is then moving heat from one area into another, with a measure of efficiency related to the amount of heat it can effectively pump compared to the amount of energy used to power the pump.

More specifically, heat pumps and air conditioners are measured based on how much heat they can move from one area into another compared to how much energy is consumed to run the pumps and blowers necessary to move the heat. This ratio of heat output/electricity input, which typically is in the range of 2.0 to 4.0, is called the *Coefficient of Performance (COP)*. In essence, recycling heat rather than heating and cooling is accomplished with a quarter to a half the energy that is required through either resistance electric heat or fossil fuel combustion.

The ability to move heat more efficiently makes subsurface geothermal heat economical for almost any region. If 50% of all energy used by humankind is for heating, then the widespread use of heat pumps, using either the air or ground as the heat source, can lower energy requirements for heating by one half, and reduce overall energy consumption by 25% to 37.5%. This energy saving potential is large, and exceeds the 25% of energy devoted to the entire transportation sector in the United States.

ENERGY FOR TRANSPORTATION

In comparison to work performed by electricity and even by direct combustion, traditional transportation energy usage is notoriously inefficient. A gasoline engine can convert only about 30% of its available chemical energy into mechanical energy, while electric cars, including charging and mechanical efficiencies, and the need to move electricity from a sustainable energy source to charging stations, are approximately 85% efficient. Conversion to electric power can then trim about 50% of our energy consumption in transportation, but will tax the grid by only an additional 15% to 25%. The electricity savings alone from large-scale conversion to heat pumps would more than cover the cost of transportation electrification.

BIOMASS AND BIOPRODUCTION

The Sun's rays produce a tremendous capacity for the Earth's flora and fauna to convert light into chemical energy through photosynthesis. Neglecting the unsustainable sequestration of such capacity from upward of a hundred million years ago in the form of sequestered fossil fuels, the energy created by biological mechanisms remains a common and sustainable source of heat around the world. Biomass concentrated in wood and peat is still used broadly, especially in Less Developed Countries in cooking and heating. Agricultural crops and agricultural and industrial waste (for instance, wood chips) can also be used to produce biofuels.

The decomposition of waste from human activities, called *biogenic waste*, that is disposed of in municipal solid waste sites and sewage plants, can be harvested to produce methane and combustible products. Ethanol is extracted from fermenting corn, sugar cane, sugar beets, and other crops to sustainably supplement or substitute for fossil fuels. Even algae can be cultivated and harvested to produce oils that can be refined in the same manner as fossil fuels to produce fuel oils.

This energy begins with the very efficient process of photosynthesis described above, so photosynthesis too derives energy from the Sun using the chemical reaction:

$$6H_2O + 6CO_2 + \text{sunlight} \rightarrow C_6H_{12}O_6 + 6O_2$$

In addition to the glucose created by photosynthesis, which is subsequently converted into other longer chain sugars and starches, cellulose, and myriad other molecules, another byproduct of this sunlight reaction is the oxygen animals breath. The photosynthesis process also scrubs the atmosphere of global warming carbon dioxide.

The biomass that is accumulated when these carbohydrate producing flora and algae die or are harvested and consumed can generate direct heat energy. They do so through their combustion, chemical conversion into fuels through fermentation, decomposition, or recent engineered processes, and biological conversions into liquid fuels, methane, propane, or others, increasingly through biological engineering.

Such sustainable fuels produced can drive heat engines and turbines that generate electricity. Fuel cells, described in Chapter 9, can also directly convert some smaller hydrocarbon molecules directly into electricity.

We are familiar with fermentation that can take sugars in biomasses such as grapes, grains, corn, and other sugar sources to produce various alcohols using yeasts as a natural catalyst. These alcohols can then be combusted in heat engines to produce warmth and electricity.

There are several other processes that concentrate biomass chemical energy. Pyrolysis is a thermochemical process that uses high temperature (400–600° Celsius) reactions in an oxygen-absent vessel to break complex molecules into simpler fuels such as methane, renewable diesel fuel, and hydrogen. In a process called hydrotreating, fast pyrolysis combines bio-oils with hydrogen via catalysts and heat to produce renewable diesel, jet fuel, heating oil, and gasoline.

Finally, gasification combines organic materials, higher heat (around 800–900° Celsius), and either steam or oxygen to produce carbon monoxide and a synthesized gas (syngas) to produce fuels to run turbines. The carbon monoxide and syngas can be further processed, for instance through the Fischer–Tropsch process, to produce sustainable liquid fuels such as diesel and gasoline.

Similarly, transesterification is a process that can convert waste vegetable oils and animal fats into fatty acid methyl esters (FAME) that can likewise be used to produce sustainable fuels, especially diesel oil.

In 2021, biomass conversions satisfied 5% of the U.S. primary energy consumption. About half of this energy production was from biofuel production, mainly ethanol production through fermentation. The remainder was predominantly derived from wood and wood waste, while about a tenth of this production was from biowaste harvesting at solid municipal disposal plants and agricultural product waste. About 10% of this production was converted to electricity, while the vast majority replaces fossil fuels in combustion.

SUMMARY

The Sun drives the mix of energy that can fuel humankind on a sustainable basis and has fueled the ecosystem since time immemorial. This solar energy has the potential to meet all the needs of humanity even if we transition away from fossil fuels very rapidly.

One aspect of this sustainable energy is geothermal. It is not technically eternal because both the Sun and the heat of our Earth will eventually decay. So too will the decay of nuclear elements and compounds that continue to warm the Earth's molten core and provide heat that we can tap near the surface. However, nuclear power, more fully described in the next chapter, can provide energy for the foreseeable life of humankind on Earth.

ESG Toolkit

Prices for sustainable power have now fallen to be lower than any newly constructed power plants.

What scale of solar power would need to be purchased through green energy credits or installed by the corporation to meet its energy needs?

Can the corporation encourage its employees to commute less or commute using electric cars, perhaps by offering low interest car loans or mass transit tokens for those who commute green?

Notes

1. https://oceanservice.noaa.gov/education/tutorial_currents/04currents1.html#:~:text=Because%20the%20Earth%20rotates%20on,is%20called%20the%20Coriolis%20effect.

2. https://www.nrel.gov/docs/fy11osti/50860.pdf, accessed September 27, 2022.

3. https://www.eia.gov/energyexplained/electricity/use-of-electricity.php, accessed September 27, 2022.

4. https://en.wikipedia.org/wiki/Geothermal_power, accessed September 27, 2022.

5. https://www.eia.gov/energyexplained/electricity/electricity-in-the-us-generation-capacity-and-sales.php#:~:text=At%20the%20end%20of%202021,solar%20photovoltaic%20electricity%20generating%20capacity, accessed September 27, 2022.

CHAPTER 8

Nuclear Power

Nuclear power has gone from salvation to scapegoat over its notorious 70-year history. It began even earlier, when Albert Einstein proposed his equation of energy mass equivalency, $E = mc^2$ in 1905. Einstein's equation was tapped by the Manhattan Project in the early 1940s and the notorious explosion of two nuclear bombs to mark the end of World War II. Peacetime brought in a promise of almost limitless and inexpensive power during the 1950s, but also spawned a Cold War based on nuclear deterrence that spanned five decades and with geopolitical echoes that still haunt us today.

Antoine Henri Becquerel, the son of Alexandre-Edmond Becquerel who had observed the photoelectric effect, discovered radioactivity in 1896. Two years later, Marie and Pierre Curie discovered two new elements, coined the term *radioactivity*, and noted the ability of atoms to give off heat and light. These three physicists jointly won the 1903 Nobel Prize in physics for their discovery of radioactivity, while, two years later, Einstein augmented their prize with a description of the precise amount of energy that can be given off by radioactive decay and nuclear fission. Research especially in Germany but also in the United States in the 1930s determined that practical nuclear fission reactions could be maintained.

Controlled fission typically begins with the transformation of Uranium 235 (U235), an atom which contains 92 electrons, and, in its nucleus, an equal number of protons and an additional 143 neutrons. A more massive form of uranium, U238 has three additional neutrons. A typical nuclear fission fuel mix is over 90% U238, with U235 typically representing 3–8% of the nuclear fuel in rods or pellets, sufficient to initiate a low enriched uranium fission reaction. When the nucleus of the uranium atom is fissioned through the onslaught

C. Read, *Understanding Sustainability Principles and ESG Policies*, https://doi.org/10.1007/978-3-031-34483-1_8

of neutrons and split into smaller atoms, it produces various fission products such as Iodine-131, Xenon-133, Cesium-137, Strontium 90, other subatomic products, and more neutrons. Some of these byproducts, when bombarded with neutrons, also split, for instance into Cesium-133.

In the process of the splitting of a U235 nucleus, some of the scattered neutrons are absorbed into the U238 that constitutes the vast bulk of the nuclear fuel source. The U238 first absorbs a neutron to become U239, which then gives off a beta particle (electron) to convert a neutron into a proton to create Neptunium-239. The emission of one more beta particle from Neptunium-239 then creates Plutonium-239.

In the end, when we compare the mass of the elements going into the reaction to the mass of the Iodine-131, Xenon-133, Cesium-137, Plutonium-239, and Cesium-134, and various particles emitted from the reaction, the difference in mass has been converted to heat energy emitted at a rate given by $E = mc^2$, where the speed of light c squared means an incredible amount of energy E is created by the conversion of only a small amount of mass m. If the plutonium can subsequently be split, the net reactions have the potential to generate a great deal more energy, to the point that the plutonium energy that is bred in such a ***breeder reactor*** could supply humanity with energy for many centuries, based on the feedstock of uranium within the Earth's crust.

GEN I PLANTS

The first generation nuclear power plants were comparably primitive. They powered an early experimental and some commercial reactors and the first nuclear-powered submarines. The first commercial plant was commissioned in the USSR in 1954 to supply about 5 MW of power. By 1960, 1 GW of power was produced globally, which rose dramatically to 100 GW a little more than a decade later as nations felt the pinch of an increasingly well-organized Organization of Petroleum Exporting Countries (OPEC) and the concomitant scarcity of fossil fuels and increase in oil prices (Fig. 8.1).

These Gen I plants were comparatively inefficient and less safe than the next iteration of plants called Generation II. The pressurized water plants in Gen I were replaced with light-water reactors with additional safety mechanisms that can be quickly implemented if necessary. The newer designs were also more cost-effective and reliable.

The early wave of commercialized nuclear power plants had a useful life of about 40 years. However, their legacy includes a large quantity of radioactive byproducts, called ***actinides***, that can last upward of tens of thousands of years. In the absence of common national waste storage sites, in the United States these byproducts remain on decommissioned nuclear plant sites, even after decommissioning. Since many of these plants were built in the 1960s and 1970s, these Gen II plants are now being decommissioned at an increasing rate.

Fig. 8.1 The 60 MWe Shippingport Atomic Power Station in Pennsylvania, the first commercial nuclear power plant in the United States (Picture courtesy of WikiCommons https:// commons.wikimedia.org/ wiki/File:Shippingport_ Reactor.jpg)

Breeder Reactors

While fission power plants provide immense amounts of power, they still only extract about 1% of the potential energy contained in nuclear fuel. Breeder reactors work by also processing the byproducts of U238 fission, most notably plutonium or thorium. The Canadian-designed CANDU reactor, and others, produces and then processes sufficient plutonium to create even more energy than the U235 and U238 consumed, which leads some to perhaps overstate that they produce more fuel than they consume.

The CANDU reactor can use natural or reprocessed uranium, thorium, plutonium, and spent fuel from light-water reactors. While the design has proven to be more expensive to build, CANDU reactors are much cheaper to fuel relative to the energy they produce. These improved designs also deter nuclear proliferation by consuming the plutonium that is sometimes processed and diverted for use in nuclear bombs.

Generation III Plants

The current state-of-the-art nuclear power plant design is called Gen III. These plants are more evolutionary rather than revolutionary. Improvements in fuel designs (pellets instead of fuel rods, for instance), improved heat capture, more self-sustaining and disaster-resistant cooling and heat transfer technology based on convection and natural flow rather than the previous generation pumps which require external power, and standardized or modular design, make for improvements in both efficiency and safety.

SMALL MODULAR REACTORS

The most recent innovations are Small Modular Reactors (SMRs). While small compared to other nuclear reactors, their size can nonetheless range from tens to hundreds of megawatts and can be assembled in combination to generate gigawatt scale power equivalent to large coal power plants. They are small and efficient, which allows them to be mass produced in factory settings and moved to facilities rather than constructed onsite.

These improvements in design and manufacturing are hoped to bring down substantially the cost of nuclear power. In addition, their compact footprint and versatility allow them to serve beyond just steam generation to power turbine generators. They can provide centralized steam and heat and are able to generate the high temperatures necessary for cracking of water into hydrogen, and other applications. This method creates what is called *pink hydrogen*.

Because about half of humans' need for energy is for combustion heating, the ability to generate centralized heat directly without first converting to electricity and then back to heat ensures enhancements in efficiency. In addition, SMR's easier transportation by railcar and truck allows them to replace and be hooked into the grid at former coal power plants. They promise to provide electricity at a levelized (amortized plus operating cost) rate of about $0.04 to $0.06 per kWh, which is less than the cost of natural gas cogeneration power, and a third to a half the levelized cost of new coal-fired fossil fuel plants. Because nuclear power generation does not produce carbon dioxide, SMRs also represent significant avoided costs of climate change mitigation and adaption.

Some SMR designs have other advantages. Leading designs rely on molten salts to move heat away from the nuclear reaction more efficiently, and can store these low-pressure salts at much higher temperatures than can safely be contained in superheated high-pressure water. The stored heat energy can then be used later to generate abundant heat.

This ability to store heat also allows such plants to essentially throttle up and down as needed by the grid or provide power when the sun does not shine. Such an ability to store and then draw down stored heat, and quickly adjust output can well-complement wind and solar generation. The stored heat can substitute for peaker natural gas electricity generation facilities and operate to fill temporary gaps in sustainable energy production. By situating SMRs at decommissioned natural gas plants, the cost of grid interconnectivity is also reduced.

GENERATION IV PLANTS

The traditional nuclear reaction has several problems beyond the production of radioactive byproducts that must be stored indefinitely. There remain significant safety issues with earlier Gen I and Gen II designs.

The shared feature of earlier designs was the need for a moderator to slow down the energetic neutrons emitted by U235 so they can readily be absorbed by U238. This typically necessitates and contaminates heavy or light water to moderate the reaction. Ideally, if this water, or neutron-augmented heavy water, is lost, the fission fizzles as the neutrons speed off before they can combine in nuclei of other atoms. This stability feature is designed to avoid the threat of a runaway nuclear reaction called a *meltdown*. However, radiated water creates a large quantity of dangerous byproducts in a form that is vulnerable to accidental high-pressure releases.

The Generation IV International Forum recognizes half a dozen design improvements that theoretically solve problems with previous generations of fission reactors. One of the most promising recently developed concepts is the sodium fast reactor. These new reactors can operate at high temperatures but low pressures by relying on heat transfer liquids, usually in the form of low-pressure molten salts, or gases rather than water. They also use other moderators, such as graphite, to slow neutrons down so they may more efficiently spur chain reactions without relying on heavy or light water for moderation.

Some Gen IV reactors can even use the molten salt heat transfer liquid to moderate the nuclear reaction. Should an accident cause the loss of the circulating molten salt used to cool the reaction and move heat to a steam turbine, neutrons are no longer slowed sufficiently to maintain the nuclear reaction. The spontaneous reaction stops and ensures that this design cannot experience the meltdown risk of older generation designs.

Beyond the greater efficiencies and ability to store very high temperature heat at pressures relatively low compared to previous generations of designs, these Gen IV designs also have a significant environmental advantage. They can process actinides such as thorium to not only extract more energy unavailable to conventional nuclear power plants, but also to reduce the radioactive footprint that is currently stored at existing and decommissioned plants. A "fast" reactor can even operate without moderation and consume almost all fissionable actinides over time. This reduces radioactive actinides to elements that degrade over centuries rather than millennia and can provide between 100 and 300 times the energy relative to previous designs, for the same amount of fuel.

Collocating Gen IV plants at present nuclear power sites then allows these facilities to take advantage of existing grid connections while they also consume the actinides that are stored at past and current nuclear power sites. Such a conversion reduces the half-life of radioactive materials from tens of thousands of years to tens or hundreds of years. In addition, like SMRs, Gen IV plants are based on passive safety circuits to be inherently safe. Such passive safety features turn potentially catastrophic failures into routine shutdowns.

While these designs do not produce global warming gases, the materials mined and processed to fuel them cause some emissions, at least until these processes can be made carbon-free. Such carbon footprints are small and comparable to the resource intensity of mining and manufacturing of solar panels and wind turbine plants. A handful of these Gen IV plants are expected to be commercialized by the 2030s.

FUSION

Fusion is the most speculative of power sources. As noted in Chapter 5, fusion has always been the holy grail of energy sustainability, but has remained a generation away for at least the past four generations. Since it is impossible to attain the pressures found near the surface or in the core of the Sun, Earth-based fusion requires containment of temperatures in excess of 150 million degree Celsius in confined space to make hydrogen atoms sufficiently energetic to collide and fuse into helium.

The latest experimental reactors have produced sufficient heat for upward of one or two minutes to sustain fusion. In 2022, one example of these devices for the first time generated more heat than the energy needed to initiate fusion, even if the facility did not produce sufficient energy for even a small share of all the ancillary needs of the apparatus. These plants are still far away from producing excess heat for long enough to overcome all the energy used in conducting the experiments.

Some advocates estimate that commercial fusion may be available by 2050, still more than a generation away. However, of late, interest and investment in fusion has flourished. Humankind is on a most promising pathway in generation to realize its potential. But by the time fusion may be commercialized, this energy source will be of little help in arresting global warming, and, even then, may not be competitive compared to solar and wind power (Fig. 8.2).

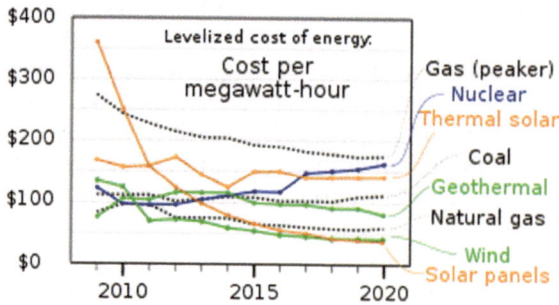

Fig. 8.2 Comparison Amortized Investment and Operating Costs of Various Energy Sources (Courtesy of WikiCommons, https://commons.wikimedia.org/wiki/File:20201019_Levelized_Cost_of_Energy_(LCOE,_Lazard)_-_renewable_energy.svg, accessed January 4, 2023)

SUMMARY

Nuclear energy has had a checkered past because of the global zeal to rapidly expand new Generation II nuclear power plants in the 1960s and 1970s. The regulation of designs is a very slow and deliberative process, for good reason, but the result has been a hiatus in the commercialization of new designs that solve many of the problems of early plants.

For instance, the newest designs that have yet to be commercialized promise inherent safety and protections from the meltdowns that were possible with earlier generations. Some of the new Generation IV designs can even consume the radioactive waste that has been left riddling decommissioned and operating nuclear power plants across the United States.

However, even if some of these new designs offer much promise on paper, history has shown that their full cost upon commercialization makes them significantly less affordable than wind and solar power. They may nonetheless be an asset in the mix of sustainable resources though because of the ability of some designs to quickly increase energy production when necessary, for instance during evening hours when the sun does not shine or when wind velocity drops.

ESG Toolkit

A corporation can purchase specific forms of energy. For instance, it can specify sustainable energy supplies from an Independent Systems Operator (ISO) off the grid that compensates clean energy providers, and it can specify sustainable diesel and fuel oil purchases.

If a corporation cannot choose its energy supplies, it may still document the mix of energy that is provided to it. Given the mix of peak natural gas power, coal, natural gas, nuclear power, and sustainable energy sources, it can then purchase offsets or lower consumption to reduce its footprint.

CHAPTER 9

Energy Storage and Efficiency

Each of the various energy sources described above is either sustainable or potentially so efficient in their use of natural capital with little alternative ecosystem value or scarcity that they can last almost indefinitely. However, not all these sustainable energy sources can meet the needs of the Sustainable Development Goals as set out by the UN Brundtland Commission without augmentation with other technologies. To meet human goals, we also need not only energy sustainability but also efficiency, resiliency, and reliability.

Resilience is the ability of our energy portfolio to adapt to changing needs or recover from challenges. One such dynamic requirement is to ensure sufficient versatility to provide energy to humans when we need it, but without contributing to global warming.

For instance, an important aspect of human quality of life is the ability to conduct our affairs after sundown. Yet, the two least expensive energy sources, in terms of levelized costs, are also intermittent. Wind power can meet our needs only when the wind is blowing. Very few locations have capacity rates, measured as the fraction of full time that a wind turbine can generate full power, in excess of even 70%. Likewise, even the best solar site can only provide the equivalent rated power of a solar cell for only six hours per twenty-four-hour day, and, of course, provide no power for almost half the day on average.

Key to the success of these lowest cost energy sources is then the ability to store power.

C. Read, *Understanding Sustainability Principles and ESG Policies*, https://doi.org/10.1007/978-3-031-34483-1_9

Battery Storage

The first mechanisms for energy storage that comes to most people's minds are batteries.

These direct current storage devices can store the potential energy of electrochemical reactions and release that energy on demand. Their performance is often quoted in terms of **Coulombic (or Faradaic) Efficiency (CE),** which is the ratio of the total energy that can be harvested from a fully charged battery compared to the amount of energy the battery required to become fully charged.

Adequate batteries for such energy storage have Coulombic Efficiencies ranging from 80 to 98%, with any such efficiency losses arising from irreversible chemical conversions or the generation of heat. Many of the most promising technologies already regularly exceed 90% efficiency, so the losses are not significant when compared to efficiency losses we often encounter when converting between various other forms of energy.

However, battery technologies remain expensive. While the technology is satisfactory for the relatively modest power and energy needs of a laptop computer or smartphone, grid and megawatt-hour scale storage are quite expensive. The needs are also significant. If solar power is hoped to meet almost half of our electricity needs by mid-century as the planet responds to the global warming challenge, the U.S. Department of Energy estimates that the United States alone would need 1600 GWh of energy storage.[1] While recent laboratory designs use abundant and inexpensive materials such as sulfur and air, most energy-dense batteries required for transportation still rely on lithium, an element not in relative abundance worldwide and which requires significant energy to extract.

Flow Batteries

While Li-Ion batteries operate in excess of 90% efficiency, their best use is for transportation, given their high energy density per unit of battery weight. This higher energy density with low weight is especially important for automobiles and aviation. However, the relatively lower abundance of lithium also makes it much more costly compared to vanadium. A vanadium flow battery is one of the most promising grid-scale storage technologies. It operates at about 60–80% efficiency, but its cost is very low, at about $25 per kilowatt-hour of storage capacity, compared to about $125 per kilowatt-hour of energy storage for lithium-ion batteries.

In addition, a vanadium battery is a **Redox-Flow** battery. Once charged, the liquids that store chemical energy are kept separate in large tanks until needed. These are weighty and cumbersome tanks, but they can store the energy indefinitely very much like a liquid fuel. A large battery station can easily be scaled up for greater capacity by building and stocking more tanks. In

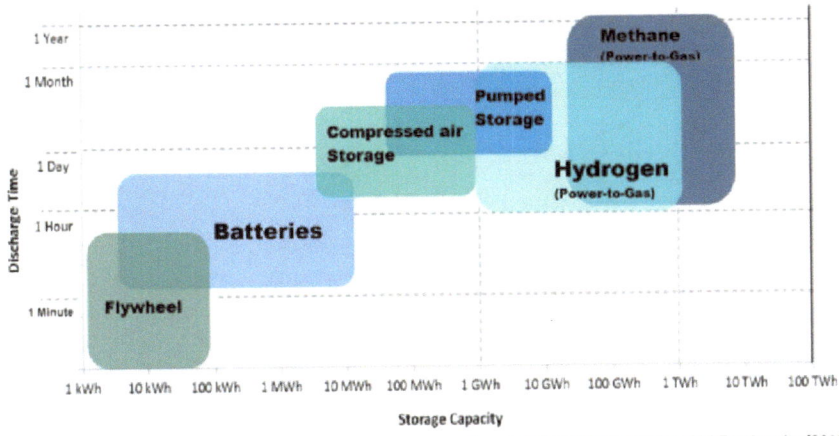

Source: School of Engineering, RMIT University (2015)

Fig. 9.1 Energy Storage—pumped hydro, latent heat, batteries (Courtesy Daniel Williams 1978, CC BY-SA 4.0. https://creativecommons.org/licenses/by-sa/4.0, via Wikimedia Commons)

some designs, these liquids can even be kept warm to store heat for subsequent release when needed (Fig. 9.1).

PUMPED HYDRO

The most extensive grid-scale energy storage system is pumped hydro. This mechanism uses extra power availability to pump water back up into a hydroelectric dam reservoir, and then uses this stored potential energy to run traditional water turbines when the energy is needed later (Fig. 9.2).

Both the capacity and the potential for inexpensive pumped hydro storage are large. Currently, China has 1,646 GW of storage capacity, while the United States has 1,074 GW of capacity. This technique is ideal in that the round-trip efficiency of pumping and generation can exceed 80%. Collocating a facility at an existing dam ensures there already exists grid infrastructure that can provide potential energy storage resources.

Floating solar panels can cover reservoirs to economize land use and reduce water evaporation. The cooler water below PV panels also increases their efficiency. Such colocation can then take advantage of a ready grid connection and the storage capacity of pumped hydro to ensure continuous power provision. One recent estimate determined that the use of such *floatovoltaics* on existing reservoirs globally could meet 240% of U.S. electricity needs through the combination of existing hydropower and new solar installations.[2]

Engineers have also proposed using shuttered mines as ideal locations for similar storage, using either pumped water or the raising and lowering of large masses to store potential energy until needed.

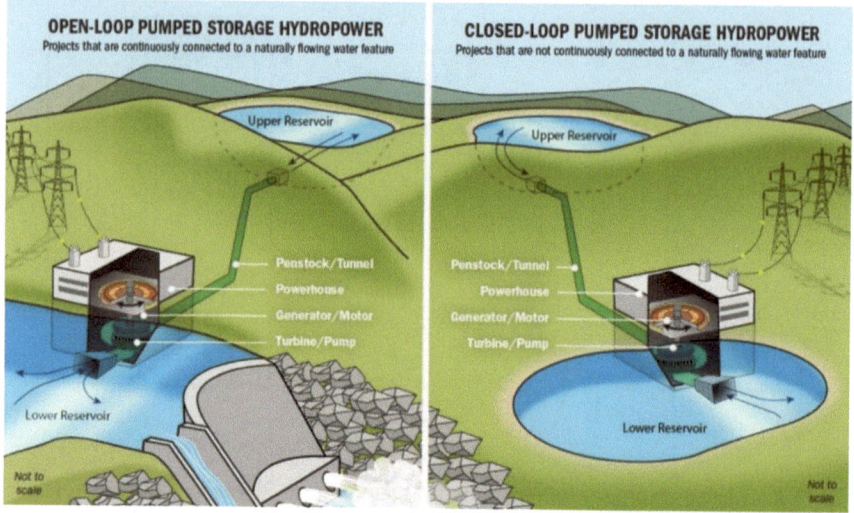

Fig. 9.2 Pumped Hydro (Courtesy U.S. Department of Energy, https://www. energy.gov/eere/water/pumped-storage-hydropower, accessed January 23, 2023)

FLYWHEELS AND COMPRESSED GASES

There are various ways to store kinetic energy so it may be used later to fill gaps in energy supply arising from darkness and cloudiness or lulls in wind that may cause solar and wind power intermittency. These techniques can also help stabilize an alternating current (AC) grid to ensure temporary loads do not cause ripples in the supply frequency and voltage that affect users.

Flywheels store energy by using temporary surplus electricity to drive a motor that spins a very heavy wheel. This energy can then be recovered very quickly or over time by tapping the spinning wheel to drive a generator. Such devices can operate with high efficiency. An energy storage system in Stephentown, NY operated by Beacon Power employed 200 flywheels to provide up to 5 MWh of energy storage. An 80 MWh facility operated by Pacific Gas and Electric (PG&E) in Fresno, California can spread the energy discharge over four hours.

Gases such as carbon dioxide or air can also be compressed with a motor and pump driven by excess electrical energy to pressurize a tank, and then drawn down to spin a generator when needed. The Hydrostor facility in California can store up to 10 GWh of energy through Compressed Air Energy Storage (CAES) at 60 percent efficiency to spread its discharge out over eight to twelve hours (Fig. 9.3).

Fig. 9.3 Flywheel Storage (Courtesy of Pjrensburg, CC BY-SA 3.0, https://creativec ommons.org/licenses/by-sa/3.0, via Wikimedia Commons)

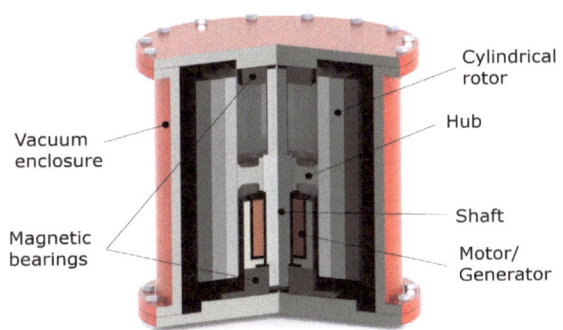

Latent Heat

About half of human energy usage is for heating and cooling. Much of these immediate needs can be met with heat pumps and geothermal energy described earlier. It is also possible to store heat that is generated intermittently from stranded or excess hydro or wind power, or from waste heat arising from industrial processes or from nuclear fission and fusion.

Often such heat is considered waste and dispersed into lakes, streams, and the atmosphere. While these sources of heat do not constitute significantly to global warming when compared to the total insolation energy worldwide, the waste heat can be tapped to replace energy required for residential and commercial heating and for industrial processes.

Already such heat regeneration is common at the most sophisticated and efficient fossil-fueled natural gas and coal-fired thermal plants that power more than 60% of global electricity needs. The best designs are cogeneration plants, often called ***combined heat and power*** (CHP) plants in the industry, that tap the wasted heat contained in steam exhausted from a steam turbine spinning a generator to provide heat for industrial processes, to prewarm water to make steam at the power plant, or to warm adjoining residential and commercial buildings. By recycling the otherwise wasted heat, a fossil-fueled power plant can recover 65–80% of the heat content of their fuel, well in excess of the 50% normally converted to mechanical power to run generators in a typical steam power plant.

This byproduct heat is typical of a temperature range of 100–180 °C, which is the ideal temperature range for commercial steam heating and for residential heating when mixed with circulating water. Large tracts of Manhattan in New York City are heated with excess steam from Consolidated Edison's power plants. This cogeneration ability is the largest of the top ten such commercial steam systems in the world and provides more heat than the next nine systems combined.[3] The Manhattan Consolidated Edison network includes 105 miles of pipe and helps heat and cool 2,000 buildings in New York City. Visitors are familiar with plumes of steam escaping sidewalk manholes that vent the system on occasion.

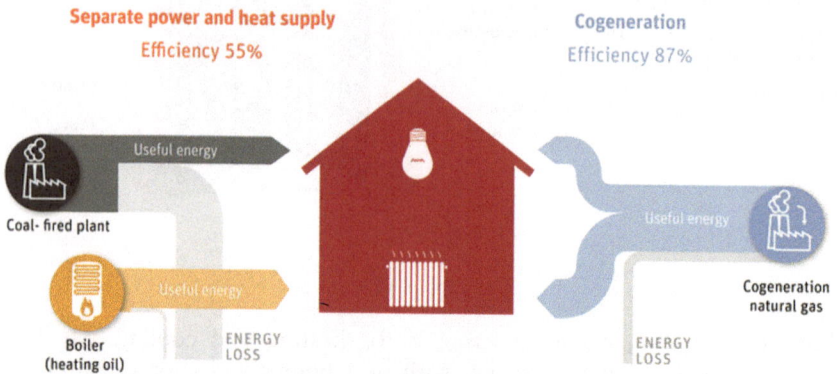

Why cogeneration is more efficient than conventional coal power plants
Comparing the energy efficiency of cogeneration with conventional coal power plant and heating system
Source: ASUE

Fig. 9.4 Cogeneration Process to Recycle Excess Heat of Combustion (Courtesy Heinrich-Böll-Stiftung, https://www.flickr.com/photos/boellstiftung/383596 36032, accessed October 12, 2022)

Such heat regeneration is most efficient when the steam generation facility is collocated near potential users of the waste heat. Recovery of waste heat can almost double the ability of a fossil-fueled power plant to efficiently use the chemical heat in methane or coal (Fig. 9.4).

THERMAL ENERGY STORAGE

Waste heat can be stored and distributed. For instance, much like the flow battery described above, waste heat can be absorbed in an endothermic (heat absorbing) chemical reaction that can subsequently be reversed when needed in an exothermic reaction to release the excess heat at another time or place. Such *thermal energy storage* (TES) systems can be designed in several ways based on carefully formulated goals.

Heat storage can satisfy two roles. The first is short-term heat shifting, from day excess heat removal to night heating requirements, or for winter usage of heat stored in the summer, called *seasonal thermal heat storage*. Highly efficient insulation that can trap heat for a long period of time, combined with storage media such as water, rock, molten salts, native Earth, sand, gravel, or other media is relatively low cost and low technology methods to store heat.

Materials differ in their ability to hold heat based on their *specific heat capacity*. Materials are chosen not only based on their specific heat but also on their optimal temperature range. For instance, a molten salt may be an appropriate storage medium for high temperature heat such as steam generation byproducts or from concentrated solar collectors driven by mirrors reflecting insolation. Ice/water phase transitions can also be a good thermal mass for cooling.

Eutectic mixtures of salts such as sodium nitrate, potassium nitrate, and calcium nitrate are ideal heat transportation and storage mechanisms for solar collectors. The salt mix melts at temperatures slightly above the temperature of low-pressure steam. The solution remains liquid in storage at temperatures of around 250–300 °C in a *cold tank* before pumped to a heat source that raises its temperature to 500–600 °C for storage in a *hot tank*. This heat can be stored easily for days or weeks, and can subsequently be used to produce steam for a steam-driven electricity generation.

As an example, a 100-megawatt turbine can be driven by a 1000 cubic meter hot tank for about sixteen hours. In this way, a combined solar and steam facility can distribute heat generated from a solar collector during the day to drive a steam-powered turbine, or use stored heat to power and warm a small city at night. This heat storage and later conversion to electricity is highly efficient and comparable to direct heat to electricity generation in the absence of storage. Any waste heat can also be used for cogeneration to warm structures at night.

Recent research into the direct conversion of heat to electricity is also being explored. These methods are believed to offer similar heat to electricity efficiencies as solar concentration and hot tank to electricity conversion but without the moving parts. These technologies allow the *decoupling* of heat supply and electricity demand. They have successfully decoupled heat generation and heat demand in centralized and distributed heat systems that power whole towns in Nordic countries.

Much more efficient methods involve phase change materials. Systems that store large amounts of heat by relying on the specific heat of phase change can be very compact and efficient. A material heated through a phase change from solid to liquid, or liquid to gaseous, can absorb a significant amount of phase-change heat, called *latent heat*, that depends on the material and the phase change. For instance, the conversion of solid to molten silicon can store more than one megawatt-hour of energy per cubic meter of silicon volume if handling and storage can accommodate temperatures in the range of 1400 °C. This volume is sufficient to accommodate the heating needs of an average home for a month. By comparison, the same cubic meter volume of ice/water phase change storage can store less than 0.1 megawatt-hours.

THERMO-CHEMICAL STORAGE

Concentrated solar energy works as an excellent heat source for a **thermo-chemical storage** (TCS) system. Certain ideal **thermo-chemical materials** (TCM) can be used to decompose potassium oxide into its constituent parts at an ideal solar collector temperature range of 300–800 °C to store 2.1 megajoules of energy per kilogram of compound. The constituent parts can be separated and then recombined later to return the heat in a way analogous to the method a flow battery can store chemical energy for subsequent conversion to electrical energy. Other molecules, such as nitrosyl chloride can be separated by photon bombardment and recombined when the heat energy is required.

One low-cost thermo-chemical storage system uses salt hydrates such as sodium hydroxide. When mixed in a solution with water, heat from a solar collector can evaporate off the water. Heat energy can be returned through an exothermic reaction with about 60% efficiency by simply adding water later. This method offers a far higher energy density than heat stored as hot water or steam alone, and the exothermic reaction can be delayed for months or years until needed without any special insulation. In essence, this reaction also works like a flow battery by keeping the two substances apart at ambient pressure and temperature and then combining them when needed.

The innovative Dutch company TNO determined that summer heat can evaporate off water to leave just a few cubic meters of sodium hydroxide salt. In winter, the salt can be combined with four to eight cubic meters of water to release about one gigajoule per cubic meter of salt storage. This energy storage capacity is sufficient to heat an energy-efficient home in many temperate climates such as that found in the Netherlands.[4]

THE HYDROGEN ECONOMY

The most hyped energy source is often described as the essential element of a **Hydrogen Economy**. Let us first describe hydrogen's advantages before its disadvantages are explored.

Hydrogen burns incredibly cleanly and gives off immense heat from the release of its chemical bonds:

$$2H_2 + O_2 \rightarrow 2H_2O + \text{heat energy}$$

On a mass basis, a kilogram of hydrogen produces the same heat energy as 2.8 kilograms of gasoline. It can also burn more efficiently in an engine, or electricity can be extracted directly in fuel cells to achieve about 50% efficiency. For transportation, 1 kilogram of hydrogen can replace about 4 kilograms of gasoline.

For heating purposes, 1 kilogram of hydrogen is equivalent to about three kilograms of gasoline and about 2.5 kilograms of fuel oil. One thousand cubic feet (1 MCF) of hydrogen weighs 2.41 kilograms, which is a fraction of the weight of an equivalent volume of natural gas. For 15% of the weight, it provides about 30% of the heat of methane, so hydrogen's energy density per unit of mass is twice that of natural gas.

In addition, hydrogen produces no carbon dioxide upon combustion, and any other inadvertent byproducts beyond water vapor can be controlled relatively easily. Finally, some believe that the huge infrastructure for the transportation of natural gas may also be adaptable to transport hydrogen.

Nonetheless, hydrogen also has several drawbacks. First, the existing natural gas infrastructure that is usually proposed to store and distribute hydrogen is often old and hence unreliable. It has been estimated that about 10% of natural gas is lost due to leaks in the production and distribution network. Hydrogen is a much smaller molecule and can squeeze through leaks at a far faster rate than can methane. Also, some pipeline distribution materials become brittle when exposed to hydrogen over time.

The existing distribution infrastructure would require significant upgrades if it is redeployed to transport gaseous hydrogen through the existing infrastructure. Should *blue hydrogen* be produced by stripping hydrogen from natural gas at the wellhead, the remaining carbon dioxide could then be transported to and sequestered in underground gas wells as the natural gas is removed. However, the existing natural gas distribution infrastructure is not optimized to return carbon dioxide from locations where hydrogen may be generated.

Hydrogen is also costly to store. Because it stores only about 30% of the heat of natural gas for a similar volume, to substitute for natural gas storage would require more than a tripling of tank volumes. Natural gas can be liquified to be much more energy dense if brought down to a temperature of about −162° Celsius. Hydrogen liquefaction requires a significantly colder temperature of −253° Celsius, which is only 33° Celsius above absolute zero. Such temperatures are difficult and expensive to maintain.

Storage is a significant issue for hydrogen in transportation which requires compact and very safe gaseous or liquid storage given the frequency of vehicular collisions and the proximity of humans. Hydrogen is not as dangerous as the Hindenburg Dirigible Disaster suggests. Much of the flames we witnessed in the notorious Hindenburg explosion video were from the ignition of the doped skin vessel that contained the hydrogen rather than the gas itself. Nonetheless, significant challenges remain to store hydrogen at a sufficiently high energy density for transportation use.

Finally, hydrogen exists almost nowhere naturally in its molecular form. It must be created, typically by stripping it from other molecules, most commonly from water or methane.

Water separation into hydrogen and oxygen occurs through electrolysis or similar electrochemical mechanisms. Electricity breaks the very tight bonds of hydrogen to water. Electrolysis is a very energy intensive process, with about 70–80% efficiency, which sacrifices about 25% of its energy at the onset. If the electricity used is green, this process creates *green hydrogen.*

As noted above, hydrogen can also be stripped from natural gas to produce *blue hydrogen.* If the resulting carbon or carbon dioxide can be recovered to prevent further global warming, this process ranges from 70 to 85% efficient. The process typically relies on steam reforming that requires steam at a temperature of 700 to 1100° Celsius to react with natural gas to form carbon monoxide and hydrogen molecules H_2, in addition to about ten tons of carbon dioxide per ton of hydrogen produced. In turn, the carbon monoxide can be used as a feedstock to synthesize other fuels. Finally, the excess heat of nuclear fission can be used to split water into its constituent components to create a supply of hydrogen gas. This *pink hydrogen* that is produced using the heat of a nuclear reaction to tip the balance of equilibrium of water toward the disassociated hydrogen and oxygen components would still need to be distributed and stored.

If hydrogen must be produced from electricity through electrolysis, and then subsequently converted back to electricity in a fuel cell, the round-trip efficiency losses from production (of 75% efficiency), electricity transportation (at 90% efficiency), and then reconversion of hydrogen to electricity in a fuel cell (50% efficiency) is then (75% × 90% × 50%) = 33.75%. In other words, hydrogen functions as an electricity storage system with only 33.75% efficiency.

As a heat source, hydrogen fares somewhat better since the combustion of hydrogen is about 95% efficient. Production (at 75% efficiency) to storage and distribution (at 90% efficiency), and combustion (with 95% efficiency) results in an efficiency of (75% × 90% × 95%) = 64%, comparable to about 85% efficiency should natural gas have been consumed for heating. If natural gas was not required as its feedstock, greenhouse gas generation is avoided. But, the round-trip efficiency for heating purposes is still substantially worse than reliance on solar electric-driven heat pumps.

Hydrogen is best regarded as an energy storage system rather than a sustainable energy source, given that it must first be produced by consuming electricity, in the case of *green* hydrogen, or derived from a fuel such as natural gas. As a battery, it compares poorly even to some of the cheapest but lower efficiency batteries.

FUEL CELLS

Many of the energy storage methods described, and especially the creation of hydrogen, require some method to convert their various forms of chemical energy into either heat or electricity. Most of these energy storage forms readily convert their chemical energy into heat. This is ideal when combustion is an essential process upon which we rely, for instance for the heating of structures, of industrial ovens, and for cooking.

We typically use a variety of *heat engines* to convert combustion heat into mechanical or electrical energy. The *internal combustion engines* in most vehicles are an example of the use of small often spark-ignited combustions to expand gases, drive pistons, and rotate crankshafts and wheels. Such heat engines are invariably inefficient, though, with about 70% of the energy lost in the translation, and only 50–60% recoverable within even the most efficient heat engines.

Fuel cells perform a much more direct conversion. They operate by performing a *redox* reaction that simultaneously removes electrons from molecules on one side of a membrane and reunites these electrons with related molecules on the other side. In doing so, it forces the electrons to follow a path of current that can drive electric motors to power the electric grid.

While such fuel cells can run on a variety of fuels such as methanol and carbonates, the most common type is hydrogen powered. In that reaction, hydrogen molecules are stripped of their electrons and permitted to reabsorb the electrons as the hydrogen protons combine with oxygen. The products of this processing of hydrogen are electricity and water. The process also produces heat and must overcome the process' activation energy. Their efficiency is about 60% at best. However, such an efficiency is about double that of an internal combustion engine.

The round-trip efficiency of hydrogen and fuel cells is not large compared to some other flow batteries as described earlier. However, the hydrogen fuel cell is an essential ingredient for the viability of hydrogen as a low weight fuel source for weight-critical applications such as airplanes, should abundant sustainable energy become available to create *green hydrogen*.

ENERGY DISTRIBUTION

Current electric energy distribution systems called electric grids need modernization. These grids were optimized almost a century ago to move electricity from a few large plants, hydroelectric near water, or coal-fired plants near coalfields, and for an economy that was far more rural than it is today. Now, modern sustainable power sources are ideally located near large tracts of sunlit fields between about 35 degrees latitude North and South, near wind-driven plains or near offshore sites, or near geothermal and hydroelectric sources.

A new electric grid design must optimize for these new energy sources. For resiliency, a redesigned grid may even connect thousands of relatively self-contained microgrids. For instance, batteries in electric cars can be charged at night and then help supplement the grid during the day. These batteries, solar panels, and most of the appliances in our home either run on direct current or could be redesigned to operate better using DC.

Innovations in modes of energy production and consumption require a redesign of electric grids. New and more efficient direct current transmission lines either must be constructed, or existing lines modified to take advantage of the ability of direct current to move electricity more efficiently than alternating current. Some of the largest new transmission lines constructed today in the United States and especially in China have used the new High Voltage Direct Current (HVDC) technology to improve capacity and efficiency. However, given the AC design from more than a century ago represents the backbone of our residential and commercial infrastructure today, such a redesign is unlikely without new and enlightened investment.

ENERGY EFFICIENCY

The final energy concept is not to use energy when we do not need it. Benjamin Franklin said "a penny saved is a penny earned." The same adage applies to energy efficiency improvements. Better insulation, windows, and building products can allow us to conserve, just as can the broad use of heat pumps, more efficient appliances, and LED lighting. Retrofitting existing homes and buildings and mandating more energy-efficient new construction carry significant investment costs but can also pay back the investment very quickly through reduced energy usage. However, since many households and businesses are constrained in their access to financial capital to make these investments, government programs that offer investment incentives are often necessary.

Such programs of tax deductions for energy efficiency improvements can especially assist low-income households that are often unable to borrow to make energy efficiency investments. Direct aid is necessary for the lower income homes or Less Developed Countries that do not have sufficient financial capacity to make such investments in energy conservation. For many households, businesses, and nations, direct financial aid may be necessary.

However, this aid pays dividends in the form of reduced need to build new power plants to fuel a Green New Deal, or in the opportunity to take offline legacy coal or natural gas power generation plants that continue to supply a large amount of power in many nations. Such aid in pursuit of greater efficiency can also permit the promotion of *economic and environmental justice* to address the concerns of the Brundtland Commission for economic development for those least advantaged.

SUMMARY

Every energy source provides a different set of characteristics. Society has come to expect reliability and resilience from the energy that heats our dwellings and structures, fuels our industrial processes, and runs our vehicles and appliances.

Some of the most promising and inexpensive energy sources rely on nature, which functions in cycles of day and night, over the seasons, and through changes in our weather. Key to energy resilience and reliability are mechanisms to store energy so we may use sustainable energy sources when we need them. Since the dawn of the Industrial Revolution, our energy infrastructure has been optimized around fossil fuels to ensure their reliability. In a sustainable world, we must create similar infrastructures to provide for the same dependability from sustainable energy sources. Energy storage is essential for this transition.

ESG Toolbox

Based on Benjamin Franklin's adage *a penny saved is a penny earned*, a corporation has many opportunities to demonstrate avoided carbon dioxide emissions through conservation.

A company can also take advantage of demand shifting to use energy at off-peak times. By permitting before or after peak hour work or by permitting work from home (WFH), a company can reduce grid load at times in which the grid must rely on natural gas peaker plants.

A company can also encourage electric car usage and mass transit for calculable reductions in a company's carbon footprint.

These policies may also have positive employee morale advantages and may enable some diverse employees to navigate workplace challenges more easily.

NOTES

1. https://www.energy.gov/sites/default/files/2021-09/Solar%20Futures%20Study.pdf, accessed September 29, 2022.
2. Yubin, J., Shijie, H., Alan, D. Z., Luke, G., Campbell, J. E., Xu, R., Chen, D., Zhu, K., Zheng, Y., Ye, B., Ye F., & Zeng Z. (2023), "Energy production and water savings from floating solar photovoltaics on global reservoirs," Nature Sustainability, 13 March.
3. Moyer, G. (2014, October 9,). "Miles of Steam Pipes Snake Beneath New York". The New York Times. Retrieved October 12, 2022.
4. https://en.wikipedia.org/wiki/Thermal_energy_storage, accessed October 13, 2022.

CHAPTER 10

The Natural History of Fossil Fuels

This chapter draws upon material I presented in a book entitled <u>BP</u> <u>and the</u> <u>Macondo Spill</u>.[1] In that book I documented how our quest for new sources of fossil fuels has induced us to take resource extraction risks that increasingly damage the environment.

Humankind's attachment to fossil fuels parallels the Industrial Revolution. New fuels powered new industries, provided profits for some of the world's largest companies and funded some of the largest sovereign wealth funds. Fossil fuels have become intertwined into economies, livelihoods, and, increasingly, our precarious energy future. The natural history of the creation of petroleum punctuates the Earth's natural history, dating back to an era not unlike our own—one of fast-rising carbon dioxide in our atmosphere.

A century ago, an article linked the death of the dinosaur to the fuel in one's car. In doing so, the article created an urban myth that is perpetuated to this day. Scientists and geologists now agree that oil formed not from the decomposition of large extinct land-based dinosaurs as was once thought, but rather from the aggregation of large stocks of some of the smallest sea-based organisms and the sequestration of the hydrocarbon molecules they contain. Indeed, one of the best predictors of oil is in the identification of geological formations that could best harbour the remnants of these marine ecosystems.

Almost all organisms are made up of carbon-based molecules that are the stuff of most life on Earth and all organic chemistry. Recall that the four most common elements in the universe are hydrogen, helium, oxygen, and carbon. The Earth has lost almost all its atmospheric elemental hydrogen and helium since its formation because these gases are relatively buoyant and have escaped our atmosphere long ago. But, while helium's nonreactivity means only a small

C. Read, *Understanding Sustainability Principles and ESG Policies*, https://doi.org/10.1007/978-3-031-34483-1_10

amount has been sequestered underground, molecules containing hydrogen, oxygen, and carbon have been combining in various other compounds since time immemorial. They are accompanied with other less abundant atoms such as nitrogen, sulphur, sodium, potassium, phosphorous, iron, magnesium, and many others, to form the building blocks of life. The most basic molecules that have given us life are simple combinations with carbon, given its strong affinity to bond broadly with other molecules, and the equally reactive atoms of hydrogen and oxygen.

Our first clues to the factors that combined to produce fossil fuels came from the 1930s discovery by the German chemist Alfred Treibs of remnants of chlorophyll in oil deposits. At first, the theory hypothesized that oil was formed from plant matter on land. Later, scientists discovered that offshore oil deposits also contained molecules that survive only on the ocean floor. These microscopic organisms are so small that a single drop of seawater may contain a million such organisms.

Now, modern science can be used to associate different microbes with different grades of oil. We can now deduce that the precursors of oil will be found where there once were prolific swamps or seas of algae during periods of high atmospheric carbon dioxide concentration. The various ages of these microbe fossils give oil geologists information about the length of time the process of heat and pressure has acted on oil deposits. The unique combination of organisms, heat, pressure, and time gives each oil field its distinct characteristics.

The combination of heat and pressure then converts these biomasses into various mixes of hydrocarbons, based on their combination of material and depth. This optimum combination of light and heat, carbon dioxide and water, created an environment in which these microbes and biomass grew and died more rapidly than the ability of the ocean to absorb them or the seabed or swamp floor to decay and redistribute them. When excessive biomass creation occurs, layers upon layers are compounded on sea beds in a form of ever-thickening black bio-mud. The theory explains why, after more than a hundred years of exploration and extraction, humans have found oil in areas of land that had once been undersea, and why, as those sources of easier-to-find oil are exhausted, we find ourselves increasingly drilling farther offshore and in deeper waters.

Over eras on Earth, organisms have evolved to create longer, even more complex, and more specialized molecules that are constituted by the same elements that make up alkanes, the primary molecules in petroleum. The simplest of these organisms are microscopic, and most of them have resided in the sea since the primordial history of our planet. The combination of abundant water, sunlight, and atmospheric carbon dioxide provided the basic building blocks for these first simple organisms.

The availability of carbon dioxide, in addition to the plentiful sunlight and water, created ideal conditions for plant life, and ultimately for oil. As these organisms lived and grew 100–300 million years ago in the Palaeozoic

Era, they combined with water and high concentrations of carbon dioxide to produce hydrocarbons and sugars. As organisms died, gravity took them to the bottom of the ocean, where trillions of dead organisms amassed to produce thick layers of decomposing organic matter. The more abundant their source of carbon, in the form of carbon dioxide, the greater this biomass production that would die off and layer the ocean floor.

When geological processes cover up this layer of organic material with layers of silt and sand, which eventually turns into rock, the ideal conditions to create oil are in place. Under this prevailing theory of oil production, we can predict the discovery of oil based on a few precursors. The initial conditions are abundant sunlight and water, combined with high concentrations of the atmospheric carbon dioxide upon which some simple organisms thrive, forces that allow the biomass to sink and form layers, and the creation of a blanketing layer of inorganic matter. A blanketing layer of sediment and sand that turned to sedimentary rock sequestered immense amounts of carbon. This blanket created the high temperature and pressure that baked the organic matter into oil.

There have been periods in the Earth's past such as the *Palaeozoic Era* that were ideal to create the concentrations of carbon dioxide so necessary for the prolific creation and eventual sequestration of organic matter. We now know from ice samples that the concentration of atmospheric carbon dioxide has cycled significantly across the millennia. The level of atmospheric carbon dioxide follows regular cycles that correspond to increased glaciation when the carbon dioxide level in the atmosphere is low, to global warming in eras with a high atmospheric carbon dioxide level.

Over these historic cycles, carbon dioxide concentrations reached lows that hover around 190 parts per million by volume (ppmv), while high levels, conducive to the creation of the algae and organisms that make oil, approach 300 parts per million by volume (ppmv). The Earth is currently on the increasing trend of one of these 100,000-year cycles, but with far higher levels of atmospheric carbon dioxide that have ever been measured or observed in the record of ice cores and fossilized trees. The rate of increase of this concentration, which now exceeds 420 ppmv, has also never been observed to now. This much higher concentration is accelerating because of our intensive use of oil since the onset of the Industrial Revolution, and its attendant release of previously sequestered carbon to the atmosphere.

This process of rising atmospheric carbon dioxide and the stimulation of microorganisms that created fossil fuels are what physicists and engineers call a negative feedback loop. The higher carbon dioxide levels stimulate plants and organisms that thrive by absorbing carbon dioxide. If these organisms become trapped in a sedimentary layer, carbon sequestration ensures the carbon dioxide cannot soon return to the atmosphere. This feedback loop reduces atmospheric carbon dioxide levels, at least until the resulting oil is burned and the sequestered carbon dioxide is returned to the atmosphere. However, once this sequestration of hydrocarbons process is reversed, it can instead trigger positive feedback loops that unleash trillions of tonnes of carbon dioxide at a rate even more rapid than sequestration trapped it.

Another factor that can accelerate the oil-making process is the abundant inflow of nutrients from rivers and streams. We now often find oil in gulfs where big rivers meet the ocean. This constant feeding of nutrients caused carbon-based organisms to amass and, subsequently, to be buried as the region is inundated with sand and silt layers. Time and pressure caused the sand, clay, and silt to turn to sandstone, entombing the primordial organic mud. Cut off from an oxygen-rich environment, and buried deeper and deeper by insulating sand and stone forming above, the organic mud cannot decay and dissipate. Instead, the mud rose in temperature from the ambient heat of the Earth's core, and was cooked and baked in a slow chemical process that consumed any remaining oxygen and produced hydrocarbons. Where the organic mud can cook at a temperature somewhere between hot and boiling water, the Earth created the ideal environment to produce oil.

Following this theory to its natural conclusion, we can surmise that future oil will be found increasingly in deeper offshore sites as more accessible pools of oil are exhausted. Cambridge Energy Associates observe that deepwater hydrocarbon extraction, defined as deeper than 2000 feet, has more than tripled in the past decade and has more than quadrupled since 2000. They also find that deepwater discoveries now make up most discoveries, and represent significantly larger fields. Increasingly, deepwater exploration and extraction are driving higher U.S. oil production. These new discoveries have allowed the United States to demonstrate greater year-to-year production of oil for the first time in almost two decades.

It has been estimated that humankind has used about one trillion barrels of oil. World crude oil reserves represented another 1.76 billion barrels, with a further 1.11 billion barrels of oil equivalent (BOE), of natural gas. Of course, the estimate of remaining reserves depends on both the price we are willing to pay and the success of new technologies to extract the remaining oil. In any event, experts believe that conventional oil reserves will last easily for another 50 years at current consumption levels. This time horizon may be shortened significantly if our pattern of economic growth and demographics shifts toward the rapidly growing economies of China and India.

Meanwhile, in the recent Anthropocene Epoch in which we have been relying on the burning of these fossil fuels, we are once again returning millions of years of sequestered carbon dioxide to the atmosphere. While we speak monolithically about oil, we actually consume hydrocarbons in a number of ways. We can differentiate between the various types of hydrocarbons produced by oil and consumed for their different energy contents.

Various Types of Hydrocarbons

As the name implies, hydrocarbons are molecules that combine the two elements hydrogen and carbon. Carbon is the basic building block of all living organisms, and is the element that defines organic chemistry. The saturated hydrocarbons that are constituted primarily of combinations of carbon and hydrogen include *paraffins*, *alkanes*, and *cycloalkanes*, the *alcohols* that combine carbon with oxygen and hydrogen in the form of a hydroxyl (OH) group, and carbohydrates that combine carbon with hydroxyls to produce sugars. All are variations of the simplest molecules that provide chemical fuel to much of what makes up our environment.

For instance, our body consumes carbohydrates, mostly in the form of sugars, and stores energy by converting carbohydrates to fats, to produce energy in our muscles when needed. The energy is used, and the sugars and fats are transformed into carbon dioxide and water. Likewise, alcohol can be burned in an oxygen-rich environment to produce energy, water, and carbon dioxide.

The simplest alcohol is methanol, with a chemical formula CH_3OH. When two molecules of methanol are ignited by combining them with three molecules of oxygen O_2, energy is given off, in addition to two molecules of carbon dioxide CO_2 and four molecules of water H_2O. We can reverse this process by combining water, carbon dioxide, and energy to produce methanol. Yeasts can create the alcohol in wine and beer, in contrast to the ability of chlorophyll to transform the energy from sunlight and carbon dioxide from the atmosphere to create sugars and oxygen that allow animals to survive.

The hydrocarbon combinations that are limited solely on carbon and hydrogen provide energy once the carbon-hydrogen bond in the molecule is oxidized, or severed, resulting in the formation of smaller molecules of carbon dioxide and water. Hydrocarbons function very much like alcohols and sugars. For instance, paraffins combine carbon atoms solely with hydrogen atoms in a ratio C_nH_{2n+2}. The slightly lighter-than-air-gas methane, which is the primary constituent of natural gas, combines one atom of carbon with four of hydrogen. The formula for this paraffin, also called an alkane, labeled C1, is CH_4.

Similarly, the next larger paraffin ethane, C_2H_6, is a heavier-than-air-gas that, with methane, makes up natural gas. Propane, also heavier than air, is labeled C3, and has the formula C_3H_8. The heaviest paraffin that is still a gas at room temperature is the C4 we know as butane, C_4H_{10}. Heavier combinations of hydrogen and carbon, from pentane C5 to C17, remain a liquid at room temperature. Still heavier paraffins run from C18 and above, and remain solid at room temperature.

Hydrocarbons are classified based on the range of carbon atoms in the various molecules. The heaviest, paraffin waxes, fall in the range of 20–40 carbon atoms per hydrocarbon molecule. However, the term *paraffin* can refer to any linear, or normal, hydrocarbon in which the carbon atoms are linked to each other in a chain using a single bond. Because carbon permits four bonds, there remain three other bonds for associated hydrogen atoms. These paraffins represent about a third of the weight of crude oil.

Alkenes are related to paraffins except that each of the carbon atoms is double-bonded to another, resulting in fewer remaining bonds for hydrogen atoms. Some commonly found alkenes include ethylene, butene, and isobutene.

Half of crude oil by weight, on average, is made up of **naphthenes**. These hydrogen-saturated carbon atoms are made up of one or more rings of carbon atoms, with the remaining bonds saturated by hydrogen in a ratio of $C_nH_{2(n+1-g)}$, where g is the number of carbon rings. The ring nature of the naphthenes is differentiated from the linear alkanes through the prefix cyclo-, for example, cyclopropane, cyclobutene, cyclopentane, and so on.

Crude oil is also made up of **aromatics** and **asphaltics**, in smaller amounts. Aromatics are similar to naphthenes, but with single and double bonds in their rings of carbon, with the remaining bonds joined to surrounding hydrogen atoms. Finally, asphaltics are molecules of carbon, hydrogen, oxygen, nitrogen, and sulphur that remain once the hydrocarbon molecules in crude oil are distilled off in the refining process. The remaining materials can be used to produce asphalt, the tar that is used for roadways.

Different grades of crude oil vary in its composition of these components of alkanes and alkenes, naphthenes, aromatics, and asphaltics. An oil that has a larger mix of low carbon number alkenes is called *light*, while *heavy* crude contains a greater proportion of the larger hydrocarbons. *Sweet* crude contains little sulphur, while *sour* crude may contain 6% sulphur or more. West Texas light sweet crude oil is ideal for gasoline production because it contained little sulphur and a higher share of the lighter, more volatile hydrocarbons needed to produce the gasoline that can combust efficiently in a spark ignition engine.

While the natural gas that fuels many homes, stoves, and power plants is predominantly methane, most of the other hydrocarbons we consume are a combination of various alkanes. For instance, hexane C6 through decane C10, in specific ratios, produces gasoline, diesel fuel, and aviation fuel. The less viscous gasoline relies on greater proportions of the lighter alkanes in the range, while increasingly thick and viscous hydrocarbons are made up with greater proportions of heavier alkanes.

The amount of energy contained in the various distillates of crude oil is related to the number of strong CH_2 bonds that can be converted to less robust bonds of the combustion byproducts carbon dioxide CO_2 and water H_2O. For instance, in the cycloalkanes that make up the bulk of crude oil the amount of energy which can be released is almost directly proportional to the number of CH_2 carbon-hydrogen combinations in the various hydro-carbon molecules (Fig. 10.1) where kcal is the number of kilocalories of energy released in combustion, and a mole is a measure of the number of molecules consumed in combustion. While the larger hydrocarbons can release

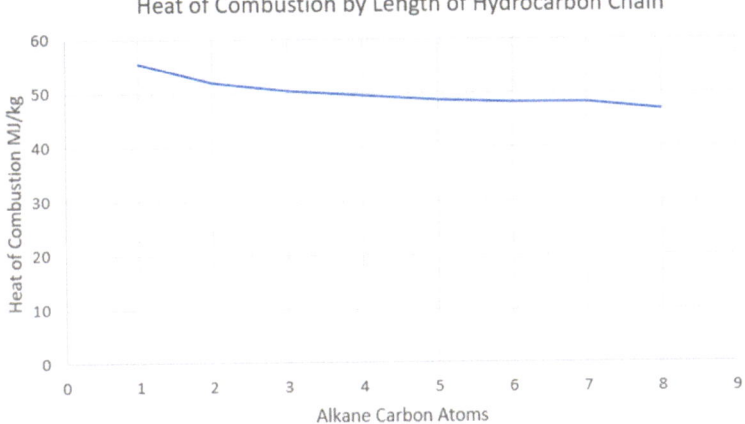

Fig. 10.1 Alkanes Ordered by Number of Bonds and Energy Density

more energy, the energy per unit of weight of hydrocarbon remains relatively constant. Hence, the energy content of hydrocarbons is differentiated primarily by the density, or weight per liter or gallon, of the fuel.

REFINING

The process of refining oil into its constituent components of alkenes and cycloalkanes relies on the fact that the lighter hydrocarbons are less dense, and hence are gaseous at lower temperatures. The traditional method to separate crude oil into these constituent molecules is through fractional distillation. When heat is gently applied to petroleum, the lightest molecules turn to a gaseous state first. The lighter molecules are progressively allowed to condense back to their pure liquid forms and are removed at various points in the distillation process. As the temperature of the petroleum rises, each molecule is distilled off until all of the constituent liquids have been separated and only residual tars remain.

Modern refining of crude oil takes advantage of other less expensive or more efficient processes to break some of the molecules up into smaller molecules. For instance, catalytic cracking uses metals or other catalysts to promote the breaking of some larger hydrocarbons into smaller ones that can be used to make more volatile products such as gasoline that commands a higher price in the market. While these catalytic reactions still require some energy, the nature of a catalytic reaction leaves the catalyst unchanged to efficiently permit further identical cracking reactions.

Once refined into its constituent parts, these liquids can be recombined to produce mixtures of fuel with the desired properties of volatility, density, and energy content. For instance, a 42-gallon barrel of light sweet crude oil can produce about 19 gallons of gasoline, made up primarily of C4 to C12, and a variety of longer alkenes. Other, heavier crudes are more amenable to the production of diesel or fuel oil, bunker oil, or other products that are heavier, denser, and less volatile.

While the science of petroleum refining is well-understood, it is not without risks. The combination of heat, reactive catalysts, chemicals used for chemical cracking, and the proximity of highly volatile and explosive products cannot be without hazard. Despite safety precautions, human error, defective valves, pipes corroded through constant contact with caustic chemicals and volatile hydrocarbons, and environmental factors create significant risks that must be managed but can never be reduced completely.

SUMMARY

The use of fossil fuels creates inherent risks and environmental costs arising from crude oil exploration, drilling, extraction, and transportation. We must also remain aware of other dangers associated with a hydrocarbon-based economy. A fossil fuel-based economy must manage the inherent risk created by volatile hydrocarbons, from refining and transportation to market, to refueling, highway risks, and the environmental consequences of burning hydrocarbons. All these factors contribute to the risks of an economy that derives a majority of its energy from volatile and highly combustible materials created a hundred million years ago and have safely sequestered carbon contained under land and under seas, until now. The world manages this risk during the most rapid release of sequestered carbon dioxide in the geological history of the Earth.

The greatest risk of fossil fuel reliance is in the existential threat to the sustainability of our atmosphere and fragile ecosystems. We will document in Section IV how the Industrial Revolution has created the greatest challenge to sustainability since the dawn of humankind. We will first develop in the next section the economic tools necessary to better understand and promote sustainability.

ESG Insights

Does your organization categorize the types of energy it uses based on its source and sustainability?

What would be the reserves necessary if every organization was required to pay the true costs of fossil fuel consumption from now and in the past?

NOTE

1. Read, C. (2011). BP and the Macondo Spill: The Complete Story, Palgrave Macmillan.

PART III

The Economics of Sustainability

The discipline of economics was contemplated as a set of tools to best manage the resources and the environment around us. Since the late nineteenth century, traditional economics has focused primarily on the positive issue of efficiency, with the more difficult normative ethics and sustainability questions receiving less attention. To accommodate fairness and equity, economists developed the concept of *Pareto Optimality* to prefer innovations that create larger gains than their *opportunity costs* and the losses of those who suffer from a change in a pattern of exchange, production, or consumption. While this Pareto criterion was designed to test efficiency, it can be reemployed in a novel way to allow us to recast the tools of economics to determine both *intergenerational efficiency* and the *economic justice* economists label *equity*. Our extension of the prevailing neoclassical model of economics into broader natural resource issues is the domain of *environmental economics*.

Sustainability is challenging. Future generations are precluded from, but also bounded by, the decisions of previous generations. Tools of economics must be extended into this new intergenerational realm by explicitly including time as an economic variable. Economists do so by first understanding intertemporal and intergenerational decision-making that is the basis for an emerging economic consensus on sustainability.

This section takes the reader through elementary concepts in the modeling of human choice at any given point in time. We will present sufficient economic theory from first principles for a reader with little economic knowledge beyond basic supply and demand. While the mathematics employed to tease our most rigorous results and intuitions may be quite advanced for some readers, we show that the mathematics simplifies considerably if our goal is to ensure perpetual intergenerational efficiency and economic justice. I also reserve the most rigorous mathematical treatments for appendices. One can comfortably skip equations in the body of these chapters and instead rely on the intuition we develop to better understand sustainable economic decision-making over time. The intuition we develop is also summarized as we draw upon it in subsequent sections following this economic analysis.

While traditional economics is generally used to treat efficiency in traditional markets, our tools can be used more extensively. A broader definition of choice in economics permits us to be as extensive as we wish it to be. For instance, the choices that give humans pleasure need not be constrained only to the goods we buy, but could also include other amenities we value, such as the existence of natural beauty, a clean environment, and a healthy ecosystem, or our regard for the future as well as the present. We show how our static decisions of today can be extended to define a path of such broad choices across generations.

We will find that humans and the ecosystem must always reckon with a resource constraint, though. We show in Chapter 12 that, even within a very simple two period model, mortal humans tend to devalue later use of a fixed natural resource compared to a previous time. We see that a fixed stock of natural capital then creates a challenging economic dilemma in that future generations are both excluded from and devalued in current decisions that draw down natural capital.

From the simple two period model, we can extend into the future to find that *abiotic natural resources* such as minerals, our atmosphere, and land will inevitably be depleted over time if managed without regard for the future. *Biotic resources* such as forests of fisheries that can reproduce are also typically managed with a preference for those who come earlier rather than later.

In Chapter 15, we present a fundamental test for economic sustainability. Is each generation of humans willing to establish *permanent funds* to replace the natural capital our generation consumes with a new form of capital on behalf of future generations? We close the section with a demonstration that we cannot solve these intergenerational *externalities* that force future generations to suffer the consequences of decisions made in earlier generations unless we also *internalize* the intragenerational externalities we impose on each other within any given generation.

As we embark on a discussion of the extension of classical economics into environmental issues that require more inclusive analytic attention, it is important to note that much of economics and finance remains decidedly market-oriented. Markets may be the appropriate domain for institutions closely tied to finances, such as the modern corporation. However, by eschewing equity and economic justice for efficiency through a central focus on market transactions, an entire and perhaps even more significant dimension is ignored. We shall see that what markets omit, in externalities and the consequences of mortal disregard for the immortality of humankind and the ecosystem, will require significant extensions of the classical economic model. We also glean a broader role for the government of peoples to ensure economic and ecosystem justice, normative issues for which the neoclassical model says little. This broader set of considerations is the domain of *ecological economics*, a subject with which we close at the end of the book.

CHAPTER 11

A Brief History of Time in Economic Decision-Making

Much of mainstream economics is oriented around decisions at a given point of time. It relies on traditional market measures such as prices, quantities, and income to describe optimal decisions humans make. Economies transform human-made factors of production to produce the goods and services households value. These goods and services are exchanged to maximize our flow of enjoyment, labeled utility, net of the costs we bear. Such traditional economic tools are less often employed to treat decisions over time rather than a point in time, or in the use of natural assets that span generations. The next task is to discover how mainstream market-oriented economics can be modified to provide us with insights into sustainability.

We are all participants in markets, and each of us has had some encounter with an interest rate. While economics has evolved over almost two hundred and fifty years, it has only been over the last century that economists have gained an understanding of how humans interact with markets and interest rates to make decisions that affect both the present and the future. Economic theory shows that humans align our regard for time to the collective regard of the marketplace.

Our modern interpretation of time in economic decision-making originated with the celebrated economist Irving Fisher at the beginning of the twentieth century. In his publication of The Rate of Interest"[1] in 1907, Fisher described the concept he labeled as *impatience*. He further elaborated on his *rate of time preference* in the 1930 The Theory of Interest.[2] Fisher showed that our degree of impatience and self-control as mortal humans has implications on how we make decisions over time. If we can first understand how each generation regards time and impatience, we can better manage resources for all generations.

© The Author(s), under exclusive license to Springer Nature Switzerland AG 2023
C. Read, *Understanding Sustainability Principles and ESG Policies*,
https://doi.org/10.1007/978-3-031-34483-1_11

Before Fisher, the interest rate was considered significant primarily for investment decisions. Fisher provided us with the framework to fully incorporate time into our decision-making by extending the interest rate to discount the future compared to the present. For instance, we recognize that an investment opportunity that can yield $1000 per year for the next ten years, but has no value beyond that, is not worth a full $10,000 in total value today.

Such future flows of income must be discounted back to today to acknowledge that a flow of income delayed precludes one from enjoying other uses of that income in the meantime. It is this *discount factor* that measures the loss of flexibility and potential reward we experience by having income tied up and its enjoyment delayed, even in the absence of an uncertain future.

We each differ in the degree to which Fisher described as our individual *economic impatience*. Some are more likely to live in the moment with little regard for the future. For others, the future weighs heavily upon their decision-making today. Those reaching the end of their mortality may hold even less regard for time in the more distant future, while those planning for the latter half of their life may sacrifice much today for greater comfort later in life.

There may also be cultural elements to our regard for time. A culture with a faith that their God shall provide for them may be less apt to feel a compelling need to provide for their own future, while others with less faith may scrimp, save, preserve, and stock up for seasons ahead because of a self-reliant spirit.

Fisher understood this very human and individual aspect of our regard for the present and a discounting of the future. He argued that those more willing to save for the future than the median are savers, while those who live more for the present than their average compatriot will be borrowers. He wrote:

> *The rates of preference among different individuals are equalized by borrowing and lending or, what amounts to the same thing, by buying and selling. An individual whose rate of preference for present enjoyment is unduly high will contrive to modify his income stream by increasing it in the present at the expense of the future. The effect will be upon society as a whole that those individuals who have an abnormally low estimate of the future and its needs will gradually part with the more durable instruments, and these will tend to gravitate into the hands of those who have the opposite trait. ... This progressive sifting, by which the spenders grow poorer and the savers richer, would go on even if there were no risk element. But it goes on far faster when as in actual life, there is risk. While savings unaided by luck will ultimately enrich the saver, the process is slow as compared with the rapid enrichment which comes from the good fortune of those few who assume risks and then happen to guess right.[3]*

Fisher's articulation of our regard for time in our economic decision-making offered an eloquent explanation for the economic motivations of each generation. His theory that describes our willingness to economize on our financial capital, or to consume it at a faster rate, was developed to better understand financial decisions. But, financial capital is not the only form of our capacity to produce and consume now and in the future. We must also manage our **natural capital**, some of which is in fixed supply. Fisher's method can be extended to help us understand why the impatience of one generation's resource extraction or regard for the environment may ill-serve later generations.

The Indifference Curve

A generation before Fisher incorporated time into human decision-making, economists had developed a tool that can be used to determine how human decisions are made at a given point in time. Formulated by Francis Ysidro Edgeworth in 1881 in his book Mathematical Psychics: An Essay on the Application of Mathematics to the Moral Sciences,[4] Edgeworth described the *indifference curve* as the collection of points of consumption of two goods that yields an identical level of human satisfaction.

Edgeworth had departed from some of his predecessors in their belief that happiness derived from our economic decisions could someday be measured objectively and compared easily across individuals. Should such comparisons be possible, the nineteenth-century economist Jeremy Bentham postulated that one could then use economic tools to construct allocations that create the *greatest good for the greatest number of people*. However, more than a century and a half after Jeremy Bentham expressed hope we could someday measure human preferences with the same rigor and cardinality as the physical sciences enjoy, we now know that the intensity of human pleasure cannot be measured on a convenient scale.

Edgeworth solved that paradox of immeasurability with a clever trick that Fisher went on to exploit a quarter of a century later. Edgeworth asked a related question that did not require such cardinality of measurement of our happiness, a measure that economists still call **utility**. Instead, Edgeworth pondered how humans are willing to trade off one good for another, all the while as they enjoy a given level of happiness.

For instance, while I cannot relate to you my current level of happiness or satisfaction with any measure that would have meaning to you, I can still compare choices according to my personal utility yardstick. I know that I may enjoy coffee cake and I may like a cup of coffee. I realize that, if I must sacrifice a bit of coffee cake consumption in a day, month, or year, I can estimate how many cups of coffee I would need to compensate me equally for that sacrifice. Such a locus of comparisons, with the number of cups of coffee on one axis of a graph and pieces of cake on another, would represent a downward sloping curve that represents the quality that, if I am deprived of one thing I enjoy, I can be compensated for it by consuming some of another item.

This tradeoff between comparable goods that maintain equal satisfaction does not require any sort of universally standardized cardinal measurement of well-being across individuals. It then offers a profound and creative way to view the satisfaction of human needs. Edgeworth was able to take the concept further, though. He expressed our utility as a function of the quantity q of various items we enjoy (called goods, for convenience), even if our example only compared two at one time. This utility function is of the form:

$$U(q_1, q_2, q_3, \ldots),$$

where each q_i represents the various quantities of the ith good one may consume. Such an indifference curve is simply the locus of points such that the utility function $U(q_1, q_2, q_3, \ldots)$ is constant. For satisfaction, or utility U, to be held constant for an individual, if the quantity of one of the elements q_i that gives us pleasure falls, the quantity of any of the other goods q_i must rise sufficiently in compensation to maintain the constant level of utility $\overline{U}$ (where the bar above the variable indicates it is held constant). For a constant level of satisfaction, this yields a negative tradeoff between items consumed.

Three additional observations can be made. First, if one were to graph a series of such curves, these downward sloping curves cannot cross. To cross would mean one identical bundle of the various goods $(q_1, q_2, q_3, \ldots)$ consumed would yield two different measures of utility, which would defy logic.

The second helpful observation flows from this requirement as well. Indifference curves farther from the origin must represent higher levels of happiness since they represent a greater amount of consumption of all the various goods. This property occurs at least up to a point when one satiates on a good such that it provides no more additional satisfaction.

Finally, the negative tradeoff between goods is not linear. If a person has little of one good and much of another, presumably the individual would be willing to trade a large amount of the abundant good for even a small amount of the dear good. In other words, our consumption exhibits **dimin-ishing marginal utility**, which describes the phenomenon that, as we have more and more of something, it provides smaller and smaller increments in happiness and hence is easier to sacrifice for something else of value.

If one accepts these observations, then Edgeworth's indifference curves are non-intersecting, downward sloping, and flatten to the right, with higher curves representing a greater level of satisfaction according to the following map of various indifference curves the yield increasing levels of utility (Fig. 11.1):

These indifference curves contain within them a wealth of information. They reveal at any point the rate an individual would trade one good for another. Consider the point in which the individual consumes two units of good 1 ($q_1 = 2$) perhaps coffee, and two units of good 2 ($q_2 = 2$) perhaps coffee cake. The slope at that point gives the (negative) tradeoff rate between the two goods, while overall satisfaction is held constant.

This rate is called the **Marginal Rate of Substitution**, and is easy to derive since we know the utility level is held constant at such a point. We can then explore how small changes is the quantity of one good must be accommodated

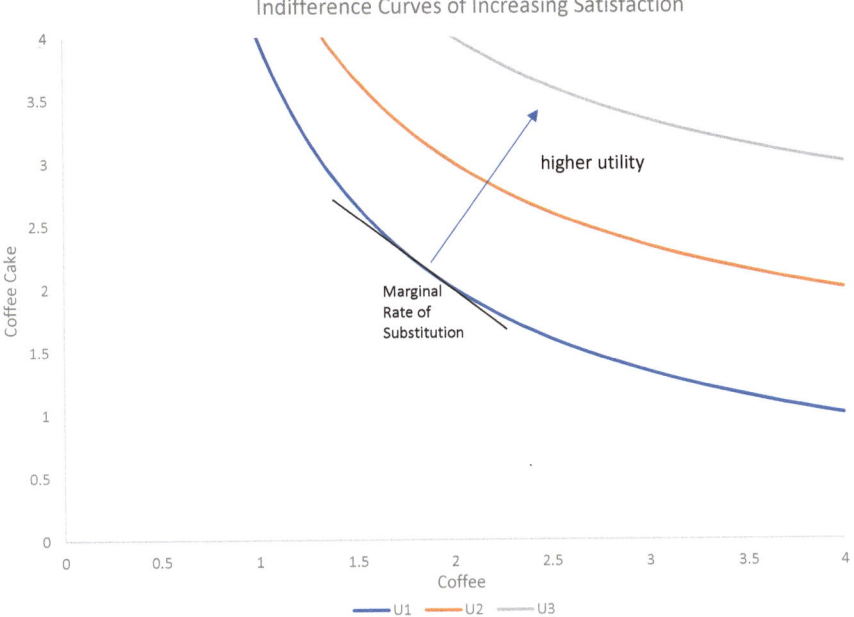

Fig. 11.1   Indifference curves representing increasing levels of satisfaction

with small adjustments in the other good to maintain utility by differentiating the utility function:

$$\overline{U}(q_1, q_2) \rightarrow d\overline{U}(q_1, q_2) = 0 \rightarrow 0 = \frac{\partial U}{\partial q_1}dq_1 + \frac{\partial U}{\partial q_2}dq_2$$

This expression contains $\frac{\partial U}{\partial q_1}$, called the marginal or incremental utility MU_1 derived from the enjoyment of a bit more of good 1, and $\frac{\partial U}{\partial q_2}$, the additional utility MU_2 derived from the consumption of more of good 2. We can rearrange the equation to find the marginal rate of substitution, which simply measures the relative contribution to our enjoyment of each good as measured by their relative marginal utilities:

$$\text{MRS} = \frac{dq_2}{dq_1} = -MU_1/MU_2$$

This expression reveals that the downward slope of our indifference curve $\frac{dq_2}{dq_1}$ is equivalent to the willingness to trade one good for another at a rate $\frac{MU_1}{MU_2}$ while overall utility is preserved. For example, if a unit of one good has half the marginal utility as the other, one would need two units of it to compensate for the loss of one unit of the other good.

We can see that this person-specific tradeoff, sometimes called a ***shadow price***, changes at different points on an indifference curve and reflects the relative scarcity or abundance of each good in an individual's consumption.

It allows us to consider how an individual values each good. The next step is to compare how individuals value goods relative to the rate the market values them.

The Budget Constraint

While Edgeworth's indifference curves are an expression of an individual's valuation of one good compared to another, it must also be combined with the ability of an individual to procure one good or another. To combine desires with economic realities, income is next introduced.

This income (or budget) line defines the combination of goods that are affordable, consistent with one's level of income I. Presumably, one could allocate all their income to purchase just one good. If the price of good 1 is p_1, then this potential maximum purchase of q_1 equals one's income divided by the good's price, represented by the ratio I/p_1. Likewise, a maximal purchasable amount of good 2 would be I/p_2 (Fig. 11.2).

However, neither extreme consumption of just one good or the other would make much sense, even if feasible, because at an extreme point one would sacrifice too much of the other good, with a high marginal utility, to permit the purchase of a good with a low marginal utility. Instead, an individual would choose some intermediate point along their budget line:

The individual's income line then acts as the constraint within which one attains the highest level of utility consistent with available income. The slope of

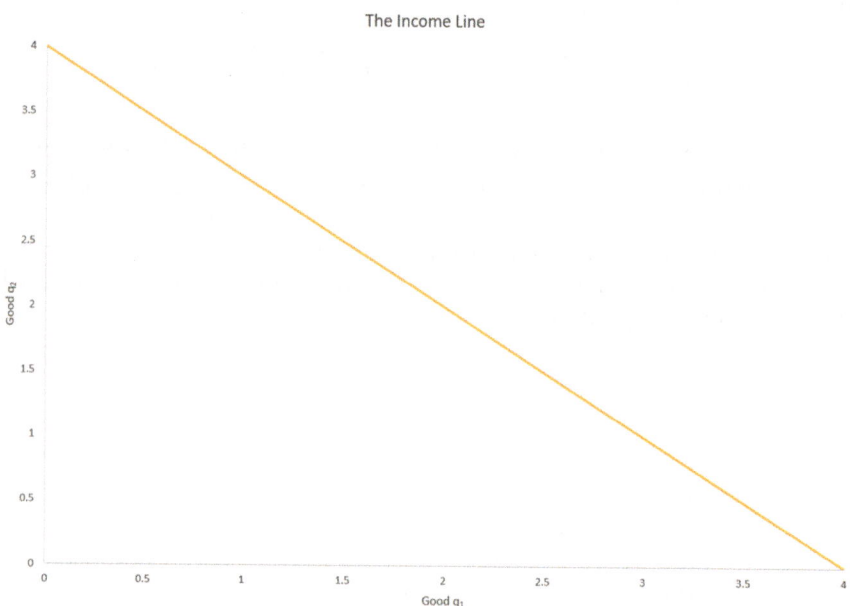

Fig. 11.2 Budget line that reflects ability to purchase a combination of two goods

such an income line corresponds to the ratio of the budget line's rise over run, or I/p_2 divided by I/p_1. Then, using some algebra, we see that the downward sloping income (or budget) line is of slope $-p_1/p_2$.

If individuals use these budget lines to attain their highest feasible utility, they translate income into happiness when their budget line hits the highest attainable indifference curve. It is easy to see that this must occur at a single tangency point. At that point, the budget line has the same slope as the highest attainable indifference curve and defines a unique combination (q_1, q_2) of goods that yields a higher level of utility U_A than any other combination attainable with the available income, as illustrated in Fig. 11.3:

Continuing with the logic that our indifference curves tell us what we want, while our budget line tells us what we can have, we attain the highest feasible indifference curve when it is just tangent to our budget line. At that point, the slope of the two lines must be equal, which allows us to make a profound observation once we equate the two slopes:

$$-p_1/p_2 = -MU_1/MU_2$$

In other words, rational individuals align their relative internal valuation of one good over another, as measured by the *marginal rate of substitution* between two goods, to their *relative market prices*. Another way to express this intuitively appealing and striking result is to compare the *bang for the buck* one

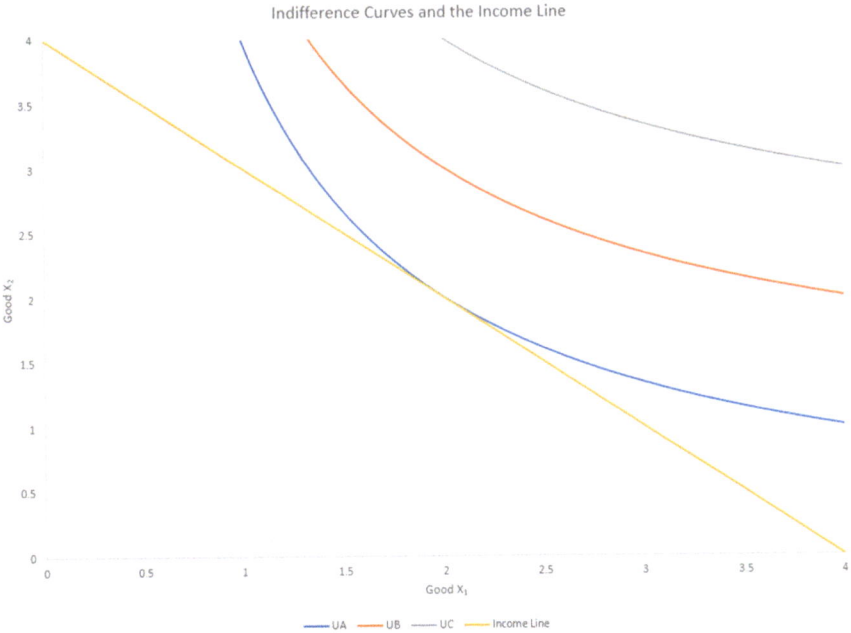

Fig. 11.3 Successively higher indifference curves and a budget line

attains by consuming each good. Rearranging the equation slightly, we can see that the bang from each good, in marginal utility MU, that the price of each good can purchase is equalized across goods:

$$MU_1/p_1 = MU_2/p_2$$

Rational individuals best leverage their ability to generate utility by allocating their income until each good yields the same bang for the buck. If the increase in utility were higher for one good than another, per dollar spent, one would purchase more such goods, and experience diminishing marginal utility, until the bang for the buck is eventually equalized across all goods.

One of the most powerful applications of Edgeworth's 1881 discovery is its extension to economic choices over time. Irving Fisher used Edgeworth's insights to shed light on human decision-making over time and provide a framework for understanding sustainability.

A New Application for an Old Technique

Fisher was interested in the dynamic aspects of consumption over time rather than the more static decision of what to consume at a given moment in time. A simple twist on indifference curves generates a powerful and insightful multi-generational tool. Let us assume instead that rational individuals can allocate their financial capital every period to optimize their consumption and maximize their utility over time. The next question to ask is how one ought to allocate their capital across periods of time or even generations.

Financial markets allow individuals to move income into future years by saving. Alternately, we can draw now upon future income by borrowing. Let us call today year 0 and denote by year 1 next year. In a two-period example, our wealth is then this year's income I_0 and the present value of next year's income I_1. We can translate next year's income into this year by obtaining a loan of size I_0 today, to be repaid with income I_1 tomorrow. If the interest rate is given by r, then one would owe $I_0(1 + r)$ in a year's time for a borrowed amount of I_0 today. Alternately, today's value of next year's income I_1 is:

$$I_1/(1+r)$$

In other words, the *present value* of a future flow of income is reduced by the *discount factor* $1 + r$, and total wealth available today for income I_0 today and I_1 in the next period is then:

Wealth today $= W_0 =$ PV of Income $= I_0 + I_1/(1 + r)$.

An individual's flow of income, suitably discounted at a rate that reflects payments necessary to determine the present value today for future income gives a measure of the present value of *lifetime wealth*. Using the interest rate r, this total wealth $W = I_0 + I_1/(1 + r)$ also represents the maximal consumption one could realize today.

Likewise, one could instead take income today and save it to generate income next year, using the same interest rate to yield maximum potential spending next year of:

$$\text{Wealth}_{nextperiod} = W_1 = I_1 + I_0 * (1 + r).$$

Economists and bankers before Fisher understood these notions of present and future values, and savings and loans. But, until Fisher, nobody had tied together a wealth today of W_0 and in the next period W_1 is analogous to Edgeworth's budget line at a given point in time. As before, we can use present and future wealth to instead allocate our potential spending between the two periods. Just as we would not use all our income to purchase solely one good or another, nor would we use all our wealth to spend solely in one period or another. Fisher argued that we allocate our financial capital to yield the greatest return over time in the purchase of intertemporal happiness.

Let us simply label as W the present value of wealth today and into the future. The economic decision then becomes whether to spend W today, $W*(1 + r)$ next year, or some combination today and next year. This can be seen on an extension of the income line to potential spending over time. This dynamic wealth line shows the tradeoff of total consumption today of W, or total consumption next year of $W*(1 + r)$, which, if graphed, looks much like the previous income line used in the static Edgeworth Indifference Curve analysis.

Maximum consumption today can be obtained by spending all wealth W today, while next period's maximum consumption would be $W(1 + r)$. We can then see the slope of the resulting intertemporal wealth line as the ratio:

$$\text{Slope} = -(1 + r)W/W = -(1 + r).$$

The *intertemporal price ratio*, given by the slope of the multiperiod income line is then $-(1 + r)$ rather than the relative prices p_1/p_2. The rate of transformation $(1 + r)$ of wealth between periods is our *discount factor* that then *prices* decisions over time.

If the price of consumption from one period to the next is given by the intertemporal discount factor $(1 + r)$, we must next define how an individual prefers one period's consumption to another. Rather than measuring goods and services each period, let us instead measure total utility in each period based on aggregate purchases and consumption C_0 and C_1 of all goods and services in periods 0 and 1 respectively. These purchases yield utility of $U(C_0)$ and $U(C_1)$ for today's period and next year's period respectively. Let us add the final dimension based on Fisher's impatience theory.

Just as markets discount future income or reward present savings based on the discount factor $(1 + r)$, an impatient mortal also discounts the future. Let this *intertemporal rate of time preference* be given by ρ. It is this impatience factor Fisher described that influences a consumer's regard for present and future consumption, as given by:

$$U_{total} = U(C_0) + U(C_1)/(1 + \rho),$$

where future enjoyment $U(C_1)$ is discounted by the factor $1/(1 + \rho)$ just as the market discounts future incomes by $(1 + r)$. Intertemporal indifference curves can be derived as before, with one modification. Just as before, more available consumption in both periods will result in greater total utility in present value terms. Let us repeat the same analysis to determine the slope of such an intertemporal indifference curve. Then,

$$d\overline{U}(C_0, C_1) = 0 = \frac{\partial U}{\partial C_0}dC_0 + \frac{\frac{\partial U}{\partial C_1}dC_1}{1 + \rho},$$

where $\frac{\partial U}{\partial C_0}$ and $\frac{\partial U}{\partial C_1}$ are once again interpreted as the marginal utility of consumption in periods 0 and 1. As before, the marginal rate of substitution between one period and the next can be derived:

$$\frac{dC_1}{dC_0} = -(1 + \rho)\frac{MU_0}{MU_1}.$$

To understand the significance of this expression for an individual's preferences of consumption between periods, through its discount factor $(1 + \rho)$, consider the circumstances that would cause one to allocate consumption evenly in each period such that $C_0 = C_1$. Then, $MU(C_0) = MU(C_1)$, and the slope $\frac{dC_1}{dC_0}$ of the intergenerational indifference curve then simplifies to $-(1 + \rho)$. The expression for intertemporal preferences can then be combined with intertemporal wealth to show the equally profound result Fisher discovered for consumption over time (Fig. 11.4):

In other words, individuals will allocate identical income and consumption across each period only if their rate of time preference ρ happens to coincide with the market interest rate r.

We have demonstrated that the expression for our preferences for consumption C_0 and C_1 can be combined with a measure of our wealth over time. Of course, not all individuals will have a rate of time preference ρ that precisely equates to the average discount rate r in the economy along the equal consumption path. Using Fisher's vocabulary, the market interest rate r is a measure of the aggregate market impatience of present versus future income. Every individual i has their own individualized level of impatience ρ_i based on their own circumstances that may involve attitudinal dimensions, demographics, measures of financial security, and even past patterns of economic justice and income redistribution. Consider various such slopes of their intertemporal indifference curves along the constant consumption ray $C_0 = C_1$:

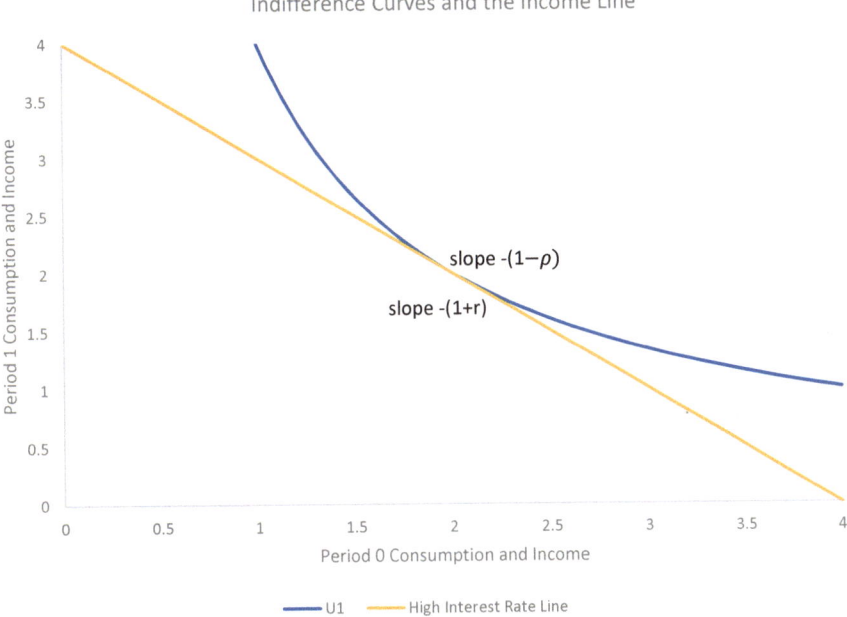

Fig. 11.4 Maximization of utility for a given budget line

Figure 11.5 shows that each of three individuals portrayed prefers a different combination of consumption over time. Only one consumer's preferences are aligned with the constant consumption ray emanating from the graph's origin. The individual indifference curve to the left has a flatter intertemporal indifference curve along the equal consumption ray because their rate of time preference ρ is smaller than the market rate of time preference r. These individuals instead choose an equilibrium to the left corresponding to lower consumption in period 0 and greater consumption in period 1. In essence, they are less impatient than the marketplace and use this difference to save in period 0 to enhance their consumption in period 1.

On the other hand, the indifference curve farthest to the right has a higher rate of time preference as evidenced by a steeper slope of their indifference curve than the market's preference along the equal consumption line. These more impatient individuals would instead then choose a higher rate of consumption in period 0, knowing that this leaves reduced consumption in period 1. These individuals borrow to enhance their spending now, at the expense of consumption later.

To see the effect of the rate of time preference another way, consider once more the intertemporal equilibrium equation:

$$(1+r) = (1+\rho)MU_0/MU_1$$

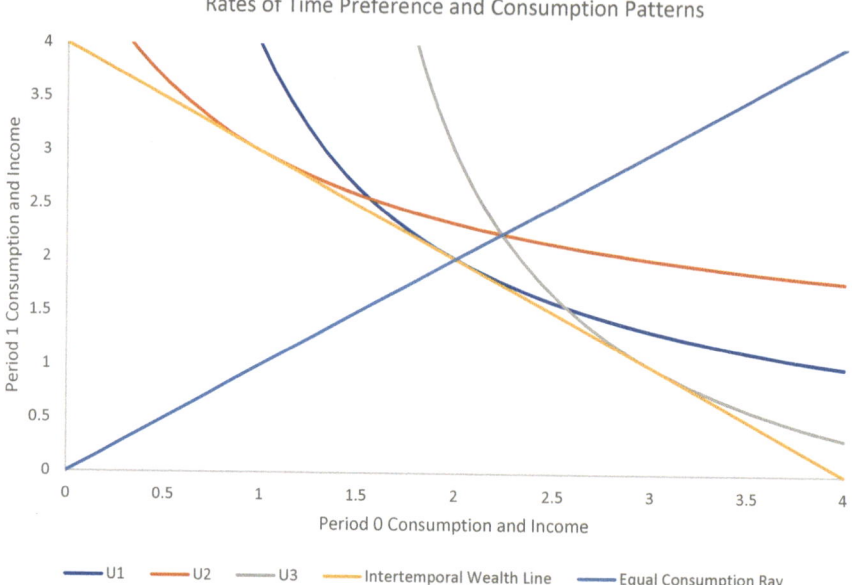

Fig. 11.5 Equilibria for individuals with differing rates of time preference

If the interest rate r rises and increases the discount factor $(1 + r)$ on the left-hand side of the equation, the right-hand side of the equation must rise likewise. In equilibrium, the ratio of MU_0 to MU_1 must then rise. The *law of diminishing utility* then implies that consumption C_0 falls, and hence C_1 rises. The higher interest rate induces a downward shift in period 0 consumption and an increase in savings in response to the higher interest rate that can then generate greater consumption C in the next period. We can see this in the following diagram (Fig. 11.6):

As the interest rate rises, our wealth today deferred to tomorrow permits us to consume more tomorrow. An individual can then consume at point E2, with lower consumption today, and greater consumption tomorrow due to the increased inducement arising from a higher interest rate.

The overall aggregate rate of impatience of society, as measured by the market interest rate, is an aggregation of our individual rates of time preference. Individuals save or borrow, either on an individual basis, or through our proxies of government, corporations, businesses, and other organizations. We pool our personal savings and borrowing, and the market computes an average *market impatience rate* to yield an **interest rate** that balances the funds of savers and borrowers.

As noted earlier, this aggregation reflects cultural and demographic nuances that may even cause prevailing aggregate interest rates to vary over time.

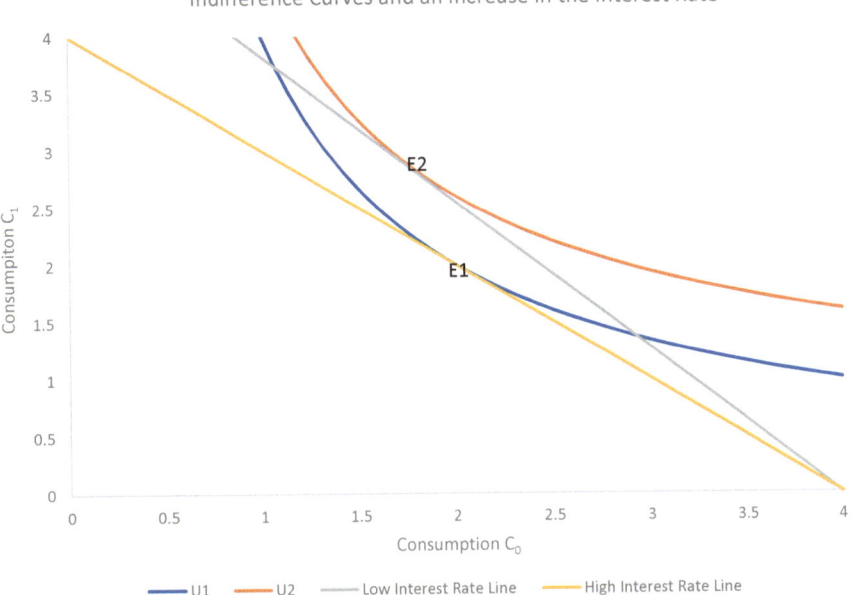

Fig. 11.6 Attainment of higher utility as the interest rate on savings rises

If a society is made up of more savers with a low personal discount rate than borrowers with greater impatience, then the equilibrium interest rate will likewise be lower. These interest rates balance the greater availability of loanable funds from those willing to defer consumption today and save for tomorrow with those interested in consuming today, investing in their human capital by borrowing for an education, or wishing to purchase a car or home so that they may spread out the consumption enjoyed from such consumer durables.

Our regard for the future tends to be individual, though. We do not use our savings to preserve the stock of *natural capital* to ensure *Earth's assets* are available for future generations. Our intertemporal decisions are instead based on personal preferences and the allocation of our personal assets for consumption over our individual lifetimes, with little regard for the allocation of natural capital over future generations.

Yet, the market interest rates we formulate because of our collective balancing of present and future consumption needs provide the signal upon which markets rely to allocate both natural and financial capital over years and generations. Individuals and society then allocate wealth and capital over time and across generations to attain such an intergenerational equilibrium that reflects society's collective discounting and our individual mortality.

INCOME STATEMENTS AND BALANCE SHEETS

Another challenge in the extension of traditional economics to intergenerational issues is that it tends to focus on income each period rather than the maintenance of intergenerational wealth. Rather than relying on a natural capital **balance sheet**, economists most often measure flows, equivalent to an **income statement**. Even the Triple Bottom Line of People, the Planet, and Prosperity often conflates social income statements and balance sheets. Indeed, measures of economic progress are often income-driven.

Typically, with some notable exceptions, prosperity is then measured along income and expenditures scales, while an aggregation of individual discount rates drives decisions rather than an intergenerational social discount rate. For instance, the destruction of human property because of increasingly dangerous storms arising from global warming is often followed with a spurt of economic activity as structures are rebuilt. Such spending spurs income for contractors and suppliers, and also results in even further spending through the multiplier effect for those who derive secondary or tertiary employment from the primary sector of construction.

Ironically, natural disasters then actually increase income and spending, and are recorded as growth in the national income accounts of *Gross Domestic Product*. Yet, when property is destroyed, crops are diminished, and lives are lost, the balance sheets of regions, countries, and the Earth's resources are unambiguously diminished. This decline in asset values through natural processes is ignored if economic indicators are biased toward the income approach rather than the asset approach.

In addition, we shall turn next to the problem that results when an income- and spending-based approach among mortals in their determination of the discount rate. This conversion between the past and present is imposed on the decisions of mortals to extract and enjoy Earth's natural capital in fixed supply, which unambiguously reduces the ability of future generations to secure the same level of usage for themselves. Individuals and entrepreneurs make intertemporal decisions that affect future generations who have no opportunity to express their preferences.

SUMMARY

We shall elaborate on the simple intuition derived in this chapter that mortals tend to make decisions over time to enhance our discounted economic enjoyment. We do not make intergenerationally fair decisions because we fail to regard equally the value of economic capital in the future, including our natural capital. This discounting of the future leaves dwindling consumption over time unless we are willing to set aside sufficient resources to ensure the natural capital stock is protected from our intertemporal shortsightedness.

This principle is the basis for our study of sustainability. As we extend these principles to explore the unique and depletable qualities of natural capital or include the effects of pollution or global warming, we find that the fundamental ethical and sustainability question always remains. To what degree do we consider the future, even in the face of dire warnings from scientists that the future is becoming increasingly precarious? This remains the sustainability question.

ESG Toolkit

Corporations must make decisions over time.

To what degree does a strategic plan balance what is enjoyed today and what is set aside for the future?

How might a corporation adjust its plans to reduce the discount rate that tends to devalue the future?

Does this tradeoff constitute a corporate ethic?

Is it material and ought this decision be communicated to stakeholders?

APPENDIX: AN EXTENSION OF FISHER'S INTERTEMPORAL MODEL TO INCLUDE PRODUCTION OVER MANY PERIODS.

To hone our intuition on optimal consumption, let us develop an effective tool that generalizes Fisher's two-period treatment of optimal consumption over time. Let our goal be to maximize the permanent flow of utility u(c) and hence consumption c forever, under the assumption that population grows at a rate n and future flows of value are discounted at a rate r. Then the present value of total aggregate utility U_0 today is given by the following intergenerational social welfare function that optimizes the path of discounted utility per person over time:

$$U_0 = \int_0^\infty u(c)e^{-rt}e^{nt}dt,$$

where U_0 is the present value of all future consumption as a discounted sum of utility u(c) each period, n is the (exponential) rate of population growth, and r is the aggregate social discount rate. This measure has the goal of maximizing utility now and into the future, but also recognizes that future utility is inevitably discounted by we mortals at the aggregated discount rate r. The relationship also explores how population growth affects our calculation.

Let production in each period be given by a function f(k) that depends of capital k that depreciates at a rate δ and must be spread out at a rate n with population growth. An economy must then decide how much production should be devoted to future capital k or present consumption c. Then, the rate of capital available for later periods and for consumption today evolves according to a differential equation:

$$\dot{k} = f(k) - (n + \delta)k - c,$$

where $\dot{k}$ is the rate of change of capital per person over time that is augmented by production according to a function f(k), and is diluted across more people at a rate n and depreciated at a rate δ, or consumed at a rate c.

The solution to this problem was developed by the applied mathematician Frank Ramsey and his mentor, John Maynard Keynes in 1928[5] and further described by two subsequent mathematical economists, David Cass and Tjalling Koopmans. The solution is easily found using a constrained optimization instrument called a **Hamiltonian**. This technique seeks to maximize the discounted value of utility over time, the $u(c)e^{-rt}$ term, but recognizes that our consumption is constrained by our available capital, its depreciation δ and the rate we must spread it over a larger population n, with any remainder reserved for our individual enjoyment. The current value Hamiltonian is:

$$\mathcal{H} = u(c) + \lambda(f(k) - (n + \delta)k - c)$$

Ramsey implicitly solved this relationship to demonstrate a path for consumption that also shows us what society must do to maintain a constant and sustainable level of consumption over time:

$$\dot{c} = \frac{-u_c(c)}{cu_{cc}(c)}(f_k(k) - n - \delta - r - c)$$

The ratio on the left-hand side of the equation describes the marginal utility of consumption u_c relative to consumption c times the rate of change of marginal utility u_{cc}. This ratio is always non-zero.

On the right-hand side of this differential equation is a constraint that measures capital we produce and leave for future generations and the amount we consume today. This capital, as measured by its marginal growth f_k net of its diminishments and dilution, represents society's return to capital, while r is the interest rate.

For consumption to remain constant across periods, and hence its rate of change $\dot{c}$ equals zero, per capita consumption must be confined to a rate that equals the increase in output over the period less the discount factor, depreciation of capital, and population growth. This **golden rule** then states that steady consumption over time requires us to consume in a manner that ensures our consumption is constrained only to the net earnings on capital, without eating into capital itself.

This path of optimal consumption is named the *Ramsey-Keynes rule* as a tribute to the collaboration the mathematician Ramsey had with his economist colleague and mentor, John Maynard Keynes. It represents the first formal model that specifies the conditions necessary to ensure that consumption can be sustained across generations.

We can also develop some intuition that will pay dividends later. If our goal is to hold the present value of consumption constant, it is not necessary to attribute a measure to utility enjoyed each period. Instead, we must only ensure that net capital is maintained across generations, with no generation exercising a privilege over another by consuming more than a sustainable share of new capital produced.

We also see that we can employ the powerful Hamiltonian tool to simplify our sustainability analyses. If society aspires to maintain a constant flow of consumption over time, it must ensure a constant and even flow of capital across generations.

NOTES

1. Fisher, I. (1907). The Rate of Interest, The MacMillan Company, New York.
2. Fisher, Irving (1930). The theory of interest, The MacMillan Company, New York.
3. Fisher, I. (1907). The Rate of Interest, The MacMillan Company, New York, at page 231.
4. Edgeworth, Francis Ysidro (1881), Mathematical Psychics: An Essay on the Application of Mathematics to the Moral Sciences, Kegan Paul, London.
5. Ramsey, Frank P. (1928). "A Mathematical Theory of Saving." Economic Journal. 38 (152): 543–559. https://doi.org/10.2307/2224098. JSTOR 2,224,098.

CHAPTER 12

Discounting and Extraction of Depletable Resources

The last chapter described some of the economic forces that determine our decisions over time. Mortals strive to consume into the future but we discount that future at the average market-wide interest rate r. We also recognize that productive capacity can be consumed in one period or conserved until a later period. We next determine how to divide capital across time to support intergenerational consumption when there is a fixed supply of natural capital. Such depletable natural capital creates a tension in the use of natural resources to support one generation at the expense of those who may follow. The intergenerational question then becomes how to allocate wealth and capital across generations.

Fisher's intertemporal choice model and Ramsey's extension to include production are helpful in that they illustrate how rates of time preference and the rates that our productive capacity depreciate affects consumption choices. Decisions for the extraction of fixed resources can also be analyzed. This chapter will describe the process by which a fixed stock of natural capital should be extracted over time. As before, we begin by analyzing first the problem over two periods and then extend the analysis into the future.

Let us begin with a simple two-period problem in which we can extract all of a fixed resource of size $\overline{S}$ today, extract the resource a year from now, or divide the resource extraction over the two periods. To frame the question as either now or next year is limiting, but the intuition that we develop will be extended later.

The principles can be demonstrated by a simple example. Assume there is a fixed stock of natural capital of a size $\overline{S}$ units. For simplicity, let the market for the extraction of this capital be defined by a linear demand curve the yields

© The Author(s), under exclusive license to Springer Nature Switzerland AG 2023
C. Read, *Understanding Sustainability Principles and ESG Policies*,
https://doi.org/10.1007/978-3-031-34483-1_12

a market price as a function of the resource extraction rate X in each of two periods 0 and 1. This simplification will demonstrate an important conclusion that applies to any sort of resource demand:

$$P_0 = p_c - m * X_0$$

$$P_1 = p_c - m * X_1$$

where $-m$ is the downward slope of the demand curve that recognizes as the quantity extracted and consumed rises, our willingness to pay falls for a bit more of the resource. This reduction in the market's willingness to pay as extraction rises is a consequence of the notion of diminishing marginal returns. The term p_c in this linear demand function (of the traditional form $y = mx + b$) is the y-intercept price that causes demand to fall to zero once a choke price p_c is achieved.

This maximal price p_c might correspond to an alternative resource we could instead consume. Such an alternative resource is called a **backstop technology**, to be described in more detail later. For instance, it could be the cost of an alternative resource such as sustainable fuel should the price of a fossil fuel rise as it becomes increasingly scarce and expensive.

In this simple two-period model of extraction of a fixed natural resource with simple linear demand, let us also assume constant marginal extraction costs c regardless of the remaining stock of the depletable resource, and there are no fixed costs of resource extraction. We can then determine the optimal level of extraction between period 0 today and period 1 in a year's time for a given interest rate r.

Extraction should be optimized to ensure that the discounted net benefits that accrue to extraction are as large as possible, on a present value basis, over the two periods. For simplicity, let these net benefits be the area under a demand curve in excess of the (assumed constant) cost c of extraction. First principles of economics tell us that no level of extraction will occur in any period if the marginal extraction costs are higher than the price the market will bear. Let us first consider an arbitrary price and quantity of extraction and a demand price at some level that exceeds marginal costs.

For example, with linear demand $p = 2500 - 10X$, and for an arbitrary extraction rate X of 110 units, we can calculate the difference in willingness to pay and costs as the sum of a triangle and a rectangle. In general, for the downward sloping demand curve defined by a price $p(X)$ equal to $p_c - m*X$, we can calculate the difference between any arbitrary price $p*$ and marginal costs c. From there, it is not difficult to calculate the net benefits accruing to users of the resource by summing up their consumers' surplus, the area between the demand curve and the price, and any producer's surplus, the area below the price that is not absorbed by extraction costs. These net benefits

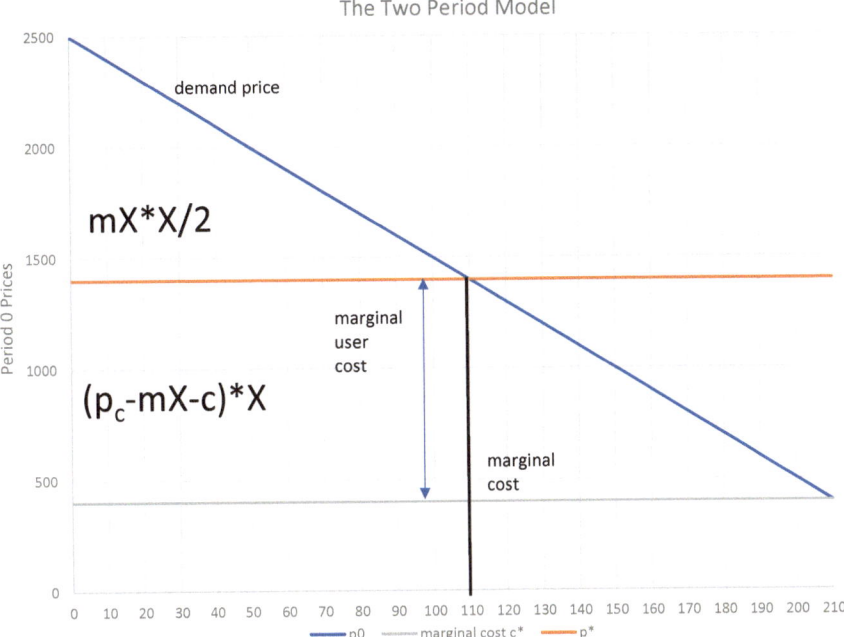

Fig. 12.1 Static resource extraction equilibrium for downward sloping demand and constant marginal extraction costs as the level of extraction X in one period rises

are simply the area of the triangle and rectangle in Fig. 12.1 for any quantity extracted and consumed X:

$$NB = (p_c - (p_c - mX)) * x/2 + (p_c - mX - c) * X$$
$$= mX^2/2 + (p_c - mX - c)X$$

The goal of optimal extraction is to maximize these net benefits.

Before we calculate how a depletable resource will be extracted over two periods, let us first derive a point of comparison. A perfectly competitive industry will continue to extract until the price p_{pc} earned for the last (marginal) bit of extraction is just equal to its marginal extraction costs c. In the absence of a second period and with sufficient available resources, this static and short-sighted industry would generate total net benefits made up of the one large triangle running to the point where the price p_{pc} is defined by:

$$p_{pc} = p_c - mX_{pc} = c$$

$$\rightarrow X_{pc} = (p_c - c)/m,$$

where b is the maximum (choke) price of the extracted resource.

In Fig. 12.1, we observe that 210 units of the fixed resource would be extracted by a perfectly competitive industry if the price p were set to marginal costs c. Let us next determine how this static market equilibrium should be modified to include a second period. Once a second period is considered, extraction should be allocated across both periods. In a two-period model, any amount of the resource that is not extracted in the first period, at a rate X_0, could instead be extracted in the next period in an amount $X_1 = \overline{S} - X_0$. However, such surpluses delayed from later extraction cause a firm to discount these future earnings.

Another way to view this principle is that a surplus enjoyed in the first period if converted to financial capital will increase in value by a factor of $(1 + r)$ if a unit of the resource is invested in a financial market. Hence, at the margin, if a unit of the resource is delayed, and hence permitted to remain in *Nature's Bank*, until the next period, it too ought to yield a higher surplus by the same factor $(1 + r)$. This intuition will be employed repeatedly in more sophisticated analyses in later chapters.

To confirm our intuition, let the width of our horizontal quantity access be given by the size of the fixed stock $\overline{S}$ of the available resource. The problem then becomes one of divvying up this width in the resource along the width of the horizontal axis to maximize the present value of net benefits.

The graph in Fig. 12.2 shows a horizontal axis of a width equal to the total stock and two demand curves. The first downward sloping line shows current period prices on the left vertical axis. On the right-hand vertical axis is the same demand curve for the next period but discounted in height because total prices and surpluses generated in the next period are of lower value today by the discount factor $(1 + r)$. Next period extraction and surpluses thus appear to the earlier period as a lower demand curve, indicated on the diagram as demand starting from the right and moving to the left of the horizontal axis below. Rather than a peak of 2500, the maximal value today of next period's price is then discounted to $2500/(1 + r)$.

The intersection of the current market value and next period's discounted value occurs where these two demand curves intersect. At that point of intersection, the marginal valuation of a unit of resource today is the same as the discounted marginal valuation in the next period. In this example, the *intertemporal equilibrium* occurs at 110 units of first period extraction and 100 units in the next period. This second period extraction of 100 units is simply the total stock of 210 units less the first period extraction of 110 units.

Our intergenerational intuition is developed by working through the mathematics of the solution. Let us begin by adding up the surpluses arising from the extraction and enjoyment of a resource across two periods. It is possible to first calculate the point where the demand price for the current period is equated to the discounted value in the next period. At the equilibrium point, society is indifferent to shifting a bit of resource extraction between one period and another.

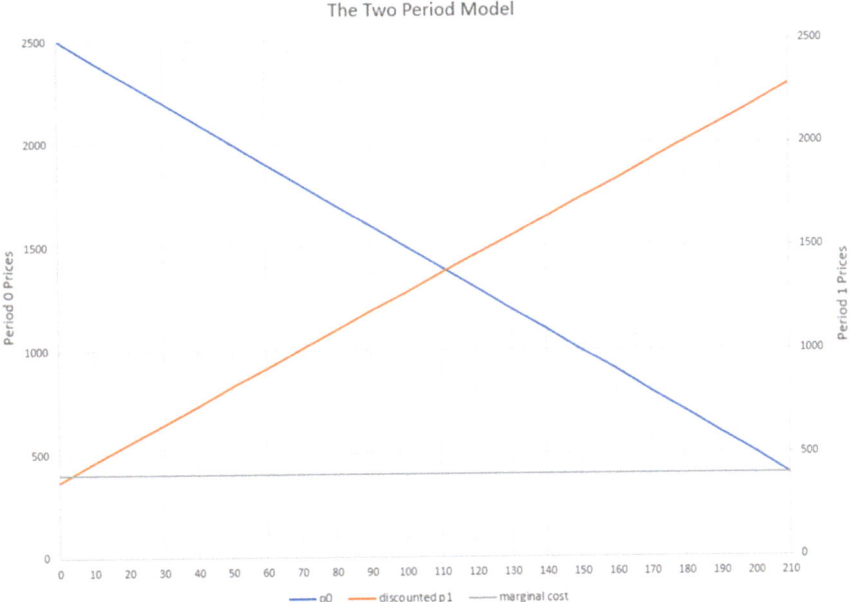

Fig. 12.2 The two-period model of extraction of a fixed resource

To find such an equilibrium over two periods, one would simply solve two equations that equate the surpluses of extraction in one period to the discounted surpluses in the next period (as one equation) and then note that these two levels of extraction must sum to the total allocation $\overline{S}$ of the resource. Let PV_0 represent the value of extraction in period zero and PV_1 the present value of extraction in period 1:

$$PV_0 = p_c - mX_0 - c$$

$$PV_1 = (p_c - mX_1 - c)/(1 + r)$$

$$PV_0 = PV_1 \rightarrow p_c - mX_0 - c = (p_c - mX_1 - c)/(1 + r),$$

and

$$\overline{S} = X_0 + X_1,$$

where $\overline{S}$, m, and p_c are all parameters and only X_0 and X_1 must be solved. Collecting terms, we find:

$$(1 + r)(p_c - c) - (p_c - c) = m(X_0 - X_1) + rmX_0.$$

We are left with two simple linear equations in the two unknown quantities q_0 and q_1:

$$r(b - p_c)/m = (1+r)X_0 - X_1$$

and

$$\overline{S} = X_0 + X_1,$$

The expression simplifies to allow us to immediately solve for present production by substituting the stock equation into the equilibrium equation. Because that $X_1 = \overline{S} - X_0$, then:

$$r(p_c - c)/m = (1+r)X_0 - (\overline{S} - X_0)$$

$$r(p_c - c)/m + \overline{S} = (2+r)X_0$$

$$X_0 = (r(p_c - c)/m + \overline{S})/(2+r).$$

From the constraint we can easily calculate next period extraction since we know that

$$X_1 = \overline{S} - X_0.$$

For instance, let us assume that the maximum demand price for the resource b is 2500, the discount rate r is 10%, the slope of the demand curve is –10, and the total stock $\overline{S}$ is 210. Then, the equilibrium quantity to extract in the current period is:

$$X_0 = (0.1 * (2500 - 400)/10 + 210)/(2 + 0.1) = 231/(2.1) = 110$$

$$X_1 = \overline{S} - X_0 = 210 - 110 = 100$$

In the first period, the price p_0 is 1400, while the marginal cost is 400. Unlike the static perfectly competitive case, we see that the price in this equilibrium is now higher than marginal costs c. In this example, the difference between price and cost in the first period is 1000. In the next period, the price p_1 becomes 1500 because of the reduced supply left to period 1. The resulting difference between price and cost has risen to 1100. Notice that the 1100 surplus enjoyed in period one, and discounted at the rate $(1 + r)$ yields the same 1000 surplus enjoyed in the original period.

THE MARGINAL USER COST

This difference between the equilibrium price $p*$ of the resource and its extraction cost c when we must ration our natural capital between periods is called the *marginal user cost*. Were the model repeated in a similar manner over additional periods, the present value of this discounted marginal user cost remains constant over additional periods as well. The marginal user cost that arises as we ration the natural capital ratchets up above the marginal extraction cost with each successive period at the same rate $(1 + r)$ to ensure that the present value of total surpluses cannot be increased by shifting a unit of extraction from any period into another.

If this marginal user cost is rising each period while the marginal extraction cost c remains constant, we see that the price is likewise rising and extraction is falling each period. This increasing price trajectory and corresponding declining extraction occurs whenever the discount rate is positive. However, if we repeat the analysis with a zero-discount rate, we see that extraction remains constant each period and the resource stock is evenly divided across all extraction periods. We obtain this important principle for any downward sloping demand curve or more complicated marginal costs.

The rising **marginal user cost** is of special significance. It is the sacrifice in value, net of marginal extraction costs, of moving a small marginal amount of extraction from one period to the next. If there is an insufficient stock of the resource to meet all demand every period, rationing of extraction is necessary, and the price is driven above marginal cost to create a positive marginal user cost. This difference in the rationed price and the extraction cost that we call the marginal user cost is then a measure of scarcity that ensures there is no advantage in additional rationing of a bit of extraction in one period to provide for the next. We are indifferent to any further adjustments in X_0 and X_1 only if the marginal user cost MUC_0 equals the discounted marginal user cost $MUC_1/(1 + r)$.

BACKSTOP TECHNOLOGIES AND SUBSTITUTES

Recall that we have assumed that there is some price beyond which an economy will no longer demand some extracted resource. This price we have called a *choke price*. For instance, in the case of fossil fuels, we expect that, should its price rise sufficiently high, the economy will instead invest in a new technology, such as electricity to replace fossil fuels and their attendant greenhouse gas emissions.

This point in which we move on from a depletable technology may be well before the resource has been completely exhausted. Instead, the choke price we discussed earlier may simply be that price for which the resource is longer economical to extract relative to our alternative. For instance, it might be the price at which the use of electricity generation from sustainable sources becomes competitive with natural gas-powered electricity plants.

Such a *backstop technology* may be known today but remains unaffordable, at least until the market price of the depletable resource rises so high that the backstop technology finally becomes affordable. It could reflect an innovation in science that may someday be commercialized, but only once competing technologies rise in price to make the innovation affordable.

The price of the backstop technology then determines the choke price p_c of an alternative technology corresponding to the cost of an alternative technology, and an expected time T when it may arrive, which we denote as a *terminal time*. This turn of phrase is drawn from mathematics to denote a *terminal condition* which pins down the end of a trajectory of the differential equation that governs the evolution of extraction.

The differential equation of interest to us is the path of the marginal user cost of a depletable resource which, when added to the marginal cost of extraction, gives the trajectory of market demand prices and extraction rates that evolve over time. Eventually, this market price $p(t)$ will rise so high that it becomes feasible to rely instead on a sustainable substitute resource, or perhaps on recycling rather than extracting the depletable resource. At that terminal time T, the price $p(T)$ coincides with the backstop technology choke price p_c.

The optimal path of sustainable resource development over time depends critically on the acknowledgment that an extraction strategy follows a dynamic and evolving path. So far, we have described the pattern of human and corporate decision-making as a series of snapshots rather than a moving picture. Our intuition remains, though. The extent that institutions can acknowledge the dynamic nature of their environment will predict their long-term success and viability.

The Schumpeter Growth Paradigm

One argument some make in favor of excessive depletable resource extraction with little regard for future generations relies on the empirical observation that science and engineering, and research and development, tend to improve human and technical productivity over time. As these innovations come to market, generations to follow may have access to a limitless and sustainable resource, and hence it is justified to live in the extraction moment. This argument is articulated in a number of ways.

First, improvements in technologies may permit extraction of fixed resources such as minerals or fossil fuels that may have proven impossible or uneconomic in the past. Resources can be categorized as proven or unproven, and economic or uneconomic. A resource endowment typically measures the amount of a fixed resource that has been measured and estimated and that can be extracted at current commodity prices. Increases in the price of the commodity over time may convert uneconomic resources into economic resources as the rising price exceeds the extraction cost. This effect can also occur if extraction technologies improve.

Additional endowments of the resource can also be discovered or inferred based on new resource finds. While these potential resources are not considered to be part of known economic resource endowments, an expectation that they may prove to be economically viable sometime in the future is at times argued as justification for greater scarce resource extraction in earlier generations (Fig. 12.3).

	Cumulative Production	Identified Resources			Undiscovered Resources	
		Demonstrated		Inferred	Hypothetical	Speculative
		Measured	Indicated			
Economic		Reserves		Inferred Reserves		
Marginally Economic		Marginal Reserves		Inferred Marginal Reserves		
Sub-Economic		Demonstrated Subeconomic Reserves		Inferred Subeconomic Reserves		
		<------ Increasing Degree of Geological Assurance				

(Left vertical axis: Increasing Degree of Feasibility ------>)

Fig. 12.3 Economic and subeconomic resources (Courtesy of U.S. Geological Survey)

Improvements in technology may also enhance our ability to reconstitute an extracted and spent resource through enhancements in recycling affordability. This implication of conservation of mass requires abundant future energy sources, but can in theory reverse previous consumption of scarce resources. Recycling will be described in greater detail later.

In addition, it is argued that the march of technological innovations over time promises that future generations shall attain an equivalent or even superior levels of consumption and happiness as they are better able to access alternative resources, both of natural and human capital. This notion argues that technology trumps scarcity.

For instance, the economist Joseph Schumpeter expanded concepts developed by Karl Marx to develop a theory of economic innovation. He argued that:

> The gale of creative destruction … (is a process of) industrial mutation that continuously revolutionizes the economic structure from within, incessantly destroying the old one, incessantly creating a new one.[1]

While Marx originally had in mind the tendency of capitalism to destroy old orders and wealth and create new ones with each major innovation, the concept of creative destruction (*Vernichtung* in German) now refers to the ability of the production process to evolve toward greater efficiency.

In fact, this notion of creative destruction was also the basis for Charles Darwin's concept of the ***survival of the fittest*** in his <u>Origin of Species</u>[2] in which "extinction of old forms is the almost inevitable consequence of the production of new forms." In the Marxian context, Werner Sombart wrote in 1913 in *Krieg und Kapitalismus*[3]:

> ... from destruction a new spirit of creation arises; the scarcity of wood and the needs of everyday life... forced the discovery or invention of substitutes for wood, forced the use of coal for heating, forced the invention of coke for the production of iron.

Schumpeter had contemplated technological innovation as an explanation for what was at the time called *long-wave cycle theory*. The theory, most often associated with Nikolai Kondratieff, and sometimes called Kondratieff Waves, argued that humankind enjoys spirts of accelerated economic development that arise from technological innovations.

For instance, the innovations of the wheel, water power, steam power, railroads, electric generation, the telegraph, the automobile and trucks, agricultural mechanization, vacuum tubes, flight, radio and television, computing, nuclear power, access to outer space, the transistor and the integrated circuit, personal computing, the information age, the smartphone, and advances in medical technology and bioengineering have each moved economic progress forward in significant, discrete, and irreversible ways.

In doing so, these waves of accelerated economic growth have created waves of destruction of technologies, wealth, and income that become obsolete in their wake. Even within the process of capitalism itself, new technologies may be accompanied by new production processes and perhaps even new ways to organize production. Sometimes, the very way that production is organized, goods are manufactured and distributed, and markets evolve may create efficiencies in themselves. For instance, supply chain management and the notion of *just-in-time* *(JIT)* have created efficiencies and conserved resources.

Despite the sometimes profound industry resistance innovations meet in the creation of a better mousetrap, these revolutions in science, technology, manufacturing, marketing, or organizing typically ultimately prevail, if not perhaps delayed at times. These phenomena are often described as *endogenous growth theory* or *evolutionary economics*.

Each of these transitions also often implies significant social adjustment, as, for instance, we see as the *social network* movement modifies how humans relate to each other as a consequence of improvements in information technology. Critics for centuries have narrated unintended negative social consequences of economic innovations, in metaphorical forms as varied as Herman Melville's *Moby Dick* and Ralph Waldo Emerson's *Walden Pond*, and the various dystopian authors from Isaac Asimov to Aldous Huxley and many in the science fiction genres such as 2001: A Space Odyssey.

Stuart Hart (2005) provides a simple example of innovation in the article Innovation, Creative Destruction, and Sustainability.[4] He described how the potential for consumption and value generation can be squandered not for any lack of ingenuity or technological improvement, but for the inability to disperse innovations for the greater good. While more than half of the world lives in LDCs with a higher level of poverty and lower economic development relative to MDCs, a willingness by MDCs to transfer technological innovations to LDCs can yield immediate returns in human happiness and unleash great potential for new and sustainable development.

Hart demonstrated that business management concepts have already developed many of the tools to enhance our collective production and consumption by better leveraging preexisting technologies. In a four quadrant diagram, he described the corporate tools for a better economic future (Fig. 12.4).

While the ESG paradigm had not yet been popularized at the time of Hart's analogy, it is apparent by his *External Factors* quadrant that an expansion in the way humans collectively think about innovations can have profound implications not only on how a lot of humankind can be improved but also on how markets can ultimately be expanded.

There are two obvious concerns that rebut the assumption of endogenous growth to excuse careful rationing of depletable resources across generations. First, with additional economic development by the expansion and dispersion of technology, demand for natural capital may actually increase. For instance, the carbon footprint of MDCs is vastly wider and deeper than for LDCs. Growth and development may occur but it may worsen rather than lessen the sustainability dilemma. To establish a more egalitarian development paradigm, then, may require sacrifices in the developed world to ensure its unsustainable footprint may not worsen. This was the political challenge of the Brundtland Commission.

Tools Today	Tools Tomorrow
• Greening efforts • Pollution mitigation • Environmental Risk management • Waste management	• Sustainable development • Community Reinvestment • Fostering of community entrepreneurship • Civic engagement • Inclusive capitalism
Internal Tools	**External Factors**
• Sustainable technologies • Eco-sensitive system and policies • Lifecycle management systems • Sustainable innovations in-house • Green innovation capabilities	• Stakeholder movements • Corporate Responsibility • Green design principles • Triple Bottom Line accounting • Transparent Corporate Governance

Fig. 12.4 Tools for more efficient intergenerational corporate strategies

Second, long-wave growth that has been observed in the past does not ensure its continuation in the future. An empirical observation does not constitute a predictive theory. In fact, technological progress may even frustrate sustainability. The vast improvements in human longevity through advances in medical research and development have not resulted in a proportional extension of human productivity. Instead, the expansion of longevity that extends retirement from a few years to a few decades creates a commensurate increased demand for the natural capital necessary to provide for goods and services to meet the consumption arising from greater longevity.

A reliance on notions of endogenous or Schumpeterian growth as a justification for either weak sustainability or an argument for reduced concern over sustainability is convenient, but not justified. With progress comes no guarantees should endogenous growth theory fail. In fact, progress may have counterintuitive and regressive aspects. In Chapter 15, we explore an approach that provides for intergenerational equity by developing and justifying the notion of *weak sustainability*. We will also derive the Hartwick Rule, which calls for public policy to optimally redistribute the fruits of economic improvements more broadly across social classes without a worsening of our sustainability footprint, while it provides for intergenerational insurance policies that protect future generations.

Summary

The two-period model is a powerful tool that allows us to develop our intuition about the role of discounting in guiding the economic decisions of mortals over time when resources are limited. It demonstrates how mortal generations divide natural capital across periods, and results in a profound observation; The nature of discounting induces us to extract more natural capital earlier, which results in greater resource scarcity and higher prices for generations that follow.

We also see that this simple two-period model is easy to extend to multiple periods once we understand the implications of the concept of a marginal user cost. This concept states that there are two costs of extraction. The first are the obvious direct costs of extraction that presumably includes the cost of negative externalities extraction imposes, as described more fully in Chapter 16. The second is an imputed cost that measures the difference between a price of a resource and its extraction cost when extraction is rationed across periods. This rationing results in a price that exceeds marginal costs, with the difference, labeled the marginal user cost, that equalizes, on a discounted basis, the cost of rationing an extracted resource across all periods.

We closed with a philosophical claim that rationalizes why concerns over intergenerational fairness are unwarranted. Some have argued that such *endogenous growth theory* assures future generations of prosperity that exceeds those who came before them. If such a hypothesis is correct, then concerns about finite resources are unnecessary. We will see in later chapters that there may be ways to ameliorate the effects on future generations from our resource extraction today. However, these approaches are based on our ability to generate verifiable intergenerational wealth rather than the assertion of situationally convenient arguments without proof, as endogenous growth theory suggests.

We end in the appendix with an extension of our mathematical sophistication by developing the powerful tool of Hamiltonian optimization that is readily adaptable to economic decisions over time and allows us to extract still more profound intuition.

In the next chapter, we further develop our economic intuition by treating natural capital that can continue to reproduce indefinitely if it is well-managed.

ESG Toolkit

Does your institution categorize available corporate resources based on their sustainability or dependence on factors that vary over time or across generations or decades?

How do the internal and external factors affecting your organization evolve?

What techniques do you use to guide the sustainability of your organizational decisions?

Appendix

A more elegant and powerful method to demonstrate the *optimal extraction of a depletable resource over time* result relies on a standard technique for optimization called the Lagrange Method. In this method, a net benefits function is calculated for each period and is then optimized across all periods. By employing this more sophisticated technique, we can accomplish three goals. First, it becomes easier to treat additional periods. Second, the technique can be extended to the Hamiltonian Method to provide even more profound insights. Finally, we are able to derive an exact measure for our **marginal user cost**, called the **Lagrange Multiplier**.

Recall the net benefits of extraction overextraction costs for linear and consistent demand and constant extraction costs is given by:

$$NB = mX^2/2 + (p_c - mX - c)X$$

In the Lagrangian Method, net benefits are set up according to the following functional form:

$$\mathcal{L} = \frac{mX_i^2/2 + (p_c - mX_i - c) * X_i}{(1+r)^i} + \lambda\left(\overline{S} - \sum_{i=0}^{N} X_i\right) \quad \forall i = 0\ldots N,$$

where N is the number of periods over which we wish to extract the fixed natural capital $\overline{S}$. This general approach allows simple inclusion of additional time periods by adding one more first order maximization equation and one more quantity to be calculated, as is demonstrated next.

It also introduces one additional variable, called the *Lagrange multiplier* λ. This multiplier easily calculates how much current value net benefit is added if one more unit of the fixed resource is made available. To see how this method works, the Lagrangian is differentiated to optimize net benefits for the choice of each quantity over as many periods N for which we may want to spread extraction of the fixed resource over time:

$$\mathcal{L}_0 = \frac{\partial \mathcal{L}}{\partial X_0} = \frac{\delta(mX_0^2/2 + (p_c - mX_i - c) * X_0)}{\delta X_0} - \lambda = 0$$

$$\mathcal{L}_1 = \frac{\partial \mathcal{L}}{\partial X_1} = \frac{\delta(mX_1^2/2 + (p_c - mX_i - c) * X_1)/(1+r)^1}{\delta X_1} - \lambda = 0$$

$$\vdots$$

$$\mathcal{L}_N = \frac{\partial \mathcal{L}}{\partial X_N} = \frac{\delta(mX_N^2/2 + (p_c - mX_N - c) * X_N)/(1+r)^N}{\delta X_n} - \lambda = 0$$

$$\mathcal{L}_\lambda = \frac{\partial \mathcal{L}}{\partial \lambda} = \overline{S} - \sum_{i=0}^{N} X_i = 0$$

These $N + 1$ equations can solve for N quantity variables $X_1 \ldots X_N$, one for each period, and the one λ Lagrangian multiplier. In the case of linear demand and constant marginal costs, these equations are also linear and hence a solution with $N + 1$ linear equations and $N + 1$ unknowns generally exists.

The Lagrangian method is employed here to determine the solution to the two-period model to ensure it generates the same results we obtained by using brute force algebra. The first order condition for the current period is determined by differentiating $\mathcal{L}_0$ with respect to the extraction rate in the first period:

$$\mathcal{L}_0 = \frac{\partial \mathcal{L}}{\partial X_0} = \frac{\delta(mX_0X_0/2 + (p_c - mX_0 - c) * X_0)}{\delta X_0} - \lambda = 0$$

Let us focus again on the two-period version of the problem. With some simplification, we see:

$$\mathcal{L}_0 = \frac{\partial \mathcal{L}}{\partial q_0} = -mX_0 + p_c - c - \lambda = 0$$

$$\mathcal{L}_1 = \frac{\partial \mathcal{L}}{\partial X_1} = \frac{(-mq_1 + p_c - c)}{(1+r)} - \lambda = 0$$

$$\mathcal{L}_\lambda = \frac{\partial \mathcal{L}}{\partial \lambda} = \overline{S} - X_0 - X_1 = 0.$$

Then, by moving λ to the right of the first two equations, we see that:

$$-mX_0 + b - c = \frac{(-mX_1 + p_c - c)}{(1+r)}$$

$$mX_0(1+r) - mX_1 = r(p_c - c)$$

$$X_0(1+r) - X_1 = r(p_c - c)/m.$$

Substituting in the third first order condition then yields:

$$X_0(1+r) - \overline{S} + X_0 = r(p_c - c)/m$$

$$X_0(2+r) = \overline{S} + r(p_c - c)/m$$

$$X_0 = -(S + r(p_c - c)/m)/(2+r).$$

Notice that this is the same expression we derived using the brute force method earlier. Using the parameters as before, we again find:

$$X_0 = \frac{210 + 0.1 * \frac{(2500-400)}{10}}{2 + 0.1} = 110$$

$$X_1 = 100$$

We can also calculate the Lagrange multiplier:

$$\lambda = -10 * 110 + 2500 - 400 = 1000$$

$$p_0 = 2500 - 10 * 110 = 1400$$

$$p_1 = 2500 - 10 * 100 = 1500$$

As expected, this more elaborate, elegant, and easily extendable technique generates the same results as the more cumbersome two-period approach derived earlier. Let us next interpret the significance of our newly introduced variable, the Lagrange Multiplier λ. It yielded the present value of the difference between the equilibrium market price and the marginal extraction cost, in this example 1000. This Lagrange Multiplier is our previously defined marginal user cost and reveals to us the equilibrium value across all periods of a shift of a small amount of resource extraction from one period to the next.

Under an efficient optimum, this Lagrange Multiplier λ then represents our *discounted Marginal User Cost (MUC)* that remains constant over time. Constancy of the discounted MUC implies a decision-maker today is indifferent to a shifting of a small amount of natural capital from one period into any other period. This constancy of the discounted marginal user cost is called the *equimarginal principle*, and provides us the general and somewhat troubling intuition that resource decisions within one generation always discount the sacrifices imposed on future generations.

We can also calculate the net benefits in the current period and the next period:

$$NB_0 = (2500 - 1400) * 110/2 + (1400 - 400) * 110$$
$$= (550 + 1000) * 110 = 170{,}500$$

$$NB_1 = (2500 - 1500) * 100/2 + (1500 - 400) * 100$$
$$= (500 + 1100) * 100 = 160{,}000,$$

while the static net benefits yield had the entire stock been extracted in the first period alone by perfect competitors would have been:

$$NB_{PC} = (2500 - 400) * 210/2 = 220{,}500.$$

We shall return to these calculations in our discussions to follow. For now, it is possible to tease out more intuition.

The Lagrange Method lends itself to optimization when net benefits are known and can be optimized. In addition, the interpretation of the Lagrange multiplier easily demonstrates its significance in allocating discounted surpluses on the margin between periods. The solutions are readily extended to consider additional periods.

Multiperiod analyses are discrete in that they separate extraction across a given number of periods. It is also possible to repeat the analysis in continuous time using an analog of the Lagrangian multiplier approach. In doing so, one can overcome the limiting factor of the Lagrange Method in that it requires a separate equation to be solved for each time period. In addition, while one can then track changes in the solution from one period to the next, for instance in prices and quantities or the observation that the discounted marginal user cost is constant under the equimarginal principle, the precise path of prices and quantities over time is not immediately apparent in the discrete time Lagrangian Method.

Extension of the optimization problem to continuous time also allows us to more easily derive the nature of challenges to sustainability and extensions to the model.

The continuous time version of *Lagrange's Method* is *Hamilton's Method*. It involves framing the problem as a *Hamiltonian*. Named after the Irish mathematician Sir William Rowan Hamilton (1805–1865) who developed the technique to better understand the mechanics and trajectories of motion in physics, the technique can also be employed to describe the trajectory of prices and quantities in the study of sustainability, among many other applications in science and economics.

Hamilton's technique is a superior method to analyze decisions that must be optimized over the continuity of time. Even a two-period discrete time model is complicated and requires solving multiple equations or matrices. Instead, Hamilton's continuous time analog instead requires the solution to one or two differential equations. In the chapters ahead we shall explore the results derived from the Hamiltonian method. We next treat the mechanics of this method.

Just as Lagrange's Method was used to describe the path of resource extraction and prices from one period to the next by optimizing the resource extraction rate, Hamilton's Method can optimize extraction subject to the same constraint that total extraction cannot exceed the available stock of natural capital.

The problem is framed in a manner similar to Lagrange's Method, but in continuous time of extraction. In general, the Hamiltonian optimizes some function $f(X(t), S(t), t)$ that depends on the extraction rate $X(t)$ and the stock $S(t)$ of a resource over time, discounted to the present by the discount factor e^{-rt}:

$$\max_{X(t)} \int_0^T e^{-rt} f(X(t), S(t), t)dt,$$

$$subject\ to:$$

$$\dot{S}(t) = g(X(t), S(t), t)$$

$$S(0) = \overline{S}$$

$$S(t) \geq 0 \quad \forall t.$$

The constraint states that the resource stock $S(t)$ declines over time as a function $g(X(t), S(t), t)$ depending on the extraction rate $X(t)$ and the stock size $S(t)$. These expressions result in differential equations to be solved, for which we know the starting point of the **state variable,** the resource stock $S(t)$, and perhaps also its endpoint or end time.

In the resource extraction case, the function we optimize is discounted at a discount rate r, $S(t)$ is the remaining stock of the fixed resource at any time t, and $f(X(t), S(t), t)$ is the utility or surplus enjoyed in any period from resource extraction $X(t)$ that is consumed in period t.

Notice the number of similarities compared to the Lagrange Method. Rather than summing the objective of each period in discrete time, the summation in continuous time uses an integral. The value of the objective at each moment of time is discounted by e^{-rt}, the continuous analog of $(1 + r)^i$ used as the discount factor when discounting in discrete time. Again, the optimization is constrained to use no more than the available resource $\overline{S}$. This stock of a resource evolves based on the rate of extraction. In the simplest case of a fixed resource,

$$\dot{S}(t) = -X(t).$$

This equation simply states that extraction and enjoyment of the resource at a rate X(t) reduces the depletable resource stock S(t). Finally, the optimization begins at the current time t = 0 and extends to any arbitrary point t = T in the future. At that point, which could be infinite, either the resource is intended to run out, or the economy moves to another backstop technology that replaces the need for extraction.

The goal remains the same, though. It is to determine the optimal level of natural capital extraction and usage X(t) over time, depending on the size of the stock $\overline{S}$, the discount rate r, how the stock evolves over time, and the net benefits of extraction embodied in the general functional form f(X(t), S(t), t).

In such an approach, the state variable S(t) of the fixed resource interacts with its extraction, the **control variable** X(t). Then, just as the Lagrange equation was formed, so can a current value Hamiltonian:

$$\mathcal{H}(X, S, \lambda, t) = f(X, S, t) + \lambda g(X, S, t)$$

This new costate variable $\lambda(t)$ is an obvious analog to the Lagrangian Multiplier and is generally positive so long as the economic problem values a positive stock of the resource stock S(t). Hamilton characterized the optimal solution as a series of differential equations rather than algebraic first order equations:

$$\frac{\delta\mathcal{R}(X(t), S(t), \lambda(t), t)}{\delta X(t)} - \lambda(t) = 0$$

$$\lambda(t) = r\lambda(t) - \frac{\delta\mathcal{R}(X(t), S(t), \lambda(t), t)}{\delta S(t)}.$$

In the fixed resource case, the objective is to optimize the extraction of the resource **state variable**, the natural capital that diminishes each period at the extraction rate X(t). Then, the system simplifies to two differential equations. One optimizes our objective function, which is typically the utility or profit derived from resource extraction with respect to resource extraction X(t):

$$\frac{\delta\mathcal{R}(X(t), S(t), \lambda(t), t)}{\delta X(t)} = 0$$

The second differential equation ensures that changes in the stock of the natural capital affects our discounted objectives at each point of time at the same rate:

$$\frac{\delta\mathcal{R}(X(t), S(t), \lambda(t), t)}{\delta S(t)} = 0$$

$$=> \dot{\lambda}(t) - r\lambda(t) = 0.$$

In optimizing the Hamiltonian, the resource is efficiently extracted, and its stock dwindles over time in a way that equates the discounted marginal user cost across periods, as before. If the objective is to optimize the value of resource extraction, net of marginal costs, then these equations simplify considerably to:

$$p(t) - c = \lambda(t)$$

$$\frac{\dot{\lambda}(t)}{\lambda(t)} = r$$

This surplus function $p(t) - c$ then equals $\lambda(t)$, which is the same *marginal user cost* we derived in the two-period model and the Lagrange Method.

This method is equivalent to the optimization of net benefits under a demand curve. To see this for linear demand, consider an arbitrary demand function that calculates price as a function of quantity:

$$\text{Price} = p(X)$$

The area between the demand curve and the supply curve that accrues to consumers and producers as surpluses, from a quantity of zero to an arbitrary q*, is represented by the following integral:

$$Net\ Benefits\ NB(X^*) = \int_0^{X*} (p(X) - c(X))dX$$

Maximization of net benefits NB(X*) requires the differentiation of net benefits NB(X*) for a change in the equilibrium extraction rate X*. Assume for now that marginal costs are constant. Applying Leibnitz's Rule to the integral yields:

$$\frac{dNB(X^*)}{dX^*} = \int_0^{X^*} (\frac{dp(X)}{dX^*})dX + p(X^*)\frac{dX^*}{dX^*} - c(X^*) = p(X^*) - c(X^*)$$

The intuition is simple. While we had formulated the **total** producer and consumer surplus when we compare the total value generated at various solutions, we need only employ marginal net benefits, which is the difference between price at the margin and the marginal cost. This equilibrium price does not contribute to consumer or producer surplus since it measures the point in which the marginal consumer of the resource is just indifferent to paying the price of the good. In other words, those *on the margin* earn no consumer surplus.

Let us compare the results to the previous case of linear demand, $p(t) = p_c - mX(t)$, we have:

$$p_c - mX(t) - c = \lambda(t)$$

$$\frac{\dot{\lambda}(t)}{\lambda(t)} = r$$

These results are the same as those derived from the less versatile Lagrange Method, but Hamilton's Method offers a richer intuition that we will use later, and is much more readily extended to renewable resources that could be managed sustainably.

As before, it is immediately obvious that the multiplier $\lambda(t)$ increases proportionally with the discount rate. In other words, the discounted present value of the marginal user cost remains constant. Often, the Lagrange Multiplier is called a *shadow price* because it functions as an important policy instrument that is not revealed directly by markets even though it influences our economic decisions.

In this application, the marginal user cost that is equivalent to the costate variable in Hamilton's Method is a shadow price that partitions the depleting resource so that its price constantly rises and extraction tapers over time until the resource runs out at the specified time. Only if the discount rate is zero would this shadow price remain constant and the resource be partitioned evenly rather than tapering over time.

This constantly rising shadow price that maintains a constant present value of future resource extraction was noted first by the Canadian economist L.C. Gray in 1913.[5] Gray's intuition was subsequently developed extensively and hence named after a later mathematician Harold Hotelling in 1931,[6] based on some of his earlier work in 1925 regarding depreciation of resource capital.[7] Our Hamiltonian confirms the intuition Gray and Hotelling postulated—that the value of an additional (marginal) unit of the fixed natural capital grows at the interest rate. As before, if the marginal user cost is rising at the interest rate, the market price of the extracted resource must also be rising, and its extraction must be falling.

From these equations, one can derive the path of extraction if the total stock is known, the time horizon is specified, and the price at the end of the time horizon is determined, perhaps based on the price p_C of some backstop technology such as sustainable recycling.

For instance, let us assume that a stock of gold can be sustainably recycled at a constant price p_C. It would then be optimal to extract any stock until its marginal extraction cost eventually rises to the choke price p_C. This backstop price and the realization that the total amount of gold extracted from the current period to then will specify when the extraction period T ends, and at a price $p(T) = p_c$.

If the demand curve is known, to determine the path of extraction, one can then work backward from the point in time in which the resource is priced out of the market. Then,

$$X(t > T) = 0,$$

Once the price of the resource has risen to the point that the economy transitions to the backstop technology. We also know that the rate of change of the market price relative to the marginal user cost equals the interest rate:

$$m\dot{X}(t)/(p_c - mX(t) - c) = -r.$$

With linear demand, these equations can then be solved to find the optimal level of extraction that preserves the equimarginal principle and hence mimics the solutions described earlier.

The Hamiltonian technique can also be easily extended to consider extraction of resources when a stock declines with extraction but can also be augmented by careful management and the addition of biomass, as is the case of *biotic resources* such as forests and fisheries. In addition, the Hamiltonian technique can easily be modified to allow marginal costs to change with the level of extraction or with the size of the remaining stock. These various extensions, to be described in the next chapter, are far more tedious using less sophisticated tools.

Introduction to the Open Access Problem

We have been able to extract some useful intuition by beginning with a simple model using basic algebra and some reasoning. Let us use these more advanced tools to expand our analysis.

Consider for a moment the open access linear demand competitive problem in which undeterred competitors extract a fixed resource. Under this benchmark, competition forces extraction to be driven to a rate in which the price converges to the constant marginal costs. We demonstrated that this solution is given by the static competitive equilibrium where extraction occurs until the price falls to marginal extraction costs without regard for future users of the resource.

The time to resource exhaustion is then the simple ratio of the fixed stock $\overline{S}$ divided by the extraction rate X where X solves for $p(X) = c$, the competitive solution where the price is driven down to marginal costs. Algebraically, with a linear demand curve $p(X) = p_c - mX$. The solution is then when $p_c - mX = c$, or $X = \frac{(p_c - c)}{m}$.

We will demonstrate later that a competitive open access natural resource stock will be over-extracted because of a phenomenon called *the tragedy of the commons*, treated more fully in Chapter 14. Let us assume a regulator knows that competitors disregard the marginal user cost and the effect of their extraction of future generations. Good public policy would then constrain resource extraction be equivalent to that obtained under the equimarginal principle. This is ensured if extraction can be allocated such that the Lagrangian multiplier grows in proportion to the discount rate:

$$\frac{\dot{\lambda}(t)}{\lambda(t)} = r$$

The intuition of the condition is that there can be no reallocation of the fixed resource between time periods that yield a higher net present benefit. Continuing with linear demand, a competitive solution that is suitably constrained follows the differential equation above:

$$m\dot{X}(t)/(p_c - mX(t) - c) = -r \quad \forall X(T) = 0$$

Rearranging this equation gives:

$$\dot{X}(t) - rX(t) = -r(p_c - c)/m \quad \forall X(T) = 0$$

It is interesting to note that this expression derives the level of extraction X_{PC} that unconstrained perfect competitors would choose. Let us use the shorthand X_{PC} for this ratio:

$$X_{PC} = \frac{b - c}{m}$$

The solution to this differential equation is given by:

$$X(t) = X_{PC} + ke^{rt} \forall X(T) = 0.$$

Invoking the terminal condition $q(T) = 0$ generates the value for the arbitrary constant k:

$$X(T) = 0 \rightarrow k = -X_{PC}e^{-rT}.$$

Then,

$$X(t) = X_{PC}(1 - e^{-r(T-t)}),$$

where T is defined by the fixed resource stock condition:

$$\overline{S} = \int_0^T X(t)dt$$

$$\overline{S} = \int_0^T X_{PC}(1 - e^{-r(T-t)})dt$$

This integral can be solved to yield an implicit solution for the terminal time T_{PC} for the extraction-constrained competitive industry:

$$\overline{S} = X_{PC}\left(T_{PC} - \frac{(1 - e^{-rT_{PC}})}{r}\right)$$

THE MONOPOLIST'S RESOURCE EXTRACTION TRAJECTORY

How does this trajectory of extraction by competitors compare with the decision of a monopolist who has full control over the fixed resource?

Note first that, unlike open access competitors, a monopolist takes full account of the **equimarginal** principle to allocate extraction each period to optimize its profits. Since the monopolist makes decisions on the margin based on marginal revenue and cost, its objective is to maximize profits rather than the difference between price and cost. Using linear demand, its revenue is:

$$\text{Total Revenue} = p(X) * X = (p_c - mX) * X$$

$$\text{Marginal Revenue} = \frac{d(p_c X - mX^2)}{dX} = p_c - 2mX.$$

Hence, the monopolist instead ensures that the allocation of one new unit of the resource stock yields the same discounted marginal user cost across each period. Recall the principle that the change in the marginal revenue relative to marginal user cost is constant and proportional to the discount rate. Their optimization then becomes:

$$2m\dot{X}(t)/(b - 2mX(t) - c) = -r.$$

The general solution then simplifies to:

$$X(t) = \left((\frac{X_{PC}}{2})(1 - e^{-r(T-t)}) \right).$$

The interaction between the resource constraint $\overline{S}$ and the extraction terminal time T is then:

$$\overline{S} = (\frac{X_{PC}}{2})\left(T_M - \frac{(1 - e^{-rT_M})}{r} \right).$$

This expression can be compared to the extraction life of the constrained competitive industry:

$$\overline{S} = X_{PC}\left(T_C - \frac{(1 - e^{-rT_{PC}})}{r} \right).$$

We can use these expressions to determine how the terminal time to the end of resource extraction changes depending on the harvesting regime. Note that:

$$2\left(T_{PC} - \frac{(1 - e^{-rT_{PC}})}{r} \right) = \left(T_M - \frac{(1 - e^{-rT_M})}{r} \right)$$

$$d\bar{S} = X_{PC}\left(1 - e^{-rTc}\right)dT_{PC}$$

$$\frac{dT_{PC}}{d\bar{S}} = \frac{1}{X_s\left(1 - e^{-rT_{PC}}\right)} > 0.$$

Quite intuitively, the optimal terminal time of extraction increases as the stock increases. Since the expression in the brackets $X_{PC}\left(T_{PC} - \frac{(1-e^{-rT})}{r}\right)$ increases in terminal time T, we see that the terminal time to the sunset of depletable resource extraction is more distant for the monopolist, even compared to competitors forced by public policy to allocate extraction based on the equimarginal principle. Since the fixed resource is stretched over a longer period for the monopolist, their price path is higher, and quantity of the resource extracted lower at each point in time.

Notice, though, that a monopolist able to perfectly price discriminate by usurping for itself all consumer surpluses will again value the resource at the margin at the market price rather than their marginal revenue, and hence would replicate the pattern of extraction that a competitive industry would choose. A perfect price discriminating monopolist extracts faster and more intensively than does a traditional monopolist.

These results are quite general. This model demonstrates that, if resource extractors are price takers who are unable to affect market dynamics and instead rely upon them, they exhaust resources quicker. The primary departure is that monopolists realize they can control the price and hence compare not prices but rather marginal revenues to marginal costs. The governing equations then evolve from:

$$\frac{(\dot{p}_t - c)}{(p_t - c)} = r,$$

to:

$$\frac{(\frac{d(p\dot{X})}{dq_t} - c)}{(\frac{d(pX)}{dq_t} - c)} = r.$$

We see that a monopolist who controls a resource will induce the difference in marginal revenue and marginal cost to rise at the market interest rate. With linear demand, the market price net of marginal cost then increases at twice the interest rate. The value of leaving a resource in the ground is then higher for the monopolist, which induces it to extract the resource more slowly, even though the higher price they charge to expand their producer surplus $p_t - c_t$ comes at the expense of *consumers' surplus*. Monopolists use a fixed exhaustible resource more sustainably.

MORE ON SURPLUSES

Each generation of fixed resource users seeks to maximize their own benefits. In the absence of some sort of market intervention, the single period static extraction rate that would occur if the resource was subject to open access with free entry results in excessive extraction that drives the resource price down to its marginal cost. Using the example described earlier in Fig. 12.1, rather than dividing extraction in period 0 and period 1 to 110 and 100 units respectively, the entire stock of 210 units would be extracted, no producer surplus would be earned, and a total *consumers' surplus* (CS) of the triangle between the demand curve and cost curve would be:

$$CS = (2500 - 400) * 210/2 = 220{,}500.$$

This simple example demonstrates some interesting features. First, discounted producers and consumers' surplus when extraction is limited by the equimarginal principle is given by:

$$NB_0 = (2500 - 1400) * 110/2 + (1400 - 400) * 110$$
$$= (550 + 1000) * 110 = 170{,}500$$

$$NB_1 = (2500 - 1500) * 100/2 + (1500 - 400) * 100$$
$$= (500 + 1100) * 100 = 160{,}000$$

$$Discounted\ NB_{Total} = 170{,}500 + 160{,}000/(1 + 0.1) = 315{,}955,$$

while competition under open access results in a lower net surplus:

$$NB_{Static} = (2500 - 400) * 210/2 = 220{,}500.$$

Allocating extraction across multiple periods enhances economic value because resources are allocated toward higher willingness-to-pay consumers over more periods. In extracting resources more slowly, all else equal, greater overall value is created, even with discounting. Still, there remain some inefficiencies in the form of deadweight losses left unenjoyed by any participant. In the absence of discounting, enjoyment of the resource would generate the following:

$$NB_0 = (2500 - 1450) * 105/2 + (1450 - 400) * 105 = 165{,}375$$

$$NB_1 = (2500 - 1450) * 105/2 + (1450 - 400) * 105 = 165{,}375$$

$$NB_{Total} = 165{,}375 + 165{,}375 = 330{,}750$$

This total is compared to 330,500 when extraction is tapered under discounting. The equimarginal principle with discounting creates minor dead-weight losses, but also introduces some ethical dimensions that will be explored later.

NOTES

1. Schumpeter, Joseph A. (1994) [1942]. Capitalism, Socialism and Democracy. London: Routledge. pp. 82–83. ISBN 978-0-415-10762-4. Retrieved October 27, 2022.
2. Darwin, Charles (1859), On the Origin of Species by Means of Natural Selection, or Preservation of Favoured Races in the Struggle for Life. London: John Murray.
3. Sombart, W. (1913). Krieg und Kapitalismus [War and Capitalism]. Leipzig: Duncker & Humblot. p. 207. ISBN 9,780,405,065,392.
4. Hart, Stuart (2005), "Innovation, Creative Destruction and Sustainability," Research-Technology Management, Vol. 48, Issue 5.
5. Gray, L. C. (1913), "The Economic Possibilities of Conservation," The Quarterly Journal of Economics, Volume 27, Issue 3, May 1913, Pages 497–519.
6. Hotelling, Harold (1931), "The Economics of Exhaustible Resources," Journal of Political Economy, Vol. 39, No. 2 (Apr., 1931), pp. 137–175.
7. Hotelling, Harold (1925), "A general mathematical theory of depreciation," J. Am. Stat. Assoc. 20:340–353.

CHAPTER 13

Discounting and Renewable Resource Extraction

To now, two classes of natural capital have been considered. The first is the class of resources that can be consumed indefinitely with no negative consequences on future generations. For instance, the Sun's insolation can create heat and electricity essentially indefinitely, at least for our human scale of existence, if not on a cosmological scale. Geothermal energy and the heat released from nuclear fuels, or from fuels that are the byproduct of fission, also fall into the category of abundance without the ability to over-extract within the realm of human needs. Even wind power falls into this category as it is fueled by the Sun's heating capacity. Likewise, tidal power depends only on the moon's rotation about the Earth and the Earth's rotation about the Sun.

At the other extreme from this class of fully sustainable resources are the forms of abiotic elemental or non-organic natural capital in fixed supply, such as gold, lithium, and other minerals that arrived when our Earth was formed, and fossil fuels that were created a hundred million years ago under unusual circumstances. These resources must be extracted, within the limitations described in the last chapter. With abundant energy, it may be possible to recycle these resources to ensure that they too can be consumed sustainably. Such recycling depends on the ability to capture, preserve, and purify our extracted natural capital, if energy is sufficiently abundant. In such cases that permit recycling, fixed resources can become sustainable once the market price rises sufficiently so that recycling emerges as an affordable backstop technology.

An intermediate class of resources is labeled renewable. These resources can be extracted sustainably if managed well, but can be depleted if overharvested. Consider a forest. It can sustainably provide firewood for heat, materials for

© The Author(s), under exclusive license to Springer Nature Switzerland AG 2023
C. Read, *Understanding Sustainability Principles and ESG Policies*,
https://doi.org/10.1007/978-3-031-34483-1_13

medicines, and lumber for construction if the resource is permitted to thrive. So long as extraction or harvesting occurs at a lower rate than the ability of a resource to reproduce and add to its biomass, and sound management need not degrade the environment, forests or fisheries can provide sustainably indefinitely.

If left unchecked to maintain their own equilibrium, such a biotic stock will grow so large that further growth is kept in check by the natural constraints of its habitat. Trees in the forest and fish eventually die. At some maximum feasible size for a resource stock, births equal deaths, or biomass loss equals gain, and the resource stock is in check by its habitat and grows no further.

On the other extreme, renewable natural capital can collapse and become extinct. If its stock is low and is subject to some sort of natural or human shock or perturbation, it can decline in size to a level that leaves the weakened population vulnerable or of an insufficient critical mass to thrive.

Between these two extremes of zero net growth upon a population crash or checked population growth once it reaches the upper bound of its supporting habitat, there is a peak which yields the greatest sustainable level of excess growth over decline over each time period. At this intermediate peak shown in Fig. 13.1, the resource is maintained at its healthiest point and can then be managed for optimal extraction. This maximal biomass growth rate that can be extracted each period without causing the natural capital to decline is labeled the ***maximum sustainable yield***.

Consider the logistic function that describes a harvested species population such as a forest or a fishery, constrained by the limits of growth of the population to a size no larger than $\overline{R}$. The population size $R(t)$ described as a logistic function yields a net growth as the rate of change $\dot{R}(t)$ obtained by differentiating $R(t)$:

$$R(t) = \frac{\overline{R}}{1 + e^{-kt}} > 0$$

$$Growth\ g(R(t)) = \dot{R}(t) = \frac{kR^2 e^{-kt}}{\overline{R}} > 0$$

Given the potential to manage resource usage when the resource may be replenished, we can modify the Hamiltonian described in the last chapter. Recall

$$\mathcal{H}(H, R, \lambda, t) = f(H, R, t) + \lambda g(H, R, t),$$

where $H(t)$ is the harvest rate for the renewable resource of size $R(t)$. The first order conditions of the Hamiltonian that characterize the optimal solution as a series of differential equations:

$$\frac{\delta \mathcal{R}(H(t), R(t), \lambda(t), t)}{\delta H(t)} - \lambda(t) = 0$$

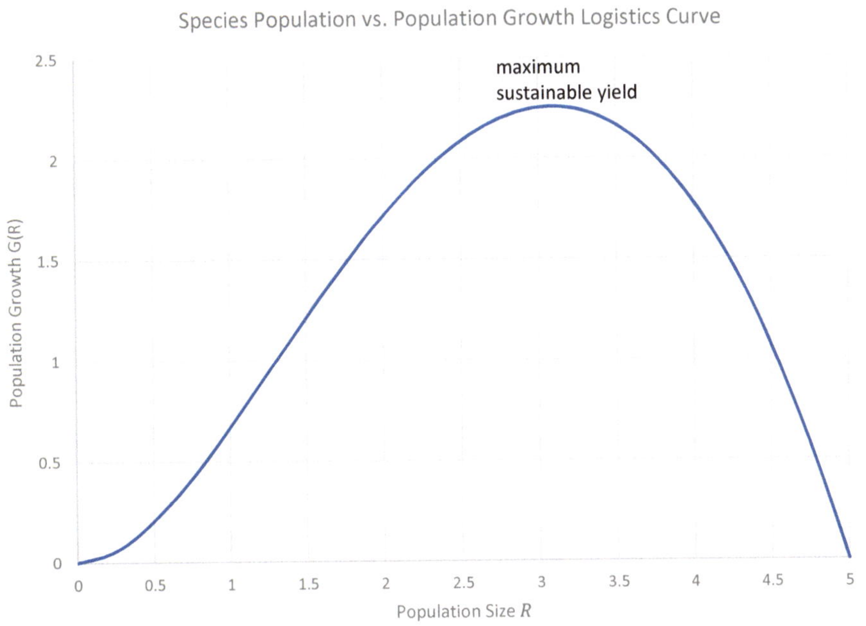

Fig. 13.1 Logistic biotic growth, the harvest function, and maximum sustainable yield

$$\lambda(t) = r\dot{\lambda}(t) - \frac{\delta\Re(H(t), R(t), \lambda(t), t)}{\delta R(t)}.$$

In the renewable resource case, the objective is to optimize the harvest rate $H(t)$ of the stock of renewable natural capital $R(t)$ to generate profits based on the corresponding harvest price $p(H(t))$, net of harvest costs and with knowledge that the harvest rate is but one aspect of the change in the stock. The other aspect of biotic natural capital is the capacity $G(R)$ for the resource to restock itself. In this case when we allow the stock to grow spontaneously in proportion to the size of the stock, the system simplifies to:

$$\Re = (p(t) - c)H(t) + \lambda(t)(G(R(t)) - H(t))$$

$$\frac{\delta\Re}{\delta H(t)} = 0,$$

$$or\ \frac{\delta((p(t) - c)H(t) + \lambda(t)(G(R(t)) - H(t)))}{\delta H(t)} = p(t) - c - \lambda(t) = 0$$

$$\frac{\delta\Re}{\delta R(t)} = -\dot{\lambda}(t) + r\lambda(t).$$

This relationship simplifies to:

$$\frac{\delta((p(t) - c)H(t) + \lambda(t)(G(R(t)) - H(t)))}{\delta R(t)} = \lambda(t)$$

$$\frac{dG(R(t))}{dR(t)} = -\dot{\lambda}(t) + r\lambda(t).$$

If the objective is to optimize the value of output, net of marginal costs, then these equations simplify considerably to:

$$p(t) - c = \lambda(t)$$

$$\frac{\lambda(t)}{\lambda(t)} = r - G_R(R(t)).$$

The first equation is our familiar result that the discounted value of the marginal user cost is given by the continuous time version of the Lagrange Multiplier. The second equation is also familiar. The marginal user cost grows in proportion to the rate we discount future user valuation, at the prevailing interest rate, except for an added growth term. This harvest growth term $G_R(R(t))$ is the benefit that arises by leaving a unit of the biotic resource unharvested so that it may grow and hence yield the potential for greater harvests in the future. This growth then contributes to a proportional relaxation of the marginal user cost.

Let us use this simple extension to our mathematics to describe an important sustainability principle.

Depending on the size of the renewable stock $R(t)$, the logistics curve shows that such growth $G(R)$ is increasing with small and vigorous population sizes, but this growth begins to decline beyond its growth peak. This peak is called *Maximum Sustainable Yield* and measures the most rapid biomass growth and hence the largest possible sustainable harvest rate of the resource.

Biotic resources that grow to their maximal size L to the right of maximum yield on the stock/growth. Figure 13.1 will grow quicker if the population is reduced. If the management goal is to maximize growth and biomass creation, a harvest can thin the population and promote more robust growth. Once the stock falls below maximum sustainable yield, additional harvest reduces the stock and population growth. A harvest to the left of maximum sustainable yield credits the financial bank by debiting the natural resource bank, while a harvest delayed credits the natural bank but reduces the potential contribution of a harvest to profits and deposits in the financial bank.

Compared to the optimization in extraction of a fixed resource in Chapter 12, delayed harvesting sacrifices a financial return r but nets an improvement in biomass growth G_R. The proportional positive marginal user cost then changes at a rate equal to $r - G_R$, rather than the rate of interest r alone. Just as in the case of a fixed resource, the marginal user cost given by the *shadow price* $\lambda(t)$ is positive, and reflects a difference between the market price of the resource $p(t)$ and its harvest cost c. This shadow price reflects the discounted price paid by future periods for one more unit of harvest from the resource stock in the current period.

Optimization of the Hamiltonian demonstrates this intuition. The proportional rate of change of the marginal user cost depends not only on the rate at which the discount rate depreciates future marginal user costs, but also on the ability of the resource to recover from a harvest. A positive discount rate increases the marginal user cost as the stock declines, and hence prices rise and harvests fall over time. Alternately, the biological growth rate $G(R(t))$ lowers future marginal user costs as the biomass harvest potential expands.

Steady state sustainability requires that the size of the stock R^* remains constant over time, which results in a constant shadow price as each period is able to enjoy the same harvest rate, at the sustainable rate $G(R^*)$ each period. If the maximization of a sustainable harvest is the appropriate public policy goal, the stock should remain at maximum sustainable yield, and every time period is able to enjoy the same maximal harvest each period. This special case would only occur if the discount rate is zero, which also requires the harvest to be peaked at maximum sustainable yield, which likewise implies $G_R(R(t)) = 0$.

To see this, let us assume for the moment that the renewable resource stock is larger than that which results in maximum sustainable yield. Harvesting above the population growth rate reduces the population and increases the harvest until maximum sustainable yield is established. At that point, a larger harvest than maximum sustainable yield would increase revenue, but the sacrifice for the resulting higher revenue is reduced harvest in the future. A positive discount rate rewards the overharvest today, and discounts the value of future consequences of the excess harvest. Once the interest earned on a marginal increase in harvest is equal to the income lost on the margin by reduced future growth, the return by converting the harvest value to financial capital, at the prevailing interest rate r, is equal to the return to the change in growth. Steady state is maintained and, at this point:

$$\frac{\dot{\lambda}(t)}{\lambda(t)} = 0 = r - G_R(R(t)) \rightarrow r = G_R(R(t))$$

This point occurs when the slope of the growth curve $G_R(R(t))$ in Fig. 13.1 is equal to the interest rate. A zero-growth rate, at maximum sustainable yield, then occurs only if the discount rate is zero. All positive discount rates push the population below the point of maximum sustainable yield.

This analogy of a unit of harvest avoided and hence redeposited in the *natural bank* must offer the same return as the prevailing interest rate in financial markets. This balancing act explains why the size of the resource stock occurs at a lower sustainable point with discounting than the maximum sustainable yield. If the stock is too high, faster extraction is temporarily advantageous as it supplements income and investment in financial markets while it also moves the renewable stock toward its optimal size. If the stock is too low, a reduced harvest allows the balance in nature's bank to be replenished and earn a higher return once the sustainable equilibrium is reestablished. The optimal harvesting rate with discounting occurs when the diminishing effect of discounting is offset by an incremental increase in the population growth rate at a harvest point to the left of maximum sustainable yield. Intergenerational equity is ensured if resource extraction reverts back to the maximum sustainable yield equilibrium.

Summary

A renewable resource can be managed in a way that depends on both the prevailing discount rate and the growth rate of natural capital that is capable of expanding its stock over time. This approach is applicable to biotic stock that can reproduce, but which, left to an ecosystem equilibrium, will expand to a maximal population constrained only by its available resources. The stock would be healthier, with natural growth greater than decay, if the stock size is reduced to the maximum sustainable yield point. At that point, there exists an interesting management prerogative. Rather than extract to the point that the resource rents earned today are equated to the resource rents tomorrow plus interest, the growth of the resource stock can be considered an asset in the natural bank. By delaying a unit of harvest, one loses financial returns but gains in resource growth. The manager will then balance these two effects. Note though, that a mortal resource owner with a positive discount rate would not maintain the stock at a level which generates maximum sustainable yield.

ESG Inquiries

Is there a resource that your corporation consumes that is depletable if managed ineffectively but sustainable if managed well?

How do your institution's decisions regarding the timing of your use of various resources respond to changes in market interest rates?

Appendix

It is also not difficult to model how greater harvest costs affect equilibrium of a renewable resource. Such variable harvest costs could have also been treated in the case of a fixed resource, but it is particularly relevant for renewable resources such as a fishery in which a thinning population typically requires more effort and cost per unit of harvest.

If the marginal harvest cost $c(t)$ depends on the stock size $R(t)$ in each period t, then the equation that governs the marginal user cost $p(t) - c(t)$

must be modified to include harvest costs $c(R)$ that depend on the size of the stock $R(t)$ over time. Then,

$$p(t) - c(R(t)) = \lambda(t)$$

$$\frac{\dot{\lambda}(t)}{\lambda(t)} = r - \frac{\delta G(R(t))}{\delta R(t)} + \frac{\frac{\delta c(R(t))}{\delta R(t)}}{\lambda(t)}$$

$$\frac{\dot{\lambda}(t)}{\lambda(t)} = r - \frac{\delta G(R(t))}{\delta R(t)} + \frac{\frac{\delta c(R(t))}{\delta R(t)}}{p(t) - \frac{\delta c(t)}{\delta R(t)}}.$$

If the stock of the resource remains constant in a sustainable equilibrium, $\frac{\dot{\lambda}(t)}{\lambda(t)} = 0$ and

$$r = \frac{\delta G(R(t))}{\delta R(t)} - \frac{\frac{\delta c(R(t))}{\delta R(t)}}{p(t) - \frac{\delta c(R(t))}{\delta R(t)}}.$$

The previous example with constant marginal costs is then simply a special case when $\frac{\delta c(R(t))}{\delta R(t)}$, the marginal harvest cost, is a constant given by c. If marginal costs rise as the stock declines, then $\frac{\delta c(R(t))}{\delta R(t)}$ will decrease, the term $p(t) - \frac{\delta c(R(t))}{\delta R(t)}$ will increase, and the magnitude of the expression on the right-hand size will decrease.

$$\frac{\delta G(R(t))}{\delta R(t)} = r + \frac{\frac{\delta c(R(t))}{\delta R(t)}}{p(t) - \frac{\delta c(R(t))}{\delta R(t)}}.$$

In equilibrium, the resource stock proportional growth rate is in equivalent to a level lower than the interest rate corresponding to a flatter part of the growth curve $G(R)$. This results in a higher renewable stock closer to the maximum sustainable yield.

The discrepancy between the harvest rate and the maximum sustainable yield offers the policymaker a tool to attain higher yields. If a policy maker can arbitrarily increase harvest costs as the stock dwindles, the equilibrium moves toward the maximum sustainable yield and hence the sustainable harvest point becomes more resilient. If these harvest costs rise sufficiently, they may even outswamp the discount rate effect and push the stock toward maximum sustainable yield. This intuition offers a tool for policymakers who wish to maximize the yield on a renewable resource but acknowledge that mortal harvesters tend to over-extract the resource.

CHAPTER 14

The Tragedy of the Commons and the Coase Theorem

The intuition of the dynamics of sustainability for a depletable or renewable resource is the interplay between extraction and consumption today and extraction delayed until tomorrow. The discount rate is a measure of how mortal generations regard those that follow them. This *rate of time preference* or, in Fisher's words, *rate of impatience*, hinges on the notion that to delay resource extraction will yield a future return sufficient to compensate mortals for sacrificed enjoyment today. The analogy we have used is that the equilibrium return to holding natural capital in nature's bank is equal to the return on the resource extraction profits that could have resulted had they been deposited in a traditional bank.

We have seen though that mortal humans typically over-extract natural capital. This excessive extraction is exacerbated when competitors race to harvest depletable or renewable resources. Such excessive harvesting occurs because these competitors do not have property right that ensure exclusive access to a resource tomorrow if harvesting is delayed. Open access resources suffer from such a lack of a property right, as we saw with the open access competitive equilibrium in Chapter 12.

In an essay in 1833,[1] William Forster Lloyd, a British political economist, pondered unregulated grazing land in Ireland and Great Britain at that time. For centuries, users of common pastoral land had exercised a sense of stewardship of public land set aside for the grazing of sheep and cows. Lloyd pondered the outcome should users have access to such common grazing land without a shared responsibility for its stewardship.

© The Author(s), under exclusive license to Springer Nature Switzerland AG 2023
C. Read, *Understanding Sustainability Principles and ESG Policies*,
https://doi.org/10.1007/978-3-031-34483-1_14

Such irresponsible land stewardship causes a breakdown of sound land management that may today be described as a *game-theoretic non-cooperative equilibrium* in which responsible resource extraction degenerates to a *catch-as-catch-can free-for-all*. Such a non-cooperative equilibrium is often described as a *prisoner's dilemma* problem within game theory. A competitive user of the resource argues that, should it extract the resource responsibly with future generations in mind, it will suffer as its good act is exploited by another more competitor who opportunistically catches as catch can.

There are many examples of careful stewardship of a common open access resource, as Nobel Prize-winning economist Elinor Ostrom noted in her 1990 book Governing the Commons.[2] In the absence of such non-binding social conventions, either regulation or the imposition of property rights may be necessary to preserve the *money in nature's bank* quality of a fixed resource when extraction is delayed.

While the notion of a *tragedy of the commons* originated in the early 1800s, such a non-cooperative equilibrium still features prominently in the price-taking behavior assumed within the competitive microeconomic model. In this paradigm, a producer does not consider the implications of its actions on the price in the marketplace. This price-taking behavior results in competition that drives the price of a good down to its long run marginal and average cost. Agents do not acknowledge that their increased rate of production results in expanded supply and hence a lower price for all producers. Instead, in the competitive model, producers assume they are *atomistic*, with only the most trivial effect on the market equilibrium.

When perfectly competitive producers share this assumption of equal and atomistic access to a market, their conjecture on the triviality of their effect on the marketplace, when combined with free entry of new producers, results in maximum long-term output at a minimum long-term price. We observe that their decisions were individually rational but collectively irrational because, unlike monopolists, no competitor realizes positive profits in equilibrium.

However, this outcome is hardly tragic. Consumers receive the lowest possible price, which results in the greatest value and surpluses of our willingness to pay over the price paid by consumers at the competitive solution. Because the traditional economic model generally assumes that we can simply make more of the necessary capital to produce competitive goods, the competitive model has emerged as the holy grail of mainstream economics. Note, however, that the desirable outcome assumes all factors of production are priced appropriately.

This efficiency hinges on the assumption that only reproduceable human resources and sustainable natural resources are employed. However, if some natural capital is irretrievably depleted and future generations are harmed in the process, an intergenerational externality is created. The competitive model is only efficient under certain limited circumstances which do not rely on unpriced or underpriced renewable natural capital or other improperly priced factors of production.

To derive the competitive result, recall the monopolist problem. A producer's decision maximizes the difference between total revenue and total costs. A producer will continue to increase output until their contribution to revenue from the last unit produced (in other words, at the margin) is just equal to its increase in costs. We can describe the increase in revenues that arises as a monopolist increases production q:

$$Marginal\,Revenue = \frac{d(p(q)q)}{dq} = p(q) + q\frac{dp(q)}{dq}$$

$$Marginal\,Costs = \frac{dC(q)}{dq} = c(q)$$

$$MR = MC \rightarrow p(q) + q\frac{dp(q)}{dq} = c(q)$$

Of course, the monopolist realizes that, to successfully sell an additional unit of output q would net it the price of $p(q)$ but would also require it to lower its price somewhat to sell the additional output. This decision requires the monopolist to weigh the increase in revenue from greater quantity sold to the loss in revenue by having to drop its price on all output to ensure the market can absorb the additional output.

Hence, the monopolist earns a price $p(q)$ but incurs a loss $q\frac{dp(q)}{dq}$ arising from the necessary price discount. The sum of $p(q) + q\frac{dp(q)}{dq}$ that represents the additional revenue if one more unit is supplied is equated to the monopolist's marginal cost in equilibrium.

The perfect competitor acting on an individual and atomistic basis does not recognize the effect of its decision on the overall collective result. It receives the revenue $p(q)$ but ignores the collective cost $q\frac{dp(q)}{dq}$ imposed on the entire market, especially as the number of competitors increases and the consequences of one individual's action are diluted across a greater number of competitors. Each individual producer can easily rationalize that their actions are insignificant.

Lloyd understood this dichotomy between the individual and the collective if instead there are resource consequences to individual actions. In his <u>Two Lectures</u> on the Checks to Population, he created an analogy to the abuse by cattle herders of the legal right to access public grazing land for their herds. Lloyd wrote:

> If a person puts more cattle into his own field, the amount of the subsistence which they consume is all deducted from that which was at the command, of his original stock; and if, before, there was no more than a sufficiency of pasture, he reaps no benefit from the additional cattle, what is gained in one way being lost in another. But if he puts more cattle on a common, the food which they consume forms a deduction which is shared between all the cattle, as well that of others as his own, in proportion to their number, and only a small part of it is

taken from his own cattle. In an enclosed pasture, there is a point of saturation, if I may so call it, (by which, I mean a barrier depending on considerations of interest,) beyond which no prudent man will add to his stock. In a common, also, there is in like manner a point of saturation. But the position of the point in the two cases is obviously different. Were a number of adjoining pastures, already fully stocked, to be at once thrown open, and converted into one vast common, the position of the point of saturation would immediately be changed.

Lloyd was an earnest proponent of the theory of marginalism as espoused by his political economist contemporary David Ricardo. He also understood the concerns of Thomas Malthus about the carrying capacity of the Earth to meet humankind's needs. If afforded a quota for the number of sheep permitted to graze, a shepherd realizes that, to exceed the quota by a single lamb, the shepherd gains all the advantages of the fattening of the additional animal, while all shepherds share in the costs of exceeding the carrying capacity of the grazing resource. This gap between individual benefits and collective costs, called the *free-rider problem*, is magnified when each shepherd makes the same self-serving calculation.

In 1968, the journal *Science* published an article authored by the ecologist Garrett Hardin. In Hardin's "Tragedy of the Commons," Lloyd's conclusion about overgrazing of an unmanaged commons is described. Hardin noted,[3]

> Therein is the tragedy. Each man is locked into a system that compels him to increase his herd without limit – in a world that is limited. Ruin is the destination toward which all men rush, each pursuing his own best interest in a society that believes in the freedom of the commons.

Hardin singled out the nature of individual decision-making rather than policies based on the collective good as the root cause of the tragedy of the commons. He added that, even if some individuals recognized the detrimental collective effect on others from their individual actions, individuals with a conscience would suffer compared to *free-riders* who act based on their selfish interest.

Economists define an externality as a consequence of an economic activity that is not reflected in its price and affects others beyond those who make the economic decision. Overuse of an unpriced, unallocated, and unmanaged resource such as an open access commons is thus an example of an externality. The need to acknowledge the *externalities* individuals impose on others will be treated in greater depth in Chapter 17.

Hardin concluded that resource management must be an essential tool to avoid the tragedy of the commons. Hardin has elevated our awareness of the *tragedy of the commons* and the need to manage an unmanaged commons. The concept of individual benefits versus collective costs was implicitly understood by Thomas Malthus in his dismal prophecies, and is also modelled in modern game theory and incorporated in the competitive economic model. Its implications on sustainability have been transformational since Hardin's influential 1968 essay.

This notion is applicable to a wide variety of problems in sustainability. Water access in the absence of treaties, forests on public lands and fisheries in international waters, and even access to some non-renewable resources in the absence of well-defined property rights may suffer from the consequences of open access. Examples of tragic market failures include the collapse of the cod stock on the Grand Banks off the coast of Newfoundland and the loss of the sturgeon stock in the Black Sea.

In our treatment of marginal user costs, recall that entities able to protect a resource respond to this gap between its market price and marginal costs because of their acknowledgment that part of the value of sacrificed extraction in the current period is compensated by extraction surpluses in the future periods, with interest. Open access guarantees no such ability to extract the resource in the future and hence makes unattainable the optimal extraction of a depletable resource over time.

THE ESTABLISHMENT OF PROPERTY RIGHTS AND THE COASE THEOREM

The Tragedy of the Commons occurs because the abusers of an open access resource catch-as-catch-can. They know that the good deed of a responsible individual agent's sacrifice of extraction in the current period will only be usurped by another's extraction in the current or the future period. Public policy can impose quotas on resource extraction to avoid such a *race-to-the-bottom* and attain the socially optimal extraction rate. By limiting output, a profit is conferred in an amount equal to the marginal user cost on any entity permitted to extract the resource under the quota. The obvious question remains. How should such an extraction advantage be conferred and upon whom?

An economics definition of socialism is that the collective public owns the fixed and shared natural capital of the Earth. If it is recognized that these fixed resources belong to us all but government does not necessarily operate commercial enterprises most efficiently, some sort of allocation mechanism must be created to ensure that these resources are developed and extracted profitably and efficiently within a free-market system, but in the interest of all humankind and the ecosystem, now and in the future.

The economist Ronald Coase won a Nobel Memorial Prize in Economic Sciences partly for his exploration of the optimal assignment of property rights to avoid such tragedies of the commons. He had been hired by the U.S. Federal Communications Commission (FCC) to propose how the airwaves ought to be allocated as radio, television, and telecommunications corporations increasingly sought frequencies as telecommunications technologies developed. Without some sort of an allocation mechanism, a frequency free-for-all would result in such problems as frequency interference and an arms race of increasing transmitter power to drown out competitors would damage the entire industry. However, Coase felt that government was in a poor position to pick broadcast frequency winners and losers should it decide to allocate frequencies based on its perception of best use of the airwaves.

The government's goal was to ensure that the airwaves resource was put to their best use, measured by the creation of maximum economic value. The FCC accepted economists' reluctance to determine who deserves more the potential for economic gain. Economic justice aside, how should the airwaves then be allocated most efficiently?

This allocation problem to ensure efficiency has obvious parallels with regard to the granting of franchises or rights to extract fixed resources. Those granted a property right to the airwaves, or to a fixed resource, may be able to parlay these grants into significant wealth. Coase determined that economic justice concerns over such endowments of wealth are secondary to the greater goal of efficient resource usage. In neoclassical economics and in markets, efficiency is concerned only with value creation, not the income distribution effects of private wealth creation.

Before we look at Coase's solution to this seemingly complex problem, note that the airwaves and depletable or renewable resources are somewhat different. Both these categories suffer from the externality that one's abuse of natural capital can diminish the enjoyment or cause the exclusion of resource enjoyment by others. But, most natural resources differ in that efficiency requires an acknowledgment of a marginal user cost for the extraction of depletable and renewable resources over time, while a radio frequency unused or abused in one period has no ill effect on a later period.

Coase extended this concept in his 1960 academic paper, The Problem of Social Cost,[4] that drew a surprising conclusion. In the absence of transaction and negotiating costs between interested parties, and in the recognition that only well-defined property rights will prevent the abuse or suboptimal use of a fixed, depletable, or congestible resource, the initial allocation of the right to a resource does not matter.

Coase claimed that *any initial allocation* of property rights to a resource will still allow a free market to discover and converge upon a *Pareto efficient allocation*. This concept of Pareto optimality simply states that, in equilibrium, there is no other allocation that could create additional benefits to some that exceed the potential losses to others. Coase was confident that free markets have a tendency to maximize value by capitalizing on the Pareto optimal solution that creates more value than any other equilibrium.

Coase showed that, in the absence of significant transactions costs, and with the ability to clearly define property rights, the most efficient allocation can be obtained without any government intervention beyond the upholding of property rights. He also showed that efficiency occurs regardless of the initial assignment of these rights. His key insight is that bargaining between economic agents will ensure a fixed resource will be secured by the entity that can create the most economic value from its use.

To see this, consider a classic example. Two companies rent adjacent rooms that share a common floor. One company performs the delicate task of diamond cutting and polishing. The process requires the steadiest of hands. The second company uses a large, noisy, and disruptive machine to shake and stir large paint containers.

Assume that the diamond polisher has been granted the right to a stable floor and a quiet work environment. If so, this right will put out of business the paint shaker. This paint company will then attempt to negotiate with the diamond polisher. If paint shaking is more lucrative than diamond polishing, the paint shaker will be able to offer a payment to the diamond polisher for their losses should the polisher discontinue business or conduct business only at hours when the paint shaker does not operate. This *side payment* will only be made, and the right exchanged, if paint shaking is more valuable than diamond polishing. In the end, the resource will go to the best use, and its opportunity cost, the alternative use, will be fully compensated.

On the other hand, if diamond polishing is more lucrative, the paint shaker will be unable to propose an offer that could profitably secure the right for itself. In this case, diamond polishing prevails. In the end, once this process of bargaining and negotiations is completed, the resource will rest in the hands of the entity that can best employ it.

Should this resource be instead allocated to the paint shaker, the alternative allocation may require side payments to entice the party holding the title to accept the reassignment of rights. Either way, under the assumptions of the Coase model, the resource will be best assigned accordingly.

In reality, such bargaining or transactions costs for the potential reassignment are not costless. If bargaining or *transactions costs* are sufficiently high, then they may render the reassignment unfeasible. This occurs when the benefits of one user are not greater than that of the other user to justify these various bargaining or transactions costs.

In addition, humans are often imperfect negotiators, especially if psychology and irrationality enter into the bargaining effort. Under these real world circumstances, it may then be the case that the initial allocation does matter. The final outcome of the most efficient use of the resource is then more likely if the most efficient and responsible resource user is offered the initial allocation.

This logic also creates a property rights dilemma. If transactions costs are indeed zero, then, in some sense, the initial assignment of property rights is unnecessary in the first place. If the Coase model suggests that a current final allocation occurs regardless of any initial property rights allocation, then initial property rights allocations serve no purpose but to dictate wealth transfers and income redistribution as these property rights are negotiated and reallocated.

The second caveat to the Coase model is in ***bargaining asymmetries***. One large polluter in a company town has significantly more power in the preservation of its status quo interest in consuming the natural capital of clean air. The thousand households that surround the polluter's mill must endure the health consequences of the pollution. In this case, each individual household has little negotiation power, and the cost of organizing all the individuals into a more potent bargaining entity is fraught with complications.

Should a community activist attempt to organize the thousand households to pool their willingness to pay to avoid the pollution and its negative health consequences by purchasing the polluter's right, some individual households may instead try to ***free-ride***. Just as with the tragedy of the commons, the household may calculate that it need not contribute to the pool but can still enjoy the benefit of the community purchase of clean air rights. The holdout household saves the membership fee based on its calculation that its non-participation will cause the other 999 households to each pay only a 0.1% higher fee to purchase back the clean air right. If this tiny increment is still attractive to the remaining households, the free-riding household receives the clean air benefit at no cost.

Obviously, if such a calculation is replicated by even a modicum of households, bargaining and representation break down and Coase's model fails. This ***holdout problem*** is worsened when there are many potential bargainers and there is an absence of compulsory participation. In such cases, a government may step in and legislate the outcome rather than leave it to the free market.

It is ironic that such coercion to participate would typically require government intervention to pre-empt free-riders given the Coase model originally argued for no need to involve government to pick winners and losers. Indeed, even in such a case where there may be significant, complicated, and potentially expensive negotiations costs, and there may be substantial differences in the sophistication and expertise of negotiators, outcomes can be unpredictable.

A better approach may be to internalize an externality such as pollution or the disruptions among an adjoining business by optimally pricing the nuisance one party imposes on another. This cost can be in the form of an appropriate tax that equals the damage an action causes. In the case of pollution, such a ***Pigouvian tax*** is typically levied on the pollution producer and is of a size sufficient to compensate for the health and quality of life consequences to nearby households. When such an optimally determined tax is imposed, any remaining bargaining will likely be far more constructive and less cumbersome and expensive. We treat the Pigouvian solution in Chapter 16. However, even this solution requires a determination of who should be taxed—the paint shaker or the diamond polisher, in this example.

SUMMARY

The tragedy of the commons has been recognized for almost two centuries as a consequence arising out of the excesses of self-interest. In the absence of well-defined property rights, humans may behave in their own self-interest without regard for the implications it may have on others or even in the long-term to themselves. This race-to-the-bottom version of competition is costly in that common resources can be rendered valueless.

However, it has been shown that the problem lies in the lack of clearly defined property rights or the failure to price and hence internalize external-ities. If property rights are clearly established, economic agents are unable to extract resources to which they do not have title. One solution to this problem, devised by Ronald Coase, merely requires society to grant rights to, for instance, extract a natural resource. Bargaining over competing uses will then ensure resources fall into the hands of entities that can best capitalize their value.

Alternately, in socialism, all forms of natural capital are assumed to belong to the people. In this case, the people can auction off the right to the resource, and can then invest the proceeds and royalties in ways that can benefit future generations. We explore that alternative next.

The internalization of externalities, the harm one decision-maker may impose on others, is most problematic when those affected do not have a voice in decisions made without them. While those affected by economic decisions within their generation can be addressed through the political or market processes, intergenerational externalities are more problematic. The next chapter formulates a solution to such intergenerational externalities, but with a great ethical onus upon the generation able to make intergenerationally significant decisions.

ESG Toolkit

A corporation often develops a pricing strategy that exercises its market power by balancing an increased price with the quantity it can sell.

In exercising some market power, are potential customers then excluded from participating?

Could an alternate strategy be developed that can somehow fold these customers into the corporation's web of stakeholders?

To what degree does the corporation benefit costlessly from community property?

If so, can corporate benevolence be tailored to incentivize community property?

Does the corporation fairly compensate present and future generations from its use of shared property?

Does the corporation unfairly exercise its greater public power to disadvantage some populations?

NOTES

1. Lloyd, William Forster, Two Lectures on the Checks to Population, 1833 lecture.
2. Ostrom, Elinor (1990). Governing the Commons: The Evolution of Institutions for Collective Action. Cambridge University Press, Cambridge, UK.
3. Hardin, Garrett, "The Tragedy of the Commons," Science, December 13, 1968.
4. Coase, Ronald (1960), "The Problem of Social Cost," Journal of Law and Economics. **3**, pp. 1–44.

CHAPTER 15

Permanent Funds and the Hartwick Rule

Ronald Coase was charged with proposing a method to allocate the valuable resource of telecommunications frequencies. His proposal was to conduct auctions that grant a property right to the resource. Assuming that an auction is able to extract all the surpluses accruing to the rights to a unit of natural capital, who should own the revenues the auction generates?

Resource auctions often resort to a more thoughtful and analytic process than employed in familiar auctions such as for art, antiques, or cattle. Called sealed bidding, a successful bidder typically formulates its bidding strategy by anticipating what it believes its competitors will bid. Such strategizing in *First Price Sealed Bid Auctions* (FPSBA) may at times result in errors and lower bidding if one tries to anticipate bids of others rather than bid according to their own valuation.

An alternate superior bidding process called *Second-Price Sealed Bid Auction* (SPSBA) may be more successful in generating higher bids and a greater likelihood the resource rights go to the entity that values the resource the highest. This auction requires the winner to only pay the best price of the second-highest bidder. In a game-theoretic analysis of sealed bidding, if the second best bid price is charged to the highest bidder, participants are more likely to bid truthfully rather than strategically and perhaps at times irrationally. Auction royalties coincide only with the second highest valuation, but strategic errors are avoided and hence shared natural resources are placed in their best use.

For instance, the popular online auction site eBay asks early bidders to enter their reservation price, which is, in essence, a sealed bid. EBay sequentially places these bids to ensure they only marginally exceed the next highest bid.

C. Read, *Understanding Sustainability Principles and ESG Policies*, https://doi.org/10.1007/978-3-031-34483-1_15

Such auctions earn a more predictable share of the discounted present value of future extraction profits that accrue to the granting of rights to extract a fixed resource.

Once granted these exclusive property rights, the auction winner then has the incentive to extract an auctioned resource over time based on the *equimarginal principle*. The successful recipient of a resource lease will act as a mortal monopoly extractor with a positive discount rate and hence will commit to harvest or extraction that yields a higher price path and lower extraction over time than would be obtained in an open access or other market equilibrium. As we discussed in Chapter 12, earlier periods experience a greater share of resource extraction and consumption, while the marginal user cost increases over time in proportion to the discount rate.

Professor John Hartwick of Queens University proposed that such resource royalties earned by governments who assign exclusive rights to the highest bidding resource extractors should be invested into a *permanent fund* that can then accrue interest at the prevailing interest rate.[1] This permanent fund accruing to a people from the sale of extraction rights to natural capital can then be used to invest in such public goods as buildings, roads, education, and other public investments.

The original concept of a permanent fund was discussed by John Bates Clark in 1899 in his The Distribution of Wealth: A Theory of Wages, Interest and Profits.[2]

> The static hypothesis that capital is not increasing means, as we have just said, that the whole net income of the capitalist class is used up daily in the form of consumers' goods. It means, also, that capital is not diminishing; and that, therefore, only the income of the capitalist, and not his *permanent fund* of productive wealth, is available to supply his wants. He has, indeed, an ultimate safeguard against starvation, which the laborer lacks; for by changing his plan of life he can use up his capital. But naturally he does not do this, and the static hypothesis requires that he shall not do it. In this condition, he needs a store of subsistence goods, if the laborers need one... It is necessary to find some term to designate the whole permanent fund of productive wealth, and the natural name for it is capital. It is also necessary to have a term for all kinds of concrete goods in which this *permanent fund* consists; and we shall call these things, including land, capital-goods. (emphasis added)

In the late nineteenth century, and before Fisher had described his impatience theory or Ramsey had produced his golden rule of steady state consumption, Clark had lamented that to corporations the term capital often refers to wealth rather than a capacity to produce. Clark was instead advocating for a definition of capital that is the collection of long-lived assets that offer sustainability of the capital stock and maintenance of production. Under such a paradigm, consumption should be limited to the annual productivity of the capital stock, which also determines the return to capital we call a rate of interest. This perpetuation of an infinitely lived stock of capital is analogous, Clark noted, to the need for laborers to provide for themselves a subsistence income that equates to the value of production on the margin.

Clark was advocating for capital to be regarded not as a flow of capacity but as a stock that can be maintained over time. These are investments that can then provide a permanent flow of benefits to society. Clark noted that fixed resources, such as land, may constitute a natural form of this sort of permanent capital.

If society ultimately owns all of Earth's natural capital, then the stock of public capital should be maintained to compensate future generations for the decreasing stock of fixed non-renewable resources that are auctioned off. Ideally, if resource rents are reinvested to offset the costs of long-term social capital investments such as education or public infrastructure, a smooth flow of net returns across periods and generations can be constructed.

The Hartwick Rule is consistent with Clark's notion of a permanent fund. It is intuitively appealing, but it also required strong assumptions that are somewhat limiting. Hartwick's premise is that, if consumption depends on the stock of available capital, and if each type of capital is substitutable for another, the depletion of natural capital can be compensated by an increase in financial, human-produced, or pubic capital.

A generalization of Hartwick's Rule permits the traditional physical capital of machines, the human capital of people, the non-renewable abiotic capital in fixed supply, and the renewable but perhaps depletable biotic capital to substitute for each other to generate a steady flow of consumption over time. Note that the optimal extraction of the fixed resource yields a return if reinvested according to the equimarginal principle. Likewise, our rate of employment of machines is also adjusted to yield the same rate of return on the margin as given by the economy-wide discount rate. We see that the return from a loss of a unit of the natural capital can then purchase a return equal to the value of production of a reproducible human-made capital. Under these assumptions, human-made and natural capital are substitutable and their value is maintained in the aggregate.

As we saw in Fisher's intertemporal model, economic agents and activities align with the shared economy-wide interest rate. As with regard to the sustainable extraction of a renewable resource, the interest rate equates the combination of the depletion of the renewable resource and its rate of resource reproduction. If the depletion of natural capital can be reinvested into human capital through education to produce the same social yield, the obvious question is whether the substitutability of natural and physical capital, each of which is governed by the same discount rate, translates into optimal consumption.

Recall Irving Fisher's intertemporal model which first motivated our discussion of intertemporal resource choice. In equilibrium, consumers respond to the prevailing economy-wide discount rate by adjusting their marginal utility from consumption in the current period to their discounted marginal utility from consumption in a future period. Individuals likewise align their discounted marginal utilities and hence their consumption to the prevailing interest rate. This interest rate then coordinates both sides of a free market—its consumption and the employment of various forms of capital in production.

These broader relationships can be compactly summarized. The various equations that govern the efficient use of all forms of capital for production result from an extension of the Hamiltonian method include human-made capital K, the rate of extraction X of non-renewable capital stock S, and the harvest rate H of the stock of renewable capital R. The equilibrium conditions are then:

$$\frac{df_K}{dt} = r + \delta, \frac{df_X}{dt} = rf_X - f_S, \frac{df_H}{dt} = (r - G_R(f_R))\dot{f}_H - f_R,$$

where f_K is the increase in the physical capital stock to compensate for the expected return r and the capital depreciation rate δ, f_X is the marginal production arising from extraction X of the non-renewable resource stock S, f_S is the loss of production as the fixed stock dwindles in size, f_H is the marginal productivity arising from the harvest H of renewable natural capital R, and $G(f_R)$ is the growth of the stock of the renewable resource.

These results are intuitive. The first identity $\frac{df_K}{dt} = r + \delta$ states that producers should continue employing capital until the last bit of human-made capital earns an amount sufficient to cover the expected rate of return and the physical capital's rate of capital depreciation.

The second expression $\frac{df_X}{dt} = rf_X - f_S$ states that this rate of extraction of the non-renewable resource over time should equal the marginal value of its production less the loss of production as the depletable resource dwindles in size.

The last identity $\frac{df_H}{dt} = (r - G_R(f_R))\dot{f}_H - f_R$ states that the rate of harvest of the renewable (harvestable) resource over time should depend on the discount rate net of the change in resource growth less the degree to which production declines as the renewable resource is permitted to dwindle in size.

These expressions are natural extensions of our rules for the optimal extraction of resources, but they are framed in a helpful way. They measure the contributions to Clark's permanent capital stock for each of the sectors of human-made capital, non-renewable natural capital, and harvestable renewable natural capital. If each of these forms of capital is employed optimally, the increased scarcity of one form can be compensated by greater productivity of another.

Capital Substitutability Revisited

These intuitively appealing results were established based on the assumption of full substitutability of the various forms of capital. Such substitutability was explicitly assumed to follow a particular functional form for the production function upon which Hartwick relied. Hartwick's approach in essence requires all forms of capital to be necessary for production, even if in small amounts. This technical detail somewhat limits the Hartwick result because of the inevitability that extraction causes a fixed resource to someday disappear. Nonetheless, Hartwick's intuition that the loss of one form of capital can, at least approximately, be made up by the purchase and production of more reproducible capital adds subtlety to our intertemporal resource extraction discussion.

The other important implication of Hartwick's analysis is that it measures intragenerational well-being based on human-made and natural capital that is maintained in the aggregate over time. Rather than an emphasis on income and Gross Domestic Product (GDP), it instead accepts Clark's criticism of an income approach and presumes well-being is based on the sustainable stock of the total capital assets available across generations.

Hartwick's rule does so by requiring the revenues from optimal usage of a depletable resource that tapers over time are reinvested into other forms of capital. These resource revenues must then be diverted into a permanent fund. Such a reinvestment to compensate for resource depletion has been labeled *genuine savings* and estimated for nations by the World Bank. These positive genuine savings then provide a pathway to economic sustainability.

A number of *petro-nations* have established permanent funds such as Hartwick's Rule recommends, often under the auspices of *natural resource funds* or *sovereign funds*.

The ability to convert resource surpluses enjoyed by one generation into long-term physical or social capital that could be enjoyed by all generations is a measure of the genuine savings of a nation. Various countries have converted the value of auctions in resource extraction into permanent funds designed to provide a constant flow of income and stock of wealth to future generations.

For instance, this financial capital is used to invest in new forms of physical capital, in more efficient transportation infrastructure, in education and health institutions, in greater physical capacity to produce, and in other public and private infrastructure; If these investments are optimal in that they can yield a return commensurate with the economy-wide discount rate, net of depreciation of the capital stock, then these interest earnings can be used to compensate each generation for their loss of access to the same abundance of natural capital enjoyed by previous generations.

Some sovereign permanent funds represent large funds capitalized into the trillions of dollars. Figure 15.1 shows that the largest commodity-based permanent funds have accumulated multiples of their annual per capita GDP.

There have also been some notorious cases of funds around other natural resources. For instance, the Republic of Nauru received funds from the Nauru Phosphate Corporation, to be reinvested into real estate and other investments that could provide a consistent and resilient flow of income to support the nation's fiscal policies. Purchased in 1970 from its former colonial power of Australia for A$21 million, the lucrative phosphate mines in Nauru generated A$100 million or more annually.

Country	Fund	Value 2021 ($US billions)	Population 2021 (millions)	Per-Capita Value	Per-Capita GDP 2021 https://data.worldbank.org/ indicator/NY.GDP.PCAP.CD	Multiple
Kuwait	Kuwait Investment Authority	712	4.329	$164,472	$24,812	6.63
Brunei	Brunei Investment Agency	60	0.442	$135,747	$31,723	4.28
United Arab Emirates	Various Funds	1363	9.991	$136,423	$36,285	3.76
Norway	Government Pension Fund	1388.2	5.408	$256,694	$89,203	2.88
Qatar	Qatar Investment Authority	445	2.931	$151,825	$61,276	2.48
Libya	Libyan Investment Authority	66	6.959	$9,484	$6,018	1.58
Saudi Arabia	Various Funds	1000	35.34	$28,297	$23,586	1.20
Alaska	Alaska Permanent Fund	54	0.733	$73,670	$74,422	0.99
Azerbaijan	State Oil Fund of the Republic of Azerbaij.	42.4	10.15	$4,177	$5,384	0.78
Kazakhstan	National Fund	133	19.003	$6,999	$10,042	0.70
Bahrain	Mumtalakat Holding Company	18.6	1.748	$10,641	$22,232	0.48
Algeria	Revenue Regulation Fund	72.6	44.62	$1,627	$3,765	0.43
Trinidad and Tobago	Heritage and Stabilization Fund	6	1.403	$4,277	$15,243	0.28
Botswana	Pula Fund	4.9	2.397	$2,044	$7,348	0.28
Iran	Oil Stabilisation Fund	62	85.03	$729	$2,757	0.26
Russia	National Welfare and Reserve Funds	191	143.4	$1,332	$12,173	0.11
Chile	Social and Economic Stabilisation Fund	19.112	19.21	$995	$16,503	0.06
Alberta	Alberta Heritage Fund	16.3	4.371	$3,729	$80,905	0.05
Colombia	Colombia Oil Stabilization Fund	12	51.27	$234	$6,132	0.04
Mexico	Mexico Budgetary Stabilization Fund	7	130.3	$54	$9,926	0.01

Fig. 15.1   The per capita value of various permanent funds

At its peak, the Nauru Phosphate Royalties Trust (NPRT) had invested A$1 billion in physical capital, in the form of hotels and properties in the Pacific Rim, South Asia, and England, as its natural capital was drawn down. Compared to its small population the size of the fund was known as the *Kuwait of the Pacific*, this fund was also drawn down to support perennial eight-digit government spending deficits. The Trust then borrowed deeply to maintain fiscal spending. With declining revenue and failing businesses within the Trust, combined with the depletion of economically accessible phosphate stocks, the Trust became illiquid. Its real estate properties were seized in default and the citizens of Nauru lost a permanent fund for which they trusted would provide for a long-term guaranteed income well after the decline of the resource.

The Nauru failure demonstrates the need for such sovereign funds to ensure that resource royalties are managed in a sustainable way that does not reduce the fund's real value over time. As such, many funds also attempt to *inflation-proof* their funds to ensure that the real value of the fund is sustained over time. As with the Hotelling Rule, each generation should then enjoy only a share of the permanent fund that is equal to the real interest accrued, net of inflation.

This ability to accumulate a substitute form of financial capital from investment assets to compensate for a loss of natural capital is an example of *weak sustainability*. This concept is designed to preserve permanent utility by allowing generations to change their production and consumption patterns over time while they preserve their overall level of well-being. Alternately, *strong sustainability* ensures that utility must be held constant without any substitution of natural and human-made capital. To attain strong sustainability, each generation must then have access to the same natural and human-made capital each period rather than the same aggregate sum of capital.

The genuine savings (GS) model measures the total stock of available capital across generations. The World Bank has been sponsoring and measuring national investments in total available capital since it adopted the genuine savings measure in 1998 under the direction of Kirk Hamilton, a collaborator with Prof. Hartwick in the Hartwick Rule. The measure provides policy makers with a *sustainability index*. If the index is positive, a nation is accumulating net capital after depreciation and hence is able to meet national sustainability needs.

The original Hartwick model must be adjusted to permit trade between nations to optimize their long-term capital stocks. In addition, the GS measure requires acceptance of the weak sustainability paradigm, and is vulnerable to any other criticisms of the Hartwick Rule. Despite these limitations, the GS model as published by the World Bank is helpful in that the data requirements are not overwhelming. Likewise, the World Bank's calculation of *sustainable development indices* provides useful data for nations as they develop public policy that enhances domestic sustainability.

Hamilton (2000)[3] described the various data needs that provide guidance on weak sustainability. The World Bank calculates differences between world prices of various depletable resources and the average costs in individual countries for their resource extraction, processing, and transportation, and their national rates of return. While some countries may distort various costs for domestic reasons, and there may exist different grades of various resources, these measures are nonetheless helpful indicators.

The World Bank also includes the depreciation of non-traded natural resources such as diminished air quality through pollution or the reduction of production arising from global warming. We treat such externalities in the next chapter, while subsequent sections in our analysis treat the deterioration of atmospheric capital arising from greenhouse gas emissions and climate change. Finally, Hamilton's approach also includes measures for the human capital stock by considering investments in education.

Under the aggregate genuine savings rate, normalized by the size of each country's Gross Domestic Product (GDP), the measure nets gross domestic savings less the consumption of capital. Hamilton included education expenditures, energy and mineral depletion rates, net forest depletion, and climate change damage to calculate their genuine savings indices for various countries.

In 2000, Hamilton and his team at the World Bank found that MDCs tend to augment Genuine Savings, while low-income developing and LDCs must often sacrifice long-term savings at times to stave off starvation. Hamilton also compares the genuine savings rate for the most resource-dependent nations, defined as nations where natural resource depletion exceeds 5% of their Gross Domestic Product, relative to their natural resource depletion rate. The dependency of genuine savings on resource depletion demonstrates the tradeoff of

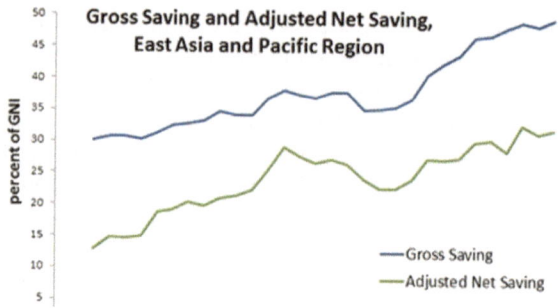

Fig. 15.2 An example of the departure from traditional savings measures once deple-tion of the natural capital stock is included (Courtesy of the World Bank, https://www.worldbank.org/en/news/feature/2013/06/05/accurate-pulse-sustainability)

reduced genuine savings as resource depletion rises. The analogy is the eating of one's seed stock (Fig. 15.2).

The resulting Genuine Savings Index (GSI) provides domestic policy makers with a better understanding of the need to incentivize savings through their fiscal and monetary policies while nations also discourage excessive natural resource consumption. The Genuine Savings approach also provides a framework for exploration of the degree to which better resource exploitation policies can foster improvements in the genuine savings rate and long-term weak sustainability.

Without such mindful investment in substitute capital as resources are extracted, a natural resource-endowed nation may actually suffer in the long run. This is known as the ***natural resource curse***.

THE NATURAL RESOURCE CURSE

A nation with abundant and valuable natural resources and with legitimate needs to provide for its population may find enticing a natural resource extrac-tion policy that emphasizes current consumption needs at the expense of future generations. All else equal, intuition would suggest that a significant permanent fund derived from natural resource extraction revenue can provide a valuable buffer to smooth the variability of income from traditional fiscal sources and enhance the human, physical, and public capital stock to improve long-term productivity and income. However, a number of countries with significant commodity-based permanent funds have performed poorly, with some exceptions.

Sometimes called the ***paradox of plenty***, or the *poverty paradox*, some nations that have strong endowments of natural resources such as crude oil or minerals, nonetheless suffer falling levels of economic development and prosperity over time. This curse is at times accompanied by lower levels of democratic attainment and social happiness. It has been likened to the misery that at times follow people after they win lotteries.

A number of countries are richly endowed with natural resources. These countries are able to generate resource exports that total more than 20% of their trade toward Gross Domestic Product or are able to fund more than 20% of fiscal spending from resource royalties. By some measures, twice as many nations perform poorly after the discovery and development of such significant natural resources as those that fare well.

The first documented case of a natural resource curse arose following a large discovery of natural gas in Groningen, the Netherlands, in 1959. The Dutch soon discovered this new and significant natural resource export caused their currency to rise in value and hence reduced the international competitiveness of the remainder of their export-oriented industries. The lack of competitiveness in an otherwise export-oriented nation resulted in a reduction in manufacturing jobs and national consumption. The decline of these other labor-intensive industries pushed the nation into a recession, and has since been labeled the *Dutch disease*.

Nations confronted with job losses arising from economic distortions in their trade sector will often compensate through additional job creation arising from fiscal stimulus funded by resource royalties. Such artificial job creation is typically of lower productivity in the service sector compared to the export-oriented manufacturing jobs lost. Hence, average wages often fall. Meanwhile, the higher exchange rate causes the cost of imports to rise, which damages both consumers and industry.

In addition, fiscal and economic reliance on a single commodity reduces a nation's diversification of economic risk. Commodity prices are often volatile, which challenges consistent funding of government-sponsored economic development programs. The level of GDP and government spending then become volatile, which induces further challenges in a nation's economic risk management.

The most significant example of the resource curse is many petro-nations' dependency on petrodollars. Oil, and most other commodities produced, generate little employment compared to the value of the commodity produced. In addition, oil and natural gas capital and infrastructure costs, especially at the onset, are large, which requires immense fixed investment. Subsequent profits returned to the multinational companies able to provide the initial development funds and expertise represent a flow of economic resources out of the country as its natural capital is extracted. The only significant rent that remains is the royalties earned, which are highly vulnerable to changes in the commodity price and prone to accounting charge-offs that are determined by the multinational companies which exploit the resource.

Production of resource exports that occur tends to rely on a small domestic workforce with only modest education and training needs. Meanwhile, multinational resource developers often reserve the most technical jobs for their own ex-patriot employees. This shift toward a modest number of domestic resource jobs and the gutting-out of manufacturing reduces demand for more sophisticated human capital and results in lower demand for and sophistication of public education.

Not only does the natural resource extraction industry tend to drain human capital, but it also diverts financial capital from other worthwhile domestic investments. Studies have shown that the net effect is a decrease in average wages and non-resource capital formation. For instance, a study of Appalachian coal mining found that a 0.5–1.0% increase in the share of coal revenues in personal income resulted in a 0.5–1.0% decrease in long run income growth.[4]

Other research has suggested that the resulting economic decisions also tend to heighten conflict and disrupt democracy. These conflicts may be internal, but sometimes also generate wars over the resource. For instance, oil resources were at the root of the Gulf War arising from competing oil extraction along the border of Iraq and Kuwait in the 1990s. In reference to fossil fuel extraction, Michael Ross (2015) concluded that "one type of resource has been consistently correlated with less democracy and worse institutions: petroleum, which is the key variable in the vast majority of the studies that identify some type of curse."[5]

In addition, political corruption may sometimes erupt as bribes to secure resource rights are sometimes employed. Meanwhile, relationships between international aid agencies and *Non-Governmental Organizations (NGOs)* are often weakened as a result of the natural resource curse.

From an environmental rather than an economic and political perspective, the resource extraction industry typically has a broad carbon footprint when compared to the manufacturing sector it often replaces directly. In turn, indirect and induced sectors also shrink because of declines in non-resource income. Jobs lost in the private and public service sectors traditionally have a low carbon footprint compared to the resource industry. The net effect is often an increase in the overall carbon footprint for a nation. This tendency is sometimes called the *carbon curse* as a corollary to the *natural resource curse*.

SUMMARY

While our positive economic conclusions are compelling, normative aspects of effective public policy and governance to ensure aggregate capital is passed to future generations are often frustrated in reality. A strong definition of sustainability that prevents any natural resource from being extracted out of concern that it would unfairly deprive a future generation may under certain conditions be replaced with weak sustainability. Under the so-called Hartwick Rule, so long as the marginal user costs that are generated from depletable resource extraction are reinvested to seed the production of alternative forms of human-made capital, future generations will have the same potential base of wealth and capital to sustain themselves as had their predecessors. But, while this notion of a permanent fund to protect future generations is intuitively appealing, the challenge is in political will. Planners often do not have sufficient tools to ensure that capital is appropriately protected and permitted to earn a rate of interest equal to the discount rate to overcome the discounting when we mortals value our generation over those who follow us.

ESG Toolkit

Sustainability depends on recognition of the links between successive stakeholders over time.

To what degree does corporate governance include a diverse set of stakeholders that can span generations?

Are corporate strategies forward-looking or reactive?

Can internal mechanism be created that articulate long-term values and effects on future generations?

Corporations must internally "price" proposed projects so the return from such projects does not depart significantly from other activities in the corporate portfolio.

What is a corporation's benchmark expected return and cost of capital?

Does the gap between these measures and a socially appropriate intertemporal discount rate cause the corporation to forsake some socially desirable long-term activities?

Can the corporation make some of these valuable long-term investments but still extract returns for shareholders?

Notes

1. Hartwick, John M. (1977). "Intergenerational Equity and the Investment of Rents from Exhaustible Resources". American Economic Review. **67**: 972–974.
2. Clark, John Bates (1899). The Distribution of Wealth: A Theory of Wages, Interest and Profits, MacMillan, New York.
3. Hamilton, Kirk (2000). "Genuine Savings as a Sustainability Indicator", World Bank Environment Department. See Hamilton, Kirk, and Atkinson, Giles (2006). Wealth, Welfare and Sustainability: Advances in Measuring Sustainable Development. Edward Elgar, Northampton, MA and Hamilton, Kirk, and Hartwick, John (2005). "Investing Exhaustible Resource Rents and the Path of Consumption". Canadian Journal of Economics. **38** (2): 615–621. https://doi.org/10.1111/j.0008-4085.2005.00295.x.
4. Douglas, Stratford, and Walker, Anne (2017). "Coal Mining and the Resource Curse in the Eastern United States (PDF)". Journal of Regional Science. **57** (4): 568–590. https://onlinelibrary.wiley.com/doi/10.1111/jors.12310, accessed October 27, 2022.
5. Ross, Michael L. (May 2015). "What Have We Learned about the Resource Curse?". Annual Review of Political Science. **18**: 239–259. https://doi.org/10.1146/annurev-polisci-052213-040359, accessed October 27, 2022.

CHAPTER 16

Environmental Externalities

A century ago, in the dozen years beginning in 1920, classical economics witnessed four important contributions to a more dynamic extension of its traditional models into the realm of consideration of time and the tendency for mortals to discount the future. At the end of that decade, Irving Fisher submitted his A Theory of Interest for publication; Harold Hotelling had described depreciation of the capital stock in 1925 and his optimal extraction rule for a fixed resource in 1931; In 1928 Frank Ramsey had described his golden rule for consumption over time as a function of a society's capacity to produce; But, the decade had begun with an influential work by Arthur Pigou in 1920, called The Economics of Welfare.[1]

These scholars each recognized that, while the tools of economics are traditionally used to describe static and independent market-based economic decisions at each point in time, they can be far better employed if economic tools are extended to sustainable decisions over time and to situations in which markets fail to properly allocate natural capital.

Henry Sidgwick was one of the first political economists to articulate the need to extend our static economic models to include a wider set of stakeholders beyond those who participate directly in a transaction. This English utilitarian economist was concerned about the ethics of economic decision-making.

During the Victorian Age, Sidgwick advocated for diversity, inclusion, and women's rights that may still be considered liberal today and consistent with aspects of the ESG Paradigm. Sidgwick also advocated for *ethical hedonism* to replace the individual self-interest advanced in *psychological hedonism* with a greater concern for the common good and for that which is objectively right. His work on what he called a *comparative methodology* inspired John Rawls to

C. Read, *Understanding Sustainability Principles and ESG Policies*, https://doi.org/10.1007/978-3-031-34483-1_16

develop a contemporary theory of an economic ethic that we will more fully describe in the next section of the book.

Sidgwick acknowledged that humans are motivated by self-interest, but he also described circumstances in which our individual self-interest in the aggregate frustrates our collective interest, just as individual acts generate the Tragedy of the Commons. First, Sidgwick argued that humans are motivated by more than the consumption that wealth and income can generate, even while our market pursuits may frustrate the happiness of others who are not considered in our decisions but are nonetheless harmed.

Such *spillover effects* had also been a concern of a Sidgwick contemporary, John Stuart Mill.[2] Mill had argued that government may need to intervene to prevent harm to society when market transactions have negative external effects. They may also include social interactions that go beyond a particular individual, such as the positive benefits that accrue to all of society when we educate our youth. These spillovers can include lower crime rates, reduced poverty, more effective democracy, and greater collective happiness.

On the resources side, *general equilibrium spillover effects* arising from greater development of alternative energy sources can result in displacements in downstream fossil fuel retailers. The most relevant spillover for sustainability occurs from the class of spillover effects called *externalities*. These are social costs or benefits imposed on other parties because of the economic decisions among private parties. To the extent that these broader costs or benefits do not appear in the consideration of private benefits and costs considerations, private decision-makers will not take these broader effects into account.

A more formal theory of such *externalities* was developed by Arthur Pigou in his 1920 The Economics of Welfare. Pigou had replaced the originator of our traditional theories of supply and demand, Alfred Marshall, as the chair of economics at Cambridge University in 1908. He went on to develop theories within economics that correct omissions by Marshall's narrow, static, and private market-focused treatment. In particular, Pigou noted in the third edition of his classic text that:

> The complicated analyses which economists endeavor to carry through are not mere gymnastic. They are instruments for the bettering of human life. The misery and squalor that surround us, the injurious luxury of some wealthy families, the terrible uncertainty overshadowing many families of the poor—these are evils too plain to be ignored. By the knowledge that our science seeks it is possible that they may be restrained. Out of the darkness light! To search for it is the task, to find it perhaps the prize, which the 'dismal science of Political Economy' offers to those who face its discipline.

His motivation was to create tools that are useful beyond the mere description of narrow individual economic decisions, but rather for the betterment of society. He argued along the ethics of Sidgwick and Mill that a broader inclusion of the values of society within our economic models enhances welfare as humans then make decisions that advance our collective happiness.

Pigou's predecessors Marshall and John Bates Clark had described the free-market private benefits to consumers and producers that arise from their private decisions. These private benefits, and their associated costs, are embedded within traditional supply and demand curves that measure the net benefits and costs of private decisions within traditional economic markets. Pigou proposed that, to ensure society makes more holistic and complete economic decisions, broader marginal social benefits and losses should also be included. He noted,

> The essence of the matter is that one person A, in the course of rendering some service, for which payment is made, to a second person B, incidentally also renders services or disservices to other persons (not producers of like services), of such a sort that payment cannot be exacted from the benefited parties or compensation enforced on behalf of the injured parties.

Pigou was particularly concerned with decisions that impose a negative social cost on society and hence decrease overall economic welfare. If there are such uncompensated losses that occur to others from decisions of private individuals, these unintended *external effects* should be *internalized* if there is no mechanism for traditional market forces to incorporate them.

These externalities do not include such inevitable effects that occur in the process of Joseph Schumpeter's **creative destruction** since Schumpeterian innovation improves overall efficiency for the greater good. Such **Pareto efficient** improvements produce value that could compensate those industries that become obsolete as a consequence of innovation, but the correct decision is to innovate even if compensation does not occur, in the interest of creative destruction. Instead, Pigou was concerned about external costs that are ignored by markets.

To see the consequences of social costs ignored, consider the classic example of pollution. When a producer pollutes in excess of the Earth's ability to cleanse the air, it in essence consumes our collective stock of clean air. In Pigou's era, such conversion of clean air into dirty air, with its intendent consequences on the welfare of others who are harmed by pollution, is a social cost that is not considered in the polluter's decisions.

Certainly, a mill prices its labor, its use of physical capital, its employment of non-renewable and renewable resources and its factors or production, and the required reward to its entrepreneurial capital. However, by not bearing the cost of the clean air it converts to polluted air, the producer does not price this cost of depletion of our clean air capital. While the degree of underpricing of such externalities is likely small for very low levels of pollution from either an individual producer or industry perspective, as production grows these social costs are magnified and distort economic decisions.

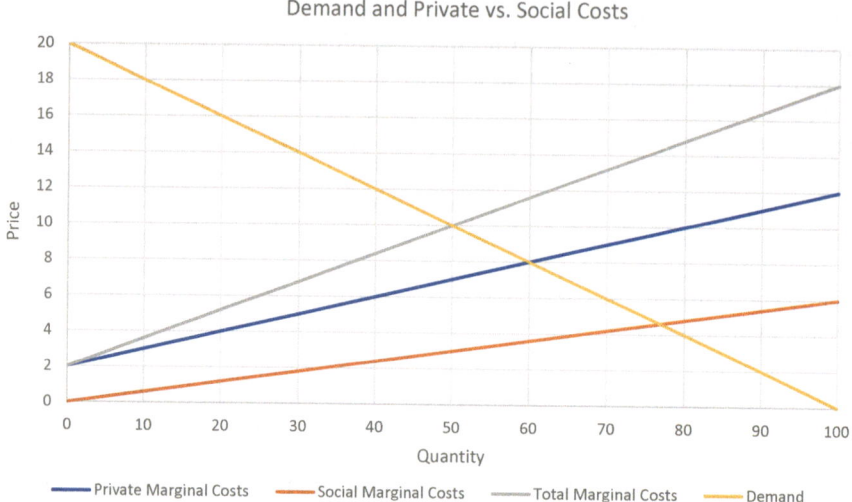

Fig. 16.1 Demand and private and social costs for a negative externality industry

These omitted social costs appear on the Fig. 16.1.

In this example, the private producer determines a production quantity of 60 that yields a market price of $8 because they are driven by their private marginal cost curve to an equilibrium point in which **Private Marginal Benefits** equals **Private Marginal Costs**. At that level of production, an additional $3.6 of marginal costs are incurred by society, even though they are ignored by the producer. If instead the producer made its production decision based on the combination of social and their private costs, only 50 units would be produced, at a cost and demand price of $10, where Private Marginal Benefits equal Total Marginal Costs, including social costs and private costs.

The failure of the producer to internalize this pollution externality results in socially undesirable overproduction by 10 units. Too much pollution is generated because it has gone unpriced. The total social cost incurred at the producer's output choice of 60 units is then the area under the social marginal cost curve with a width of 60 and a height that rises from 0 to 3.6. This total area of the triangle that represents the social cost of pollution is then 0.5*60*3.6 = $108. Had the producer instead adopted the marginal cost curve that includes its externality, the total social costs incurred would fall to 0.5*50*3 = $75.

Pigou was concerned about this greater social damage that occurs when producers do not incur the full costs they impose on society. He noted that, if an omnipotent planner-regulator imposed a fee equivalent to the social cost at the social optimum, then these higher marginal costs would induce the producer to make a production choice consistent with the maximization of overall social welfare.

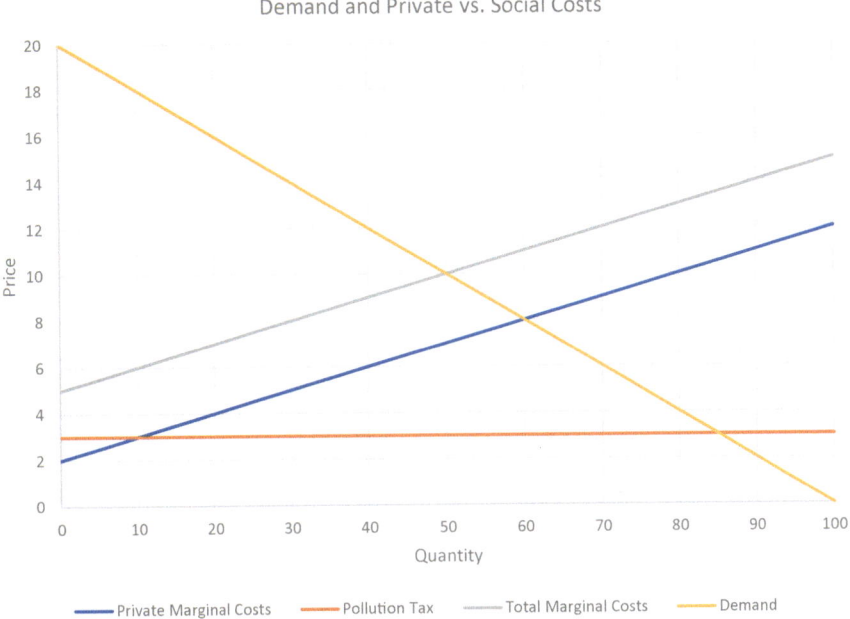

Fig. 16.2 Private and socially optimal equilibrium when a negative externality exists

To see this, note that, at the socially optimal level of production of 50 units, private marginal costs are $7 and social marginal costs are $3. If a regulator imposes a tax of $3, the private marginal cost to the producer is $3 greater at every level of production. Then, the producer would instead act as below (Fig. 16.2).

If these social costs are internalized accordingly, the producer will then choose a level of production of 50 units and incur a tax of $3. This solution results in a total internalized pollution cost of $150. This internalized pollution tax has induced the producer to operate at the socially optimal level. The planning solution also avoids some costs by producing 10 units less, at an average cost of $7.5. The producer saves $75 in production costs and society saves $75 in social damage, while $150 in tax revenue is raised.

The same results could be obtained if regulators imposed a graduated pollution fee that precisely mimics the increasing harm rate described by the social marginal cost curve. Such a graduated fee would be more complicated to calculate and regulate compared with the simple marginal social cost at the social optimum when producers internalize their true costs, including the costs to society for otherwise unpriced resource consumption. Either path would induce optimal production, though.

The regulator could also simply limit production to 50 units. If a regulator sets a maximum production level, the correct level of production would be obtained but society would still not be compensated for the costs the producer imposes. Instead, in the example of pollution, a lower quality of life, life expectancy, and other health and aesthetic costs would be absorbed by consumers rather than producers. The additional profit the firm would earn by retaining the potential $300 pollution fund would benefit the owners of production rather than society as a whole.

If production is owned by society as a whole, or if the welfare of the owners is considered to be equally valuable as the welfare of society, then society may be indifferent to these redistributive effects. It is clear, though, that there are social justice aspects to consider, depending on our assignment of pollution rights to producers over the right to clean air for citizens.

The Effect of Pollution on Future Generations

Pigou's extension was static in that it did not model the evolution of environmental degradation over time. He had modeled externalities as if they were costs alike other factors of production. Ultimately, any factor of production is either a form of human-made or natural capital, or it is produced with human-made or natural capital.

A greater problem arises when a more chronic form of pollution accumulates to cause permanent damage or is sufficiently long-lived that it should be treated in the context of the depletion of our atmospheric stock. To properly treat the long-term and dynamic aspects of such externalities, the economic model of sustainability must be extended.

The first step to attain efficiency is then to impose an appropriate Pigouvian tax to remedy any externalities at a given time. Such a tax is equal to the cost of damages arising from pollution. Then, with our tools of dynamic optimization of national wealth and income at hand, the long-term implications of pollution and environmental degradation can be treated from an intergenerational perspective.

In 2000, Kirk Hamilton of the World Bank published an article entitled Genuine Saving as a Sustainability Indicator[3] that modeled environmental degradation from an intergenerational perspective as a way to measure how such negative externalities ultimately subtract from a nation's Genuine Savings. It also provides us with some sustainability intuition as we treat the cost of very long-term depletion of our atmosphere through the release of anthropogenic greenhouse gases.

Hamilton produced a model described in the chapter appendix that derives an intergenerational economic well-being function that can be enhanced by earned interest r on the stock of society's permanent wealth W, net of our consumption U. Let us define a conversion factor U_C that is the marginal contribution to utility from the value of additional consumption. Then, Genuine Savings in utility terms is given by $U_C GS$ which should be limited to the increase in permanent welfare $\dot{W}$ each year. Hamilton concludes that:

$$U_C GS = \dot{W} = rW - U.$$

Hamilton includes pollution externalities that detract from Genuine Savings each period. His now-familiar prescription retains the familiar concept that a society should not draw down permanent funds at a rate faster than interest accrues, once various externalities are internalized each year. To yield a constant level of consumption and utility then, a planner must first ensure that externalities are resolved within each generation. Hamilton then shows that only a constant level of Genuine Savings can ensure smooth consumption and economic welfare across generations. In doing so, we wed the static traditional economic world with Pigou's prescription of the need to internalize externalities and the intergenerational need to ensure one generation does not consume resources in a way that deprives future generations.

Hamilton's conclusions bring into mainstream economic measures of wealth the cost of actions of a generation that harms future generations. The example above used pollution as the social nuisance. We will see in the next section that this example can be easily extended to treat our current emissions of greenhouse gases that contribute to global warming. We shall employ the results we have obtained and the intuition that our growth of social wealth must equal the rate we discount the value of future generations.

Public policy should be optimized to maintain intergenerational welfare and happiness. This requires sufficient regard to the accumulation of a stock of wealth to indemnify future generations for their losses of access to natural resources and a clean environment. Failure to internalize externalities creates incentives for overextraction and for an underappreciation of pollution abatement. If so, intergenerational economic justice is rendered impossible.

SUMMARY

We have now developed the economic tools necessary to properly analyze sustainability in terms of familiar economic concepts. We can move on to explore how to include ethics into our sustainability discussion. As we do so, we see that our economic decisions have two parts. If we are to ensure sustainability we must first ensure that sound economic decisions are made within a generation, which requires us to properly internalize externalities. This is intragenerational efficiency. Then, we must connect the generations to create intergenerational efficiency by properly incorporating the effects of our decisions on future generations.

ESG Toolkit
A corporation balances its usage of resources at its disposal.

As a corporation contributes to the depletion of underpriced fixed resources, is it substituting for these losses with augmented production of physical and human capital?

> Does the corporate strategy recognize its carbon footprint must at some level be mitigated through sustainable investment elsewhere?

Appendix

Hamilton's extension of the concept of Genuine Savings can explicitly model negative externalities such as pollution. Consider the accumulated environmental degradation from pollution as denoted by D. This stock of pollution that undermines natural capital is worsened by emissions e. Let societal production follow a production function $F(K, L, H)$, where K is our human-produced capital flow, H is the rate of extraction of our renewable natural capital stock R, and L is our use of human capital. Our production $F(K, L, H)$ from the use of these resources can be used for consumption, to abate pollution, and to produce and improve our human capital. Any production left over after consumption, abatement, and education can then generate greater capacity to produce in the future.

Our addition to future productive capacity $\dot{K}$ is given by:

$$\dot{K} = F(K, L, H) - C - A - M,$$

where C is the amount of productive capital that is diverted from the capital stock and which we consume, A is the capital cost of pollution abatement, and M is the measure of capital diverted to education. Given this ability to expand future opportunities and capacities, our nation's discounted present value of present and future wealth of utility is given by the sum of discounted utility derived from consumption of a human-produced composite good C and the enjoyment of environmental amenities B:

$$W = \int_0^\infty U(C, B)e^{-rt}dt.$$

This intergenerational goal to optimize our collective utility through our consumption C and enjoyment of natural amenities B over time is discounted at an appropriate interest rate and must also follow some constraints.

The first constraint is that production combines our physical capital K, our human capital L, and the harvest level of natural capital H according to the production function $F(K, L, H)$. This level of production is devoted to the sum of consumption C, additions $\dot{K}$ to the physical capital stock, pollution abatement A and investments in human capital (education) M:

$$F(K, L, H) = C + \dot{K} + A + M.$$

The rate that such education translates into additional human capital L is a function $l(M)$ of a society's investment of capital into education:

$$\dot{L} = l(M).$$

A share of the capital produced must also be devoted to pollution abatement. Hamilton modeled the change in the level of pollution D as depending on emissions e, as a function of the level of production and the degree of abatement A, less the ability d of nature to cleanse some pollution from the environment:

$$\dot{D} = e(F(K, L, H), A) - d.$$

This level of pollution damage D also degrades our ability B to enjoy nature. Hamilton models this enjoyment of natural amenities B as a function of the pollution level X such that $B = b(D)$, with our consumption of nature's amenities B that falls as pollution D rises.

Finally, the stock of natural capital R falls as resource extraction H rises, but Hamilton allows for the possibility that our renewable natural stock can also grow, for instance in the case of a biotic resource such as a fishery:

$$\dot{R} = -H + g$$

The exercise can then be expressed as a current value Hamiltonian to be optimized over time with our now-familiar first order conditions that include changes in produced capital K, the stock of pollution damage D, the natural resource stock R, and the level of human capital L. The Hamiltonian maximizes utility arising from consumption C and enjoyment of the environment B:

$$\max \mathcal{H} = U(C, B) + \lambda_K \dot{K} + \lambda_D \dot{D} + \lambda_R \dot{R} + \lambda_L \dot{L},$$

subject to our various capital constraints, where the various costate variables (equivalent to Lagrange Multipliers) $\lambda_K, \lambda_D, \lambda_R, and \lambda_L$ are the sensitivities of our current level of utility $U(C, B)$ to changes in capital K, the level of pollution D, the size of the natural capital stock R, and human capital L respectively.

Based on the intuition we developed in our previous chapters, we do not need to actually solve the Hamiltonian if our goal is instead to ensure a constant level of enjoyment over time, including the degree to which we value ecosystem values and environmental amenities. Hamilton was able to draw some immediate conclusions arising from his inclusion of the level of pollution into our intergenerational optimization.

Hamilton defined a Genuine Savings function that includes additions to our capital stock but which also depreciates because of pollution. He noted that inclusion of the various constraints into the Hamiltonian creates the following expression:

$$\mathcal{H} = U(C, B) + U_C$$

$$\{\dot{K} - \left(1 - b\frac{\delta e}{\delta F(K, L, R)}\right)\frac{\delta F(K, L, R)}{\delta R}(R - g) - b\left(e(F(K, L, R) - d) + \frac{l}{l_M}\right)\},$$

where U_C is once again the rate a society translates the value of permanent capital into utility. Our interest is to ensure that the current value of consumption remains constant over time. Hence, intuition we developed earlier shows that the elements in the parenthesis must be offsetting. He drew this conclusion based on some intuition.

Hamilton constructed his expression based on the rate the capital stock can contribute to consumption. The first element in the parentheses is the contribution to utility from improvements in the capital stock $\dot{K}$.

To understand the second term $\left(1 - b\frac{\delta e}{\delta F(K,L,R)}\right)\frac{\delta F(K,L,R)}{\delta R}(R - g)$, note that $b(D)$, the cost to environmental amenities B we suffer in environmental amenities damage from the aggregate pollution level D, should be used to determine the correct internalization of these pollution externalities as Pigou suggested. Its value is the true marginal social cost of pollution. Such pollution emissions are in an amount $e(F(K, L, R), A)$, net of the degree to which nature can cleanse pollution at a rate d. Then, the product of the marginal cost of pollution and the marginal level of pollution emissions as the level of production rises gives us the share of production that must be taxed to internalize and offset the pollution externality.

The net reduction to capital accumulation net of these pollution expenses because of a drawing down of our natural capital at a rate R net of its growth g is then given by:

$$-\left(1 - b\frac{\delta e}{\delta F(K, L, R)}\right)\frac{\delta F(K, L, R)}{\delta R}(R - g).$$

The third element, $-b(e(F(K, L, R) - d)$, measures the damage pollution emissions $e(F(K, L, R)$ inflict on our natural capital stock, net of the ability of the Earth to heal itself at a rate d, and priced at the correct Pigouvian marginal emissions cost on our enjoyment of natural amenities.

Finally, the last term l/l_M is the addition to human capital l priced at the marginal cost l_M of creating one more unit of human capital by investing in education M.

Hence, the long expression in the parentheses simply sums the additions to and subtractions from our intergenerational genuine capital base over time from production, resource extraction, pollution, and education. If our goal is to maintain a constant level of utility across generations, the sum of these credits and debits to our permanent stock of various forms of capital should remain constant.

Through his calculations, Hamilton provided us with a more elaborate statement of Genuine Savings. Using some shorthand for the marginal increase in emissions arising from production $\frac{\delta e}{\delta F(K,L,R)}$ as e_F and the increase in production $\frac{\delta F(K,L,R)}{\delta R}$ arising from additional resource extraction as F_R, we can express our intergenerational assets as:

$$Genuine\,Savings = GS = \dot{K} - (1 - b e_F) F_R (R - g) - b(e - d) + \frac{l}{l_M}.$$

This expression simplifies somewhat if the pollution nuisance cannot be naturally cleansed, which implies the term d equals zero. If natural capital is unable to replenish itself reasonably quickly, our environment and atmosphere in essence behaves as a depletable resource.

From this expression, Hamilton is able to tie together human-made and natural capital, the diminishment of natural capital and well-being through pollution, and our investment in education. Policymakers then have insights into a broader set of economic measures that determine intergenerational economic justice beyond traditional measures of income and investment.

With this intuition at hand, let us take a step back. Recall our goal is to maximize that total discounted utility across generations is given by the aggregate intergenerational welfare function:

$$W = \int_0^\infty U(C, B) e^{-rt} dt.$$

We see then that the addition to utility at each point in time, our intergenerational welfare evolution $\dot{W}$ is constant if the utility U each period is equal to its share of intergenerational utility, based on the discount rate r. The change in Genuine Savings GS over time, with the marginal utility of consumption U_C the conversion factor between assets and utility, is then equal to the value of steady state utility over time:

$$U_C GS = \dot{W} = rW - U.$$

It is comforting that our familiar rule still applies so long as we properly internalize pollution externalities at each point in time. If so, generations can maintain a level of Genuine Savings that provides for smooth enjoyment of consumption and environmental amenities across generations. This then requires us to not only internalize intragenerational pollution externalities. We must also internalize the externality of intergenerational natural capital depletion. To do so, each generation must set sums aside to compensate future generations for the deterioration caused by earlier generations of our atmospheric natural capital.

NOTES

1. Pigou, Arthur (1920). The Economics of Welfare, MacMillan, London.
2. Mill, John Stuart (1859). The Collected Works of John Stuart Mill, Volume XVIII—Essays on Politics and Society Part I.
3. Hamilton, Kirk (2000). Genuine Saving as a Sustainability Indicator. Environment Department papers; No. 77. Environmental economics series. World Bank, Washington, DC. © World Bank. https://openknowledge.worldbank.org/handle/10986/18301. License: CC BY 3.0 IGO.

PART IV

Sustainability and Public Policy

An extension of the traditional economic model to the efficient and sustainable use of natural capital creates powerful insights that can then guide good public policy. Given the preferences of humankind and the need for ecosystem resiliency, these tools can prescribe a sustainable path for resource usage on our planet. The fundamental issue then comes down to a political choice. If we know how an omnipotent planner can determine resource usage for subsequent generations, from an intergenerational perspective, we must still resolve some questions to ensure intergenerational efficiency. To what degree do early generations regard all generations as equivalent? Is each generation prepared to treat natural capital as a public good to enhance the present and future generations? Is each generation willing to set aside resource royalties to compensate future generations for our depletion of natural capital? Within a generation, are nations and peoples treated equivalently or do some secure a preferential position? Finally, do we support values and wealth in some peoples over others? These are questions public policy must resolve.

We have discovered generations benefit by investing in the production of human-produced goods and education. However, each generation is prone to divert natural capital to itself. Sustainability requires a generation to set aside an equal amount of wealth into a permanent fund that acts as a reserve to compensate future generations for depleted natural capital. If generations fail to do so, such a permanent fund shortchanges all subsequent generations in a compounding way since theory shows that such permanent funds should grow at the rate of interest. Clearly, what is missing from our political discourse each generation is a frank conversation of our accumulating errors of permanent fund omission. Next the implications of economic justice are explored.

CHAPTER 17

The Philosophy of the Social Discount Rate

The degree to which generations make sustainable economic decisions depends critically on how each generation internalizes its externalities. Sustainability first requires negative externalities within a generation to be internalized. Then, remaining intergenerational externalities must be remedied to indemnify future generations. If these tandem objectives are met, one remaining issue is an assurance that each generation does not inappropriately discount economic values of subsequent generations. Another issue, under the purview of *ecological economics*, is whether ecological and other broader normative values are properly considered in the economic solution.

Let us step back for a moment to understand the motivations for increased recognition of future generations in our economic awareness in sustainability. In a revolution in our understanding of social optima, Pigou, Ramsey, Fisher, and Hotelling collectively established our understanding of the importance of time and the internalization of externalities. John Maynard Keynes weighed in further on the incompleteness of static neoclassical economics when he described the reasons and cures for depressions in his seminal 1936 book The General Theory of Employment, Interest and Money.[1]

The next wave of appreciation in sustainability arose when fossil fuel consuming countries were facing rapidly rising oil prices in the 1970s. An oil embargo on the United States by the Organization of Petroleum Exporting Countries (OPEC) in retaliation for U.S. support of Israel in its seizure of land in Palestine during the Yom Kippur War. The resulting fossil fuel shortage caused an increase in the price of oil from $25 to a peak of $140 over the 1970s. This *oil shock* caused economists to reflect on the sustainability of fossil fuels and on issues related to energy self-sufficiency.

© The Author(s), under exclusive license to Springer Nature Switzerland AG 2023
C. Read, *Understanding Sustainability Principles and ESG Policies*,
https://doi.org/10.1007/978-3-031-34483-1_17

Motivated by this growing vulnerability to the finiteness and fragility of nature's capital, Massachusetts Institute of Technology (MIT) economist Robert Solow produced the **Solow Growth Model** in 1974 that studied the growth path of an economy dependent on a fixed supply of a resource such as oil.[2] He asked the question of whether a country could sustain its consumption over time if it invested in human-made capital as a substitute for resource capital. Solow drew upon conventional economic tools to do so, but employed them in a novel way.

The economic concept of an *isoquant* is a locus of possible points that compare combinations of different types of capital that can maintain the same quantity of production (and hence consumption). For instance, in the classic example of the interplay between human labor and machines, one could maintain one quantity (isoquant) of production by employing a lot of labor and few machines or a lot of machines and few laborers. If there exists diminishing marginal productivity from either form of capital, machines and capital may be optimal in moderate amounts.

In 1817, the economics pioneer David Ricardo provided the first known model of natural capital in his exploration of agricultural value and rents in a publication titled On The Principles of Political Economy and Taxation.[3] For instance, consider the use of land and machines to produce a crop. One could use a large amount of land with a modest amount of equipment, or instead convert to highly capital-intensive indoor vertical farming with more modest land needs but with a much greater capital requirement. More intensive land use then typically requires more extensive human-made capital use, and vice-versa, according to Fig. 17.1.

The diagram shows that greater quantities of each form of capital are represented by higher isoquants that in turn create a greater amount of production. In addition, the straight line denotes the total factor budget devoted to either purchasing only human-made capital K, in an amount I/P_K for a given budget I, or employing only extracted natural capital X in an amount I/p_X, or by some combination in between. The slope of this line is given by the relative price P_X/P_K of natural and human capital respectively. Hence, our choice of the combination of competing forms of capital depends on their relative prices.

Solow's insight is that consumption can be maintained if the scarcity of one resource forces an economy to move along an isoquant in favor of a more abundant resource. Solow asserted that the unsustainable resource, in his case a fossil fuel, if extracted according to Hotelling's depletable resource equation, could be substituted by a human-made resource if society were sufficiently willing to save in the present to expand the production of the human-made resource in the future.

A social savings function may then be constructed and funded with royalties from the efficient depletion of oil as described by Hotelling's equation and Hartwick's Rule. Solow's use of a relatively simple and easy to solve idealized

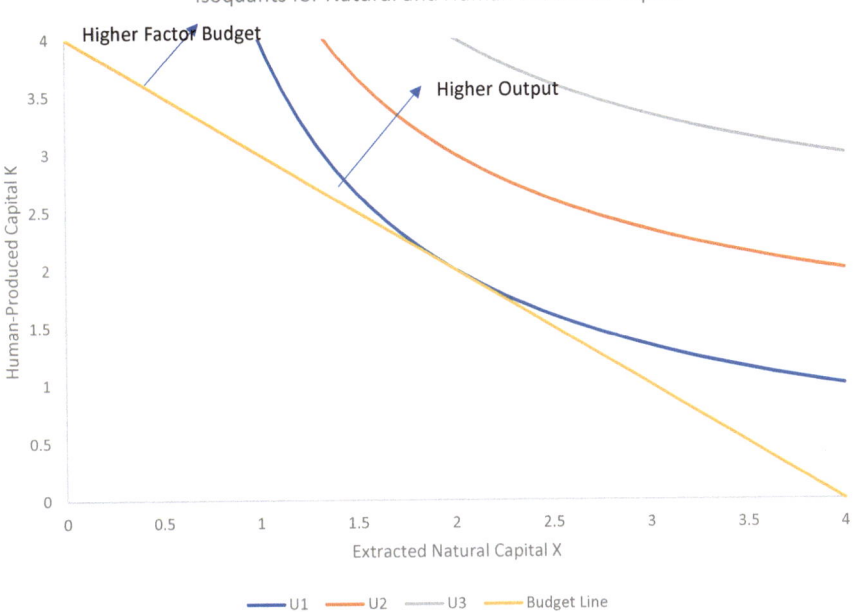

Fig. 17.1 Preferences over the enjoyment of natural and human-produced capital

production function, called a Cobb–Douglas function, allowed him to make a couple of interesting conclusions within his highly stylized model.

Solow was able to derive an optimal saving function sufficient to enhance sufficient human-made capital that could offset the loss of natural capital. He was also able to solve for a permanent and constant value of consumption across generations if the natural resource is depleted only slowly and at a diminishing rate over time. By converting physical units of capital into financial capital, Solow successfully determined the path of resource depletion and consumption.

Note, though, that asymptotic reductions in natural capital mean there remains forever some natural resource extraction. In Solow's model, a fixed resource is never depleted entirely. Its extraction is merely whittled away asymptotically. As such, his model never anticipated the need to prevent further extraction of natural capital to prevent environmental catastrophe or atmospheric degradation that can result in a permanent loss of natural capacities, at least with respect to the planning horizon of humankind. Nor did he treat population growth in his basic model.

However, given a particular functional form for the production function upon which he relied, he was able to show that the total value of Gross Domestic Product (GDP) can be viewed in two equivalent ways. One is that output equals production $F(K, X)$, where K is the use of human-produced capital and X is the extraction of natural capital. Production can be viewed to either provide consumption C or augment growth of the human capital stock $\dot{K}$:

$$GDP = F(K, X) = C + \dot{K}.$$

If the production function has a particular property called **constant returns to scale**, then we find that:

$$GDP = K\frac{\delta F}{\delta K} + X\frac{\delta F}{\delta X} = K F_K + X F_X,$$

where F_K and F_X are the respective marginal productivities of human and natural capital. Then, ***Net Domestic Production (NDP)*** defined as GDP less the value of resource depletion $X\frac{\delta F}{\delta X}$, remains constant each period:

$$NDP = GDP - X\frac{\delta F}{\delta X} = C + \dot{K} - X\frac{\delta F}{\delta X} = K\frac{\delta F}{\delta K}.$$

This is a fascinating and troubling conclusion on its surface. It states that the current value of national income is unconstrained by the loss of natural capital. Rather, total usable production only depends on our human-made capital choice. However, this conclusion is based on the presumption that resource rents are constantly set aside in a permanent fund to improve the stock of human-made capital over time in an amount equal to the decline of the natural capital stock $X\frac{\delta F}{\delta X}$. It also assumes that human-made capital is always substitutable for natural capital, and that there always remains some natural capital in nature's bank.

We saw in Chapter 15 that his model begged the question of permanent funds to substitute for the loss of resource capital. If resource rents are constantly set aside for the benefit of future generations, they can use this financial capital to enhance their reproduceable human capital and enjoy the same level of consumption as past generations. John Hartwick subsequently elaborated on Solow's conclusion by describing the nature of permanent funds, and, in his 1990 Natural Resources, National Accounting and Economic Depreciation,[4] the inclusion of pollution. In addition, commentators have noted that human consumption can actually increase if the rate of formation of human-made capital exceeds the loss in the value of natural capital, i.e., $\dot{K} - X\frac{\delta F}{\delta X} > 0$. As such, consumption falls if our capacity to produce using human-made capital does not keep up with the value of the loss of natural capital.

Contributions by Hamilton[5] (1995) and Hamilton and Hartwick[6] (2005) showed that, under a more complete analysis of factors that represent a human capacity to produce and consume, so long as a society's Genuine Savings (GS) earns a sufficient return r to replenish any drawing down of this balance of societal capital, the path of consumption C can be maintained:

$$C = rGS - \dot{G}S$$

This relationship has become the basis of the World Bank's Genuine Savings Index.

SOLOW GROWTH AND ECONOMIC JUSTICE

Such insights into issues of *economic justice* and sustainability have arrived in waves. They culminated in the inclusion of the important development of time and externalities in the 1920s and the discussion of economic growth in the 1970s and 1980s. These approaches have allowed us to formulate criteria for intergenerational equity. They have also been controversial in that, by modeling growth, these approaches provide an olive branch to those who argue that technological innovation can correct the problems that unsustainable natural resource extraction creates. If so, economic theory then prescribes a path of smooth consumption and hence the ability to provide for economic justice. The next question is then one of an intergenerational ethic. Do generations have the will to preserve the human and natural ecosystem?

At the center of these discussions and controversies are three Nobel Prize-winning economists, Robert Solow, his student William Nordhaus, and John Rawls.

Solow is most associated with his research in optimal growth theory. Using the tools that are also employed in the construction of paths of optimal consumption without abusing the Earth's stock of natural capital or the environment, optimal growth theory demonstrated how economic progress, for instance through research and development, can maintain or expand economic output as measured through Gross Domestic Product (GDP) and other measures. Of course, these same tools could arguably provide a scaffold upon which some could hang arguments that unspecified future technologies can indemnify future generations from the harmful effects of resource overextraction or environmental degradation.

Solow was somewhat sympathetic to innovation and economic growth early in his career. He produced his earliest model of technological growth in 1956 when the United States was experiencing a wave of optimism in business, engineering, and science in the wake of huge economic gains following World War II. Even then, his model would look familiar to our method to describe the conditions necessary to attain sustainability, both in the techniques and in quickly highlighting some essential missing aspects.

Solow's goal was to show how improvements in our understanding of the black box of neoclassical economics we call the production function can yield greater insights into the capacities arising from the combination of humans and machines. These forms of human-made capital are enhanced through the realization of technical progress. Indeed, scholars have estimated that about 80% of the economic progress in the middle of the twentieth century arose because of technological improvements.

One cannot hold responsible Ramsey, Hotelling, Solow, and Nordhaus for the inconvenient misapplication of the tools they developed. There is nothing in the modeling of past technological progress that ensures future progress will be either easy or politically palatable. Sometimes lost in the observation regarding innovation is that such economic growth often expands demand for other forms of capital. For instance, the carbon footprint and hence the depletion of atmospheric capital by more developed countries is far greater than its depletion by less developed counterparts, despite vastly more abundant human-produced capital potential of MDCs.

Solow won the Nobel Memorial Prize in Economic Sciences for his work on growth theory and his extension to matters of public policy as he incorporated natural capital into his 1974 extension of his work from the 1950s. In his Intergenerational Equity and Exhaustible Resources,[7] Solow (1974) showed what must be done to ensure that sustained economic prosperity does not come at the expense of the environment. More telling though is another paper he gave that year; the Richard T. Ely Lecture entitled The Economics of Resources or the Resources of Economics.[8]

In the 1974 address, Solow initiated his speech in his description of the trajectory of sustainability with a quote:

> Contemplation of the world's disappearing supplies of minerals, forests, and other exhaustible assets has led to demands for regulation of their exploitation. The feeling that these products are now too cheap for the good of future generations, that they are selfishly exploited at too rapid a rate, and that in consequence of their excessive cheapness, they are being produced and consumed wastefully has given rise to the conservation movement.

Solow was not reflecting on his own work in these prescient words. Rather, he quoted Harold Hotelling from 1931, almost half a century earlier. While he spent the first half of his lecture describing extensions of his past modeling of growth to include the 1970s reality of resource constraints, especially in fossil fuels, he shifted in the latter half of the paper with his most consequential conclusions. He noted the gulf in incentives between private actors within a generation and the needs of policy makers to not shortchange future generations.

Solow outlined a number of reasons why private agents in one generation generally discount the aspirations of future generations. He noted that private discount rates may be consistently higher than the socially justifiable discount rate. These include:

- Private individuals and entities discount the future because it is risky and uncertain, and hence is less valuable than the better known and certain present. In the absence of effective and complete insurance from these risks, agents tend to be generation-centric.

- Some of these risks are individual and not societal, and hence there are asymmetries between individual decision-making and the optimal society responses.
- Individuals plan for a time horizon that is always shorter than that of society as a whole. Some individuals may plan for a very short remaining time horizon, while the market interest rate averages the rate of time preference of mortals of sometimes short but always finite length of life. Yet, we presume society is infinitely lived.
- Individuals are motivated to make decisions based on the value of their investments. Since income taxes on capital investments remove a share of returns, an individual may always demand a greater return than society as a whole. This taxation premium biases personal discount rates up further if individual saving for the future is taxed.
- Frank Ramsey, in his 1928 seminal paper on intergenerational consumption concluded that the discounting of future utilities was unethical. Solow agreed that individuals may lack imagination about the effects of their discounting, or they acknowledge their own mortality. Ramsey and Solow asserted that to treat future generations unequally is ethically bankrupt even if it is individually compelling.
- Solow endorsed the ethical justification of a zero-discount rate, but noted that he would accept a discounting of future consumption if it could be demonstrated that future generations may realize more prosperity than the present. This observation is an extension of his modeling of technologically driven growth theory, but he notes that no theory ensures such future prosperity. He added that "the future may be too important to be left to the accident of mistaken expectations" and the vagaries of an affluent society's work ethic.

In the wake of Limits to Growth and the Blue Marble, Solow noted that the gap between a generation's decision-making and a perpetual society's planning needs are even more critical once one considers the fragility and finiteness of Earth's resources. He cited theoretical conclusions that show any optimal economic path for a society will eventually result in zero future consumption and exhaustion of non-renewable Earth capital if the employed social discount rate is positive.

When future generations are discounted, some generation will eventually prefer to exhaust any remaining Earth resources because of the increasing drag of ever-dear resources on each generation, at the permanent expense of all generations that may follow them.

Solow argued that only a dispassionate public policy entity unswayed by the preferences of its current generation could make the proper determinations of natural capital extraction and environmental degradation by employing a zero-discount rate that does not devalue future generations. He harbored faith in the various tools at the intergenerational planners' disposal. These include conservation subsidies and resource severance taxes. He admitted, though, that such policies must be promulgated through traditional political channels.

However, if the average corporation is incapable of taking a long view, and if members of corporations are drawn, at least in part, into government, Solow was less than confident that "a government bureaucracy" could be transformed into "a guardian of the far future's interests." Solow remained critical of faith that our collective institutions could better safeguard the future than well-constructed policies even-handedly applied to protect future generations.

Instead, Solow held more faith in the role of our institutions to promulgate and disseminate information about Earth's reserves, new technologies, and the nature of consumer demand. A population educated about the expectations of future generations and resources could create greater regard to the needs of future generations.

An Organization's Cost of Capital

If production and consumption align based on the cost of capital, some significant challenges are immediately obvious.

First, neglecting technological innovation, population growth, or changes in societal tastes and preferences, a constant flow of consumption and utility across every generation will only occur if the interest rate and rate of time preference we impose on intertemporal decision-making is zero. We shall return to this ethical issue in turn.

Another challenge is the universality of the interest rate upon which we depend for sustainability. For instance, in economic models, it is typically assumed that all economic agents have access to both savings and borrowing at the same prevailing interest rate. Such *perfect capital markets* do not exist. Individuals cannot borrow at the same low-interest rate offered to savers. Indeed, some households that could most benefit from investments that provide long-term sustainability are precluded from borrowing at all, which then constrains their consumption to current income.

One aspect of a more holistic *lifecycle approach to consumption* is that young people may find it advantageous to invest in their own human capital by borrowing for education. If these investments are resilient and do not appreciate rapidly, the earlier this investment can be made, the greater the life-long return on investment. However, young people are often precluded from sufficient borrowing that would permit them to invest optimally in education.

Society has created institutions to overcome such constraints, through public education for children and either subsidized higher education or subsidized and guaranteed loans for higher education. Despite these important institutions, the assumption of identical and symmetric access to lending and borrowing is a strong requirement of intergenerational equity and is likely unrealized in practice.

On the production side, nor do businesses have equal access to capital at the prevailing interest rate. If an extractor of a finite resource cannot borrow or save toward future extraction earnings at the prevailing interest rate, it may not have the appropriate incentive to defer extraction.

Corporations regularly calculate their cost of capital. A corporation is willing to invest in new projects based on a predictable interest rate commensurate with its typical level of corporate risk. If stock markets are efficient, such an investment in itself, perhaps through borrowing or the issuance of new stock, can be made at their risk-adjusted rate of return. To the degree that financial markets appropriately judge the risk-adjusted rate of return of corporations, a business is able to borrow and save at its prevailing interest rate.

Corporations can borrow in three ways. It can issue corporate bonds or secure credit from banks and financial institutions. Markets formulate the requisite interest rate r_D for debt to compensate for corporate risk. A firm can also issue preferred stock by providing an expected yield denoted as r_P, which an efficient market also adjusts for risk. Finally, a firm can issue new equity in the form of common stock, with a market-adjusted rate of return r_E.

Each of these methods has various transactions costs. Bank loans require a spread over the bank's cost of capital to cover bank expenses and profits. A new issue of bonds requires the creation of a prospectus and the willingness of a broker or investment banker to market the bonds for a fee. Likewise, investment bankers may facilitate a new issue of preferred or common stock. The various fees for these issues represent significant transactions costs.

In addition, producers regard these funding sources differently because of tax policies. Interest paid to debtholders is typically regarded as a business expense which allows these payments to reduce a producer's tax obligation. Then, for a producer with a marginal tax rate T, their **weighted average cost of capital** (WACC) is given by:

$$WACC = W_D r_D (1 - T) + W_P r_P + W_E r_E$$

where these various weights WD, WP, and WE sum to one and represent the typical weights an individual corporation chooses for funding new projects.

The resulting cost of capital that firms then employ as their corporate discount rate they apply to future projects is always positive and significantly higher than the zero optimal social discount rate. This implies that corporations fail to meet the public policy goal of optimal resource extraction over time based on a generationally neutral zero-discount rate.

These various strategies idiosyncratic to individuals or producers, the myriad imperfections in capital markets, and the impossibility of households and producers to access perfect capital markets call into question the practicality of equilibria between households and producers across generations. However, these market failures introduce various public policy avenues to correct such imperfections and attain optimal resource usage over time. Yet, most resource extraction decisions are made by corporations.

The Rawls Veil of Ignorance

We know that corporations invest and extract according to their interests. It is impossible for them to set aside their original position and substitute instead a social perspective. Yet, central to a sustainable path for the economies of both MDCs and LDCs is an ability to properly balance current interests with intergenerational interests. Implicit in much of Solow's critique is that our current economic reality influences our regards to such an extent that concerns of future generations are muted or perhaps even irrelevant.

Our ability to place decision-making on a higher and intergenerationally agnostic plane above self-interest is one that has interested philosophers for centuries. Social philosophers, from Thomas Hobbes to John Locke and Jean-Jacques Rousseau, have proposed various thought experiments in an attempt to place mortals in a neutral position unwed from their personal and generational interest in the outcome of a deliberation.

A contemporary philosopher, John Rawls, is credited with a rebirth in the discussion of *economic ethics and justice* in modern society. The son of a prominent attorney, Rawls went on to create a theory for which the legal profession would surely embrace, and for which thousands of courtroom discourses have since quoted. This notion is inherent in the motto that justice should be blind.

In 1971, John Rawls, then a philosophy professor at Harvard University, published what a peer, Will Kymlicka, deemed had led to "the recent rebirth of normative political philosophy." Rawls' A Theory of Justice[9] sought to establish a principle of justice as fairness that commends society to confer maximal benefits on the least advantaged members of society.

In this sense, he developed a theory of ethics and justice that operationalizes a concept that had been articulated across cultures and centuries. Even religious prophets have aspired to a higher societal ethic removed from self-interest, as in the Christian Bible, King James Version, Matthew 25:40, which states "Inasmuch as ye have done it unto one of the least of these my brethren, ye have done it unto me."

In the intra- and intergenerational economic awakening of the 1920s and early 1930s, John von Neumann had developed a similar concept as a solution to the theory of games. In 1928 Neumann had proposed an equilibrium concept which, if pursued by each of two competitors in a zero-sum game, would result in a stable and determinable equilibrium which is superior to any other potential equilibrium.

Neumann's concept was the *minimax solution*. In this solution, two rational actors strive for an outcome that minimizes the maximum loss each may receive in equilibrium. In turn, this solution yields the best among all bad outcomes, or equivalently, equals the worst among best outcomes, and hence is sometimes called a saddle point, or a minimax.

The same notion applies to the equilibrium concept sometimes advocated in dividing a literal rather than an economic pie. As siblings squabble over a fair division of dessert, parents may proclaim—*One cuts the pie, and the other picks the piece*. In the economic and societal analogy, in a deliberation over an ethically justifiable social or economic issue, one must place aside their position, privilege, and advantage in society and instead place themselves in the position of the least advantaged. In doing so, economic justice is framed as an issue of fairness and equity.

Rawls recognized the difficulty self-interested mortals may have in putting aside our *original position* to make choices consistent with the greater good. To do so, he developed his *veil of ignorance*. Rawls' thought experiment of a veil of ignorance over an original position argued that fairer economic and social outcomes would occur and economic and social justice would prevail if people ignored their position in the social and economic structure with regard to race and gender, age and intelligence, wealth and religion, and skills and education.

To this list one could add generation. Deliberators know they are empowered with the ability to fully engage in political, social, and economic institutions, but economically just results occur only if we can disregard the privilege and position we bring to our deliberations.

To see how such a removal from position can influence one's outlook, first consider your own personal regard to such economic and social institutions as access to universal healthcare, the desirability of free prenatal care, access to universal daycare and pre-Kindergarten programs, the ease in which young people can invest in their human capital by accessing high-quality public education, and the degree to which racism, classicism, or elitism is tolerated in society.

When surveyed or asked to vote regarding such issues, most people opt for institutions that are not unfamiliar to those that have served us individually reasonably well, regardless of where these institutions fall on the spectrum of experiences of all humankind. It is of course difficult to deliberate on the fairest outcome of each human and economic dimension.

The political process is wrought with the difficulties of achieving a societal consensus on these important issues for humanity. Each of us tends to be wed to a set of beliefs that are incorporated into our individual social fabrics, influenced perhaps by our upbringing and position in society, and we all vote accordingly in a democracy.

Instead of attempting to solve for optimal social and economic conventions in each of these dimensions based on our individual standing, consider the following contrived circumstance that I call a *baby lottery*.

In a period of your life, you and your spouse have decided to parent a child. You have provided and cared for your unborn baby by making good health choices and investing in prenatal care. You are as confident as any new parent that you can provide opportunities for your child to at least enjoy the quality of life that you have been able to forge for you and your spouse.

On the day of birth, you and your spouse have made your way to the hospital and successfully delivered a child late in the evening. The newborn baby is removed by the maternity ward staff to the neonatal intensive care unit (NICU) for washing, the taking of blood samples routine testing, and preparation to be taken home while the mother rests. The nurses assure you both that there is no reason for concern.

In the morning, you both proceed to the NICU to pick up your newborn on your way home. You poke your head in and ask for your baby. The staff give you the baby nearest the door. Somewhat confused by the lack of names on the baby cribs, you ask if this baby is yours. Their response is "it is now."

In this baby lottery, on average every couple conceives an average baby and the baby the couple receives at the NICU door is also, on average, well, average. There is no ethical basis for a couple to claim that the baby they receive is either better or worse than the one they conceived. Assuming you willingly accept this removal of your original position, how would you determine the optimal level of the various societal dimensions described above, but now under such a veil of ignorance?

Most people placed in such circumstances under a veil of ignorance propose social and economic institutions that more fairly protect the interests of the least advantaged. Were a couple to know that they would return home with a random baby, they may prefer a medical system that permits affordable or free access to prenatal care for the baby's natural mother, and may even be willing to adopt rules that deter dangerous substance consumption while pregnant. They may advocate for affordable or free access to hospitals so mothers do not consider a hospital birth as one only afforded in desperation.

Knowing that a couple's genetic stock may have gone home with another parent, their parental instincts may hope their natural child also has access to universal daycare and a good public education. Presumably, one would prefer their genetic child and the child they brought home to be free of racial or gender discrimination, knowing that the child they bring home may not be of their same race, preferred gender, or biological lineage.

In short, people may opt for a set of social conventions that ensure even the most disadvantaged child must endure a minimal set of social and economic risks and be afforded every opportunity to thrive. They opt for a minimax solution.

This veil of ignorance can also be applied to the quandary of intergenerational justice as well. Using Rawls' principle, society ought to construct a path for economic fulfillment that benefits no one generation more than another. If such were the case, we may be willing to accept the notion of weak sustainability with the knowledge that, while the circumstances for each generation may not be identical, any shortcoming in one dimension of capital is fully compensated in another so that each generation has an equal opportunity to thrive. A dearth of one resource is compensated by additional endowments set aside of a new resource by a past generation on behalf of a new generation.

Rawls concluded that exploration of a *just savings* fund may be warranted. However, he remained concerned about such a permanent fund. His just saving principle was described in section **49.2** of his book as:

> The principle of just savings holds between generations, while the difference principle holds within generations... (This difference principle states that) Societal and economic inequalities are to be arranged so that they are both a) to the greatest benefit of the least advantaged, consistent with the just savings principle, and b) attached to offices and positions open to all under conditions of fair equality of opportunity.

Rawls actually harbored concern about a true Genuine Savings approach. He argued for a level playing field of opportunity, but he distanced himself from a moral prerogative of intergenerational equity if one condition is not satisfied. If past generations have not provided fairly and justly for the present generation, the present generation should no longer be obliged to indemnify future generations from excessive resource extraction. This rationale makes sustainability elusive.

THE SUSTAINABILITY RESPONSIBILITY

If one rejects Rawl's qualification of intergenerational equity, the veil of ignorance is a profound tool that argues for a path of smooth consumption for all generations over time. This conclusion is not without complications. Beyond Rawls' concerns, social philosophers must still resolve a number of questions:

1. What role will technological gains play in this prescription?
2. Does smooth consumption translate into smooth utility?
3. Are resources and wealth fairly distributed within and between generations?
4. If the population of subsequent generations grows, is the available capital and consumption on a per capita basis or total basis?
5. Have the various externalities on natural capital and ecosystems been adequately remedied?
6. If a principle for sustainability can be constructed for the Earth, are individual nations willing to share in that consumption?
7. If Less Developed Countries (LDCs) cannot develop in ways More Developed Countries (MDCs) can, will they have little choice but to sell their resources at prices below their true shadow prices and also consume their marginal user costs rather than reinvest them?

Partha Dasgupta tackles some of these dilemmas in the 2008 article The Welfare Economic Theory of Green National Accounts.[10] He began by acknowledging the need to distinguish between the measure of livelihood governments and business prefers, the **Gross Domestic Product** *(GDP)*, and a more appropriate measure of the Earth's natural and human-made assets, net of their depreciation over time. One cannot construct a path of optimal sustainable consumption without properly measuring the natural and human-made wealth that sustains it.

As a rubric that demonstrates misplaced economic policies, Dasgupta in turn asked four questions and then described whether these questions are addressed through various approaches. He asked:

a. how is the economy performing at any point in time,
b. how has it performed too now,
c. will it perform better under improved management, and
d. what policies should we then adopt?

As Dasgupta delves into these questions, he also noted a fundamental measurement problem. Even in *triple bottom line accounting*, which advocates for the construction of a dynamic balance sheet for Nature, humans typically add up assets, much like what we might see in an accounting *balance sheet*. But, in reality, this linear approach to wealth is made up of contributions from many assets, each of which incurs costs and consumes resources decidedly differently and non-linearly. Indeed, the differential ability of one resource to contribute often depends on complex relationships with other resources as well. While the linearity and uni-dimensionality of an economic value system reduced to money as the primary measure of financial capital is simple and somewhat appealing, it cannot adequately represent the complexities of the economic elements it hopes to embody.

It is then difficult to explore the intricacies of the conversion of resources to wealth and its effective translation into consumption and happiness. We take highly complex and non-linear economic and ecosystem relationships and convert them into simple linear sums on a balance sheet. All we can really say for sure is that, if these natural and human-made assets decline, so too will happiness, assuming we are at all times managing the economy for optimal efficiency in the conversion of resources to happiness. In essence, money and wealth may not predictably buy happiness or ecosystem health. For instance, significant wealth may not compensate for the values many may hold for a strong and sustainable ecosystem sacrificed in the interest of progress.

Dasgupta noted other problems as well. First, our models tend to assume constant population and typically do not speculate on the crystal ball of technological improvements. Second, implicit in our predictions for the future is that the present is managed optimally, with fully internalized externalities and with foresight of the nature of the future. Yet, we know that many activities of natural resource relevance do not enjoy well-functioning markets, if any at all. Finally, economists often speak of convex sets of possibilities to convert our available resources into production and wealth. Yet, the world is not the smooth and predictable set of circumstances economic equilibrium typically requires. Non-convexities create discontinuities. Resources can disappear entirely and hence bring some processes to a screeching halt.

Dasgupta offered a partial solution to the long-term problem that at least incorporates intergenerational values into a monetary measure. If we could construct an accurate wealth index, and if the contents of that wealth of natural and human-made assets could be efficiently employed, at the correct prices and rewards, and with all externalities internalized, then the change in wealth $\dot{W}$ will equal the return $rW(t)$ for each generation less its consumption $C(t)$:

$$\dot{W}(t) = rW(t) - C(t).$$

At the most basic level, and not unlike the equations we have developed, a constant Earth balance sheet requires that a generation's consumption must not exceed the appropriate rate of return on its permanent wealth. Society should not eat its own seed. Included in our consumption may also be our appreciation and valuation of the natural world. Also implicit in this formulation is that the loss of an increment of Nature's assets, properly priced, must and can be compensated by an increase in human-made assets.

Dasgupta asserted that our total capital stock $K(t)$ evolves as a function of our ability to convert factors of production into product. He included a factor $A(t)$ to represent an index of society's shared knowledge and the effectiveness of our institutions to generate productivity. This total factor productivity may evolve and is multiplied by a generation's production function.

From this production Dasgupta subtracts off our investments in human capital $J(t)$, our consumption $C(t)$ and the depreciation of total capital assets $K(t)$ at a depreciation rate δ. Then,

$$\dot{K}(t) = A(t)F(K(t), L(t), H(t)) - J(t) - C(t) - \delta K(t),$$

where $L(t)$ is the stock of human capital, which depreciates at a rate μ (i.e., $L(t) = J(t) - \mu L(t)$, and reflects the loss of human capital as people expire at the death rate μ), and where $H(t)$ is the extraction rate of natural capital. Denoted by $R(t)$ our renewable natural capital, which likewise can, if managed well, potentially grow as some function $M(R(t) - H(t))$:

$$\dot{R}(t) = M(R(t) - H(t)).$$

Dasgupta had observed that our renewable natural capital stock $R(t)$ is a complicated non-linear expression since it grows in some manner related to its size $R(t)$, net of its harvest rate $H(t)$. It could then be the case that the resource could reach a tipping point and disappear, which leaves us with a production function $F(K(t), L(t), 0)$. Should resources fall below a tipping point, it could jeopardize our capacity to produce. We see such stark concern in the plethora of science fiction novels where humankind is forced to search for resources on other celestial bodies to sustain life on Earth, or even in the reasons why Elon Musk argues we must populate Mars.

Dasgupta also takes issue with the conversion of consumption, fraught with complications as it is, into utility. Obviously, utility is not a linear function of consumption. The enjoyment of a meal for one sated with food is certainly far less than that of one starving. In addition, the utility we derive may even depend on our regard for the consumption of those around us, as Thorstein Veblen commented in his *theory of conspicuous consumption.*[11]

Finally, embedded in our utility measure is our intrinsic enjoyment of natural capital merely from the comfort that they exist, such as the degree we value the existence of Yosemite Park or the Grand Canyon, even if we never travel to enjoy them. Such values are considered by many to be a shared part of the human condition.

We also know from the Tragedy of the Commons and the practical difficulties in establishing property rights based on the Coase Theorem that all societies do not fully and completely employ institutions that can efficiently manage our resources. If, for instance, our capital stock K and our labor stock L are embodied with property rights, but our natural stock R suffers from the tragedy of the commons, natural resources may be extracted at prices well below their true social marginal user costs. Distorted prices anywhere mean somewhat distorted prices everywhere even according to classical economic theory.

Even within one factor class, distortions can exist. It is possible that one piece of capital may be more fully utilized in one industry than another, and hence aggregation of factors and their rewards may induce other problems. Indeed, the market likely abounds with mispricing and incompleteness, and hence the shadow prices we depend upon for sound resource management are equally spurious.

These shadow prices become more questionable if a resource comes too close to non-linearities and discontinuities such as tipping points. Ecological economists measure resilience as the distance away from such points of discontinuities or inflection points. Resilience acts as a buffer that provides a margin from collapse or a buffer from these discontinuities and non-linearities in an uncertain environment.

To move us forward to some sort of measurement paradigm and partial resolution, Dasgupta postulated a *comprehensive investment criterion* that assumes a policy maker is able to correct for the various mispricing described above and hence can measure a society's comprehensive investment $I(t)$ as:

$$I(t) = p(t)\dot{K}(t) + q(t)\dot{L}(t) + n(t)\dot{R}(t).$$

To do so requires a correct and evolving price or value $p(t)$ of the employment of physical capital K, the true value $q(t)$ of the use of labor $L(t)$, and the proper value $n(t)$ of natural capital $R(t)$. He then demonstrates that the true present value of utility $W(R(t))$ of an intergenerational economy, as a function of the state of the economy $W(t)$ at each time in the future to ensure an intergenerational welfare level W, evolves over time according to:

$$W(\dot{R}(t)) = I(t).$$

In other words, intergenerational welfare should evolve over time as a function of our willingness to invest in the future generations. As a corollary, this comprehensive investment rate is a measure of its effect on future consumption.

Dasgupta challenged policy makers to the task of evaluating individual projects, and society's investments in the aggregate, to the extent that, from an intergenerational sustainability perspective, projects should enhance comprehensive investment and permanent wealth. He demonstrated that a current value Hamiltonian that optimizes consumption and utility over time is then equivalent to the construction of a set of policies that ensure constant happiness over time. The solution gives:

$$\mathcal{H}^*(t) = U'\big(C^*(t)\big)C(t) + I(t),$$

where the policy maker must then construct an appropriate path of comprehensive investment over time. The Brundtland Commission's notion of sustainable development then requires comprehensive wealth to not decrease over time. To do so, a society must ensure that the value of consumption by each generation does not exceed its **Net National Product (NNP)**, properly adjusted for depreciation of natural and human-made assets.

Another way of stating this solution at an intuitive level is that the rate of change of the comprehensive value is the sum of the rate of change of utility as a function of consumption and the rate of change of comprehensive investment over time:

$$\dot{W}(t) = U'(C(t))\dot{C}(t) + \dot{I}(t).$$

As before, we see that we maintain a steady state level of properly priced social value for humankind only if the amount removed by each generation, represented by the first term $U'(C(t))\dot{C}(t)$ on the right-hand side of the equation, is offset by the amount of social investment $\dot{I}(t)$ each generation adds, to be enjoyed by subsequent generations.

Finally, investors often calculate in their profits the capital gains that occur as an asset becomes more valuable, such as may occur when an asset experiences increasing demand. This is in some sense a paper wealth that should not allow too easy claim of one generation that they have adequately invested in the next. These proper prices of assets must then have subtracted from them such capital gains one generation may try to capitalize financially even though the future usefulness of their capital stock remains unchanged.

There is one complication that is more difficult to treat from a Rawlsian perspective. If population grows, does that create a responsibility of earlier generations to leave even more resources for future generations so that average *per capita* consumption can be preserved? This would then require us to estimate a shadow price that evaluates today's expected population growth in the future. Similarly, should these prices be adjusted so that more is set aside to compensate for uncertainties and population choices of future generation?

PERMANENT FUNDS AND POPULATION

The issue of population growth induces two complications. The first is whether the correct weak sustainability ethic requires each generation to pass to the next a smooth consumption path or a constant per capita consumption capacity. If population grows, do earlier generations have the responsibility to anticipate population growth rates chosen by later generations, and leave greater permanent fund contributions to these later generations to ensure the permanent fund indemnifies the choice of future generations to expand the population?

A second ethical question also arises from population growth. With increased population growth comes greater demand for resources, including sustainable resources. The higher prices that fixed or limited factors of production then command means that the share of the GDP of subsequent generations must devote to consumption of these increasingly scarce resources rises because of their higher prices. These higher prices then compound the difficulty for future generations to enjoy the same level of consumption of previous generations, even with the optimally set permanent funds that have, ideally, been left for them. Such increased resource prices relative to demand of other capital goods over time further distort demand of human-made production capacity as well.

The ethical question remains whether a generation must anticipate the choice of additional population growth and challenges facing future generations. These ethical questions of fixed resource usage and future population growth and economic development are most apparent in the management of our shared atmosphere. But, such an ethic may commend population control, and hence may disadvantage LDCs that are not afforded the same opportunities to grow their populations and cultures as MDCs were afforded in previous generations. Economic justice issues then emerge.

SUMMARY

The ramifications on policy-making with regard to the setting aside of original position and the imposition of the veil of ignorance are obvious, even if our decision-makers would remain challenged to fairly and consistently meet such an intergenerational standard. Permanent funds and sufficiently well-funded Genuine Savings may assist policy makers in this exercise, but informational requirements and the incorporation of new opportunities and risks will remain challenging.

The next set of topics will describe how these various principles that have been developed can be applied to address and correct various sustainability challenges.

The theoretical models of sustainability all arrive at a similar conclusion from their various perspectives. Each generation has a responsibility to add to the permanent social well-being at a rate equal to what it removes. This is equivalent to weak sustainability in which as natural capital is removed, some permanent fund must be augmented in its place, not merely to provide for wealth, but instead to ensure no decrement to well-being.

Rarely is such a responsibility acknowledged by nations of peoples though. There is no easy answer to these daunting qualifications to sustainability except to say that inaction is no solution. We perhaps do not yet have all the tools and measures to do a proper accounting of natural capital. Nor do we have the ability to indemnify those developing economies that can least afford doing the right thing for the future. Finally, there are no universal political yardsticks to come up with global solutions in our globally connected world.

While challenges facing mortal generations in an immortal world are profound, we can't afford to not try.

ESG Toolkit

Decisions of organizations are often history-dependent and their leaders are drawn from that corporate culture. The measure of Net Natural Product better places natural resources as a shared asset rather than a consumption item.

How can a triple bottom line balance sheet be constructed at the corporate level to include these values?

To what degree does the corporate history and culture constrain diverse ideas and frustrate socially optimal decisions?

How might a corporate board put aside original position at times in the interest of stimulation of diverse thought?

NOTES

1. Keynes, John Maynard (1936). The General Theory of Employment, Interest and Money, Palgrave Macmillan, London.
2. Solow, Robert M. (1974). "Intergenerational Equity and Exhaustible Resources". Review of Economic Studies. **41** (Symposium): 29–46. https://doi.org/10.2307/2296370.hdl:1721.1/63764. JSTOR 2296370.
3. Ricardo, David (1817). On the Principles of Political Economy and Taxation (1 ed.). John Murray, London.
4. Hartwick, John M. (1990, December). "Natural Resources, National Accounting and Economic Depreciation". Journal of Public Economics. **43** (3): 291–304.
5. Hamilton, K. (1995). "Sustainable Development and Green National Accounts", unpublished Ph.D. thesis, University College, London.

6. Hamilton, K., and J.M. Hartwick (2005). "Investing Exhaustible Resource Rents and the Path of Consumption,". Canadian Journal of Economics. **38** (2): 615–621.
7. Solow, Robert M. (1974). "Intergenerational Equity and Exhaustible Resources". The Review of Economic Studies Vol. 41, Symposium on the Economics of Exhaustible Resources (1974), pp. 29–45. Published By: Oxford University Press.
8. Solow, Robert (1974). "The Economics of Resources or the Resources of Economics". The American Economic Review. **64**, No. 2, Papers and Proceedings of the Eighty-sixth Annual Meeting of the American Economic Association (May, 1974), pp. 1–14.
9. Rawls, John (1971). A Theory of Justice, Harvard University Press, Cambridge, MA.
10. Dasgupta, Partha (2008). "The Welfare Economic Theory of Green National Accounts". Environmental and Resource Economics https://doi.org/10.1007/s10640-008-9223-y.
11. Veblen, T. (1899). The Theory of the Leisure Class, Macmillan, New York, NY.

CHAPTER 18

Application of Public Policy to Attain Sustainability

Public policy is informed by the insights described by our economic models of sustainability. The resulting public policy challenges can be divided into two categories. The first includes challenges raised by externalities created within a generation. If resources are squandered within a generation, optimized resource decision cannot be made between generations. Then, economic justice requires intragenerational efficiency. This second category ensures that the proper intergenerational transfers are made so every generation can attain the well-being of previous generations able to access our shared natural capital. The first section of the chapter describes public policy for the attainment of *intragenerational efficiency*. Then, the more vexing *intergenerational efficiency* challenges are described. This latter category is much more difficult from a public process perspective because they involve the caretaking of interests for stakeholders who do not yet have a seat at the economic table.

RESILIENCY FOR RENEWABLE RESOURCES

Effective management of the risks of natural capital depletion or degradation is well-understood and is widely used in environmental economics to prescribe the management of depletable and sustainable resources. These management protocols are first designed to ensure that the use of natural capital remains within the boundaries of sustainability. Sufficient resiliency also ensures these resources do not cross the threshold in which they collapse or remain unusable for an extended period of time. Examples of renewable resources include those replenishable biotic resources that are able to grow at some sort of natural rate $G(R)$ as a function of the size of the renewable resource stock R. Should the

C. Read, *Understanding Sustainability Principles and ESG Policies*, https://doi.org/10.1007/978-3-031-34483-1_18

resource size R decline too far, a tipping point can push it below a sustainable threshold that frustrates the ability of Nature to reproduce.

This point beyond which recovery cannot reasonably occur can be labeled the *crash threshold*. In the case of biological resources that can repopulate, an animal species may become so geographically thin and dispersed that matching between males and females becomes improbable. Alternately, the species may be unable to form herds of sufficient size to protect its young and most vulnerable. For instance, marine mammals such as whales can suffer from geographical dilution to make procreation difficult, while herds such as caribou can be so thinned that a depleted healthy and strong outer perimeter of a herd cannot protect the vulnerable inside the herd from predators.

These phenomena arise either from the need to maintain a critical mass to propagate, or the need for strength in numbers to protect against a hostile environment. An example was provided in Chapter 2 of two interrelated species that interact in ways that can push a prey species to a crash point based on its relationship with a predator species. If species are interdependent, a crash in a prey population can induce a crash in a predator population, even if elements of our ecosystem are related in simple linear ways.

In addition, some potentially sustainable resource stocks can be pushed into extinction because the unique habitat upon which they depend may be threatened. The dramatic escalation in species loss arising from global warming or the disappearance of geographically specific species due to the increased need for agricultural land or from excessive use of pesticides are examples of such geographical encroachment and species decline.

Once a species falls below such a crash threshold, recovery may be difficult or impossible. The goal of public policy, then, is to both maintain the population of a biological species at an optimal level that maintains a sufficient distance from the crash threshold. This requires management for both sustainability and resiliency in the face of natural population fluctuations.

Recall the logistics population growth relationship for a biotic resource. Public policy must determine the optimal population size to maximize the value of population growth and a sustainable harvest but must also ensure sufficient resiliency to keep such an equilibrium well away from the crash threshold. Policy should certainly ensure that the population is maintained sufficiently above a threshold that provides for a critical population mass. For instance, the population size below 0.5 in Fig. 18.1 results in a degenerative population that then crashes spontaneously over time:

While resiliency is dependent on population dynamics, it is also a function of human management of the population. Recall that the anthropocentric management of a biotic population sets a rate of change of growth that corresponds to the discount factor r. Only in a zero discount rate regime would this point coincide with *maximum sustainable yield*. The higher the discount rate, the lower the sustained population size. Resiliency is reduced and species crash probabilities are increased with a higher discount rate or corporate weighted average cost of capital. This low sustained population size arising from a

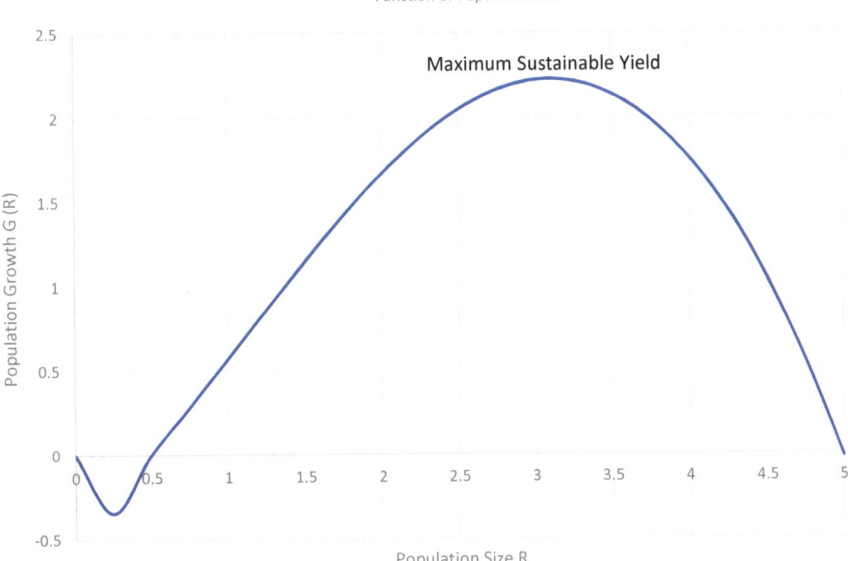

Fig. 18.1 Harvestable rate of growth of a biotic population such as a fishery or forest

high discount rate then pushes the population closer to the *crash threshold*. The resource becomes more vulnerable to exogenous forces such as disease or human activity that may randomly pressure the buffer. Such a managed resource is more prone to extinction.

Effective policy must then meet a few criteria. First, it must ensure that property rights or quotas exist to avoid the tragedy of the commons that arises when a race-to-the-bottom occurs as extractors catch-as-catch-can. Second, public policy must ensure that the optimal stock size be maintained to ensure a healthy population and a higher sustainable yield to act as a buffer against resource crashes. This policy may require a lower imputed discount rate and hence a stock size higher than a privately managed resource may determine. Finally, the policy must be adjustable over time to ensure that a safe buffer is maintained to ensure resiliency and avoid a resource crash.

These criteria require a careful application of the tools of public policy such as quotas, taxes, and monitoring. These techniques are not done to ensure intergenerational equity, although they do allow for a population to be sustainable and thus intergenerationally fair. Rather, the techniques are primarily designed to ensure intragenerational efficiency in the use of the sustainable resource.

Fisheries and ITQs

The classic example of such a regime is in the management of a fishery.

Let us first differentiate a commercial fishery from a fish farm. Aquaculture is equivalent to sustainable farming if all externalities of the activity are properly internalized. Like agriculture, these externalities may include the underpricing or the unpricing of pollution arising from fish farms, any depletion of the growing medium and other factors of production, and environmental costs arising from monocrop cultivation or the use of factors of production such as feed and fertilizers. Assuming any such environmental externalities are properly internalized, both the farmer and the fish farmer have the incentive to manage the resource to sustain a yield as described with the optimal management of a potentially depletable resource.

This sustainable equilibrium is in contrast to a natural fishery. We know that fisheries cannot operate in open access without suffering from the *tragedy of the commons*. Regulation is essential.

There are a number of possible regulatory regimes. The first is to permit open access to a fishery but impose some sort of a quota to limit over-fishing. Fisheries often do this by limiting the duration for which open access is permitted. Hunting and fishing seasons, timed to avoid natural reproduction cycles, can be set to ensure that overextraction is made difficult. In addition, to limit potential extraction, either licenses can be sold or rationed and the extraction technique be constrained.

We saw earlier that an artificial increase in extraction costs will result in a decrease in effort and extraction. If extraction costs, measured in effort costs per unit of extraction, tend to increase as the stock thins, either such a public policy or nature's automatic equilibrators can help to limit extraction. Managed fisheries may do so by artificially limiting the amount or type of equipment used in the harvest. Nets of a certain size, limits to the duration of an opening, the monitoring of catches, and other artificial regulatory tools can limit the catch to that determined by policy makers.

However, these methods are inefficient. The restriction of fishing technologies unnecessarily increases harvest costs. Permitting free or licensed entry within a narrow harvest window can limit the total catch, and may also be dangerous as harvesters race to extract intensively within a narrow time frame. It also requires significant and specialized fish boats to be used for only a small slice of time, which results in wasted physical capital at other times.

Individual Transferable Quotas

The solution to inefficiencies arising from harvest externalities is the imposition of a fish quota. The quota specifies the number of participants times the size of each quota to sum to the chosen harvest level $H^* = \sum_{i=1}^{i=n} h_i^*$, where h_i^* is the quota assigned to each of n licensees. Under such a licensing system, individual harvesters can determine their most efficient harvesting technologies. Hence, this system does not suffer from artificial limits placed upon the harvest under alternative management regimes.

A question remains though. How should these licenses be granted? If a fishery is considered a resource held in the public trust, to grant a license is to offer a potentially profitable resource to interested harvesters. This opportunity for a harvester to extract this natural capital is not unlike the auctioning of airwaves Coase confronted. His proposed solution, adopted in some fisheries management regimes, is to auction these property rights to earn a harvest up to the quota amount, in return for payment of a winning bid at auction. These bids would ideally be proportional to the risk-adjusted future flow of expected profits beyond the costs and required return to each potential harvester.

As with airwaves, the greatest social and permanent fund value results if these individual quotas are transferable. If quotas are not transferable, the amount potential participants may be willing to bid would be constrained by the expected time horizon of their participation. By making *individual transferable quotas* marketable, potential ITQ purchaser is then expected to bid a greater amount P_0:

$$Auction\ Present\ Value\ P_0 = \sum_0^T \frac{(p_t - c_t)h_t}{(1+r)^t} + \frac{P_T}{(1+r)^T},$$

where p_t is the unit price of fish in period t, c_t is its (assumed) constant marginal cost, including all necessary factor payments, h_t is the granted fish quota, r is the bidder's discount rate, T is the expected duration the bidder expects to harvest, and P_T is the market value of the quota at time T when the ITQ is subsequently sold. If an auctioned quota is not transferable, the discounted quota resale term $\frac{P_T}{(1+r)^T}$ is not capitalized into auction bids.

In steady state, and with a well-managed fishery, an upper bound to royalty revenues is then given by:

$$P_0 = \frac{(p_t - c_t)h_t}{r}.$$

The best bid auction captures the entire present value of an indefinitely sustainable fishery that imposes control to ensure the maximum harvest is sustained. At any time in the future, another set of participants is able to purchase the right to the resource according to the same present value of the ITQ.

This system of ITQs has been used for the past few decades to great effect. First proposed by Christy (1973),[1] they have been employed by nations since the 1970s and now control about 10% of all fisheries. They still challenge the ability of small operators to access sufficient capital at the prevailing interest rate to compete effectively in ITQ auctions. Such auctions favor those who have superior access to capital markets, and hence tends to favor corporate entities.

This approach also still depends critically on the discount rate of participating corporations or individuals. On an individual level, their weighted average cost of capital, including the inherent risk premium of their entities that raises their interest rate still further, limits total bids compared to society's optimal zero discount rate.

In addition, there are other public policy values beyond profits. A generation may have legitimate concerns that disadvantaged communities or countries are not afforded the same access to capital markets or resources because of past disadvantages, an inability to raise capital or expertise, or because of a system tilted toward the powerful or the wealthy. For instance, an LDC may have been precluded from auctions in the past because of a lack of extraction infrastructure. As a matter of public policy, regulators can reserve some quotas for small and traditional operators, or for indigenous peoples, or those who have traditionally depended on the fishery. The best and most socially responsible bid may not be the highest bid. In addition, regulators can even subsidize access to financial capital for certain preferred bidders.

Should the stock be challenged in future years, there may be provisions to adjust the size of the quota or buy back quotas to reduce the quantity of outstanding individual quotas. While one may object that such buybacks by the public of a public resource can be costly, note that significant revenue is generated from a public forest land lease or fishery auction that should presumably be set aside as a permanent fund. Recall too that the purpose of ITQs or forest leases is not solely to earn a profit, even though the revenue of a well-run royalty system should in theory raise sufficient revenue to act as a permanent fund for the public use of this sustainable resource. The goal is to ensure the resource is harvested efficiently and safely. This auction system is typically superior to other more constrained and less efficient management regimes.

One flaw of such a solution is that future generations may be deprived of the full use of its natural capital. Revenues devoted to the permanent funds are lower than the full surpluses enjoyed by those who bid for resources. Higher than socially optimal discount rates, risk premiums, *monopsony rents* resulting if few harvesters bid all result in royalties that underprice benefits, and hence shortchange future generations. Differential access to capital markets and allocation of quotes also confers benefits unequally across individuals, corporations, and society. Such irreversible decisions ultimately affect social and economic justice, even if efficiency may be obtained.

FORESTRY AND CARBON TAXES

A forest is not unlike a fishery. Often forest land is owned by the public, although there are privately held forests that are managed effectively, much like management of privately owned farmland must also be cognizant of ecosystem concerns over monoculture cultivation. As with a fishery, should a forest be operated open access and go unmanaged, there is little incentive for replanting and the forest may well suffer from the tragedy of the commons. In addition, competition among timber extractors may result in wasteful procedures, such as the destruction of access roads to prevent rival harvesters from free-riding on their road building.

Forest rights can also be auctioned or offered under concessions that specify the amount and species of various trees to be harvested, obligations to keep approved access roads in place to serve the public interest, and requirements that a harvested forest will be replanted and remediated in specified and monitored ways.

One problem that arises is that those offered extended duration concessions may prefer to harvest in the least cost manner without concern for other social or ecosystem values, which often results in monoculture cultivation, clearcutting, and replanting of select varieties that grow most profitably. These practices are costly in terms of land degradation and forest health that remains vulnerable to fire and disease if a stand is of only one species and age.

The granting or auctioning of concessions can be designed to avoid such monoculture externalities with their resulting disease and fire vulnerability. A policymaker can determine a pattern of selective logging that describes how trees should be harvested in a stand. Such selective logging allows a planner to optimize harvest maturity. Such a management strategy promotes a healthier and more sustainable forest stand. Forests can also be harvested to mimic the pattern of natural disturbances such as from forest fires, or to limit the harvest to ensure the forest reestablishes its aesthetic qualities. These goals can be built into the management regime for optimal timber extraction.

By designing a selective forest harvest plan, a stand can also be maintained to sequester carbon in an optimal biomass each period. A well-designed timber harvest can remove a maximal amount of biomass that sequesters carbon dioxide as logs are milled into lumber that is then used to build structures.

To manage for maximum biomass production, a healthy and well-managed forest stand can function better than a natural forest from the perspective of carbon sequestration. An overly mature forest is more prone to disease and decay. When a tree dies and decays, it releases much of its decaying biomass in the form of carbon dioxide and methane. Harvested timber sequestered into human dwellings and other uses removes a potential source of carbon dioxide and methane emissions. Proper forest management techniques can then be designed to also minimize global warming through optimal carbon sequestration.

A well-managed forest can then be an effective tool to optimize forest health and optimize the sequestration of global warming gases. The design of a management regime with these goals in mind is not the form of timber harvesting typically preferred by industry, but investments in active forest management accomplish other important social goals.

Greenhouse gases can also be sequestered through the prevention of clearcutting of forests. The purchase of forest resources to maintain sequestration and prevent conversion to grassland constitutes an essential ingredient of global carbon markets. Such markets offer an opportunity for a carbon dioxide emitter to purchase offsetting reductions in emissions by auctioning off covenants that prevent a forest from being harvested. Such carbon markets will be discussed subsequently.

Forest preservation does not optimize the ability of a forest to most effectively absorb and sequester carbon dioxide and prevent methane and carbon dioxide emissions. Sophisticated management practices can be more effective from a carbon removal perspective as they also maintain a healthier and more diverse forest, preserve the value of medicinal plants and trees, and protect forest aesthetics and our implicit contingent valuation of a forest.

AIR POLLUTION

While the consequences of emitting gases into the atmosphere have been described and increasingly understood since the 1800s, awareness of the chronic consequences of pollution has accelerated in the latter half of the twentieth century. The long-term depletion of our atmospheric stock through the emission of greenhouse gases can take millennia to rectify and will be treated in the next chapter. This chapter instead treats the flow of pollution that imposes more transient intragenerational consequences to the environment.

Global effects of emissions such as greenhouse and ozone-depleting gases aside, air pollutants are problematic in that they tend to be localized, even if they may span borders. Air pollution beyond the threshold for which Nature can cleanse may accumulate rapidly and create chronic health and aesthetic consequences. *Fund pollutants or flow pollutants* are cleansed by Nature up to a threshold. Any emissions beyond the *absorptive capacity* of Nature can result in persistent or permanent environmental damage and are called *stock pollutants.*

A variety of pollutants occur from the combustion of fossil fuels, especially contained in the effluent that arises as coal is burned in power plants. Acid rain that results when sulfur-laden coal is combusted and allows sulfur-dioxide (SO_2) molecules to combine with water vapor to disperse sulfuric acid into the atmosphere. The acid vapor can travel for hundreds of miles before it returns to Earth in the form of acid rain which can then contaminate lakes and streams and kill wildlife. For example, regional pollutants from Midwest U.S. coal plants form acid rain that falls on Ontario, Canada, while the steel plants in Canada create acid rain that contaminates lakes in the Adirondacks of New York State in the United States.

Most problematic are *stock pollutants* that are only slowly cleansed by nature, if at all. These include heavy metals, some toxic chemicals, and the most persistent greenhouse gases, especially carbon dioxide. For instance, the lead that was once mixed with the gasoline to improve the fuel's octane rating was emitted into the environment and concentrated near roadways and adjacent to heavily traveled urban streets. Such metals may even find themselves drawn into the food chain and ingested.

The policy issue is to find the correct balance between the level of flow pollution emissions and their consequences. Necessary emissions abatement increases as the tolerated level of emissions is reduced. Meanwhile, the damage to emissions increases with increased emissions. The optimal level of pollution balances the marginal costs of abatement with the benefits of additional abatement. This tradeoff is shown in Fig. 18.2.

In the example above, the optimal pollution level is five units, and the costs imposed for additional abatement or for damage from the last unit of emissions are both equal to $3. Assuming that the public policy prerogative is toward clean air, the optimal Pigouvian tax to internalize the pollution externality is then $3. The triangle below the marginal damage curve to the left of the equilibrium emissions level of 5 units gives total harm to society, while the triangle to the right and below the marginal abatement costs gives the total cost to reduce unabated pollution of 10 units to the socially optimal level of pollution of 5 units.

Policymakers can arrive at this most efficient level of a flow pollutant through *command-and-control* or through market mechanisms. A command-and-control approach requires policy makers to impose a pollution quota, carefully monitor emissions, and impose fees accordingly. Alternately, by imposing a pollution fee of $3 per unit of emissions, the policy maker induces the industry to abate pollution from 10 to 5 units.

With the U.S. Clean Air Act of the 1990s, various important pollution markets were subject to a *cap-and-trade* regime. Under such a system, each

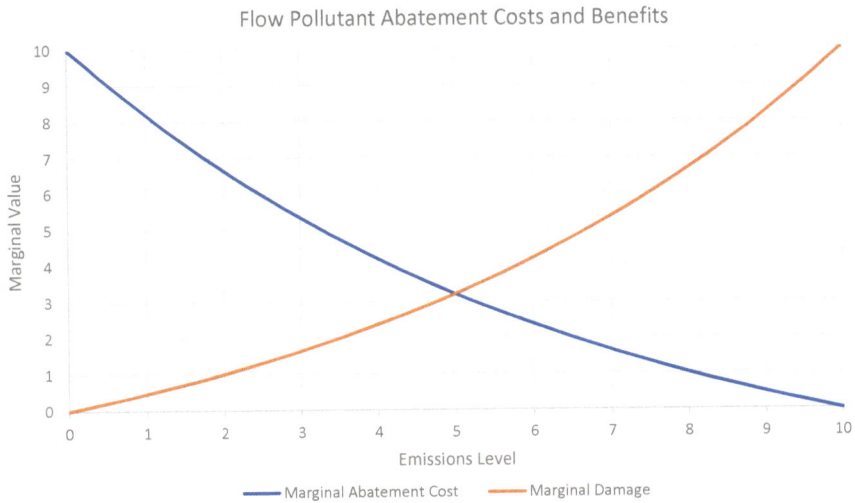

Fig. 18.2 Determination of the equilibrium level of pollution abatement

polluter can purchase the right to pollute up to a certain level. In this example, let us assume that each of two polluters is granted a right to pollute 2.5 units, at a per unit fee of $3. If such permits are issued, the equilibrium remains at five units, but the producer who is technologically able to pollute less than its permit by incurring a lower than $3 cost can sell some of its pollution allowance to the other polluter who may be unable to meet the $3 abatement cost. The level of pollution remains 5 units, the clean producer has an incentive to invest to become cleaner yet so it may sell more of its permits, and the other producer may also explore new technologies if such investments cost less than the cost of permits they negotiate with a cleaner producer.

On the surface, this solution appears to create the same solution as in the diagram. However, under cap-and-trade, a polluter has an incentive to discover a cheaper way to abate their pollution. A polluter facing command-and-control quotas has little incentive to clean up further if it were granted an emission allowance.

Such a market approach can also be modified over time to reduce the pollution emission allocations. This can be done by regulation or by the purchase of allocations, as is done with ITQs. Policy makers can use past auction revenue to repurchase emission rights as conditions change, with the central goal of an optimal level of pollution attained through market mechanisms that does not exceed the Earth's carrying capacity. While permitting pollution appears on the surface to not be in the public interest, recall that the public policy goal is to set the efficient level of pollution abatement. In this market solution, the industry itself may even be self-enforcing because innovators have an incentive to monitor the pollution of others to be sure innovators receive the highest price for these pollution credits.

WATER POLLUTION

Societies became acutely aware of the costs of air pollution not long after the urbanization that followed coal-fueled industrialization. While urban areas first began to take note of links between air pollution and health problems in the nineteenth century, the cost of waterborne diseases, or damage to water supplies was more difficult to detect. Rachel Carson's Silent Spring and the Cuyahoga River fire of 1969, displayed prominently on the cover of Time magazine, all helped galvanize public attention of threats to the environment. Increased social awareness culminated in the first Earth Day on April of 1970. This growing awareness fomented a movement and was an impetus for the formation of the *Environmental Protection Agency* in January of 1970.

Water is essential for organic life. It manifests in the atmosphere as water vapor and in rain, on the surface as groundwater, lakes, rivers, and streams, in subsurface aquifers, and in the ocean. Its presence in habitats affects the ecosystem in various ways. Human-made disturbances to the *water cycle* and the ecosystem may arise from point sources. These include the acid rain caused by coal-fired plants and infiltrating lakes and streams, or can be broadly distributed from sources such as poor handling of human waste, drainage of storm water that scrubs roads and fields of metals pollution or fertilizers, or plastics found in landfills. The disposal of solid waste that only slowly disintegrate into microplastics and are digested by humans and especially water-based animals can cause profound health consequences up the food chain over time.

Navigable waters act as a primary transportation network and are subject to statutes that prevent illegal flushing of bunker oil or contaminants from holds of ships, or even from maritime disasters that can release millions of gallons of crude or refined oil. As the price of oil increases with demand and with the exhaustion of onshore crude oil reservoirs, a greater reliance on expensive and riskier offshore oil drilling results imposes *operational risks* for calamities described in Chapter 10, such as British Petroleum's *Deepwater Horizon Macondo Spill* in the Gulf of Mexico in 2010.

As with air pollution, Nature can tolerate and process a limited quantity of organic pollutants that flow into its rivers, lakes, and oceans, up to a threshold. These *flow pollutants* become problematic when the scale of pollution exceeds Nature's carrying capacity. Nature can even compound the problem. Some pollutants become more toxic, such as mercury emissions that make their way through the sea life food chain and eventually concentrate in such marine mammals as albacore tuna that can average concentrations as high as one part per million of this toxic heavy metal.

Unlike air pollution, which is often visible and easier to detect, water pollution is often hidden and ubiquitous. Almost every part of the world is regularly cleansed with rainwater that ultimately makes its way from the surface or groundwater into streams, the ocean, and our reservoirs and irrigation canals, or into underground aquifers and wells. Precipitation can carry with it contaminants from the atmosphere, land, rivers, and lakes into the oceans. Concentration of airborne and surface contaminants can well exceed the ecosystem's ability to cleanse itself and may even concentrate to chronic levels that threaten the ecosystem in ways more profound than air pollution and concentrated solid waste.

Monitoring and control of degradation through the water cycle is difficult given the myriad and distributed sources and types of effluents and their difficulty to easily observe. Comprehensive sampling and testing regimes can help trace sources, but at considerable cost and sometimes inconsistent conclusions. Regulatory attention in the United States occurred as early as the 1899 Refuse Act that was concerned primarily about navigable waters. A series of acts in the 1970s and 1980s tackled industrial discharges with increasing effectiveness as they mandated the use of *best available technologies* that are *economically achievable (BAT)*.

Humankind's biggest polluters are often urban areas which, before the wave of U.S. *Environmental Protection Agency (EPA)* activity in the early 1970s, often discharged sewage directly into rivers and the ocean. Billions of dollars were invested to subsidize the construction of urban sewage treatment plants in urban areas that could rely upon centralized sewage systems. This approach that specified a best available technology was replaced in the 1990s with a **total maximum daily load** approach not unlike the regulatory regime that limit air pollution discharges.

Regulators have imposed emission standards for various solid pollutants such as phosphate and phosphorous, heavy metals, biological contaminants, and nitrogen-rich materials. These standards were effective when pollutants could be easily monitored. In the United States, regulatory authorities also encouraged **citizen civil suits** that permit whistleblowers and concerned citizens to detect and sue illegal dischargers. Knowing that civil suits have uncertain outcomes, even to meet the *more-likely-than-not preponderance-of-the-evidence* court standard, and that attorneys may work on a contingency as a share of the judgment if successful, legislation allowed citizens who prevail to collect *treble damages*. Successful and highly publicized civil suits against *Pacific Gas and Electric (PG&E)* in California and *WR Grace* in Massachusetts created a cottage industry for concerned citizens motivated to clean up their own backyards.

The regulatory analysis for the optimal deterrence of such pollution is a little different from air pollution. Low effluent thresholds are often appropriate, given the persistence and potential concentration of damage within waterways of much lower volume than the atmosphere, and hence less opportunity for dilution. Regulatory vigilance is more difficult given challenges of detection. Finally, since water pollution tends to eventually drain into the ocean, solutions to water pollution often require global treaty negotiations, a subject that will be treated in a later section.

AGRICULTURAL PRACTICES

Modern agriculture is highly efficient. In LDCs, many people live close to the land and are primarily employed in agriculture. MDCs have moved from economies in which fully three quarters of its citizens were employed to feed their nations to less than two percent. Modern farms are outdoor factories in which fertile land is no longer the predominant input it was in David Ricardo's days. In addition, efficient transportation means that land need not be collocated with the population that depends on it.

Instead, land is increasingly becoming but one agricultural factor. Agriculture takes advantage of good insolation, a ready supply of water either through rain or irrigation from human-made reservoirs, aquifers, or water control projects, high technology farm machinery that are increasingly autonomously operated, and extensive use of fertilizer and **genetically modified organism (GMO)** seeds that are highly optimized and prolific creators of food biomass. Indoor farming in which crops are stacked vertically in multi-level buildings and nourished hydroponically and with the optimal mix of nutrients further intensifies land use and increases its productivity (Fig. 18.3).

Rachel Carson described the dangers of uninformed and irresponsible use of pesticides, with their attendant effects on surrounding organisms and those

Fig. 18.3 An example of indoor vertical farming (Courtesy Mos.ru, CC BY 4.0 https://creativecommons.org/licenses/by/4.0, via Wikimedia Commons)

that feed upon them. MDCs have developed somewhat effective regulatory regimes following the awareness that developed from Carson's exposé. However, the damage of fertilizers, pesticides, and modified seeds, once considered miracle enhancements to productivity, has necessitated large investments in agricultural oversight and legislation to limit the damage poor practices can inflict, even as we all depend on the bounty agriculture provides.

Genetically modified seeds are another example of the perceived triumph of technology. These seeds often quickly outcompete more traditional seeds in the marketplace and in fields, and provide for greater crop yields, but also often impose a greater need for artificial nutrients. They are typically sold on a proprietary basis, with a farmer no longer able to save some of its crop to act as seed in the next year. Instead, the right to use these seeds must be repurchased annually.

The economics of the use of such engineered crops are compelling, but their effects on the ecosystem balance are potentially profound. Nature functions well from the advantages of diversity that creates natural resiliency as all metaphorical eggs are not placed in one basket. The complex web functions in a sustainable balance that can be disrupted by single crop engineered farming. In addition, the pressures of the marketplace that demands increasingly productive crops gives rise to unintended consequences that may not become evident for years or decades to come.

Two essential nutrients that are accelerating in demand are fertilizers and water. Fertilizers are rarely recycled effectively. The energy and fixed natural resources necessary for fertilizer production requires minerals such as phosphorous to be mined. Such extracted elements eventually disperse back into the environment in ways difficult to recycle, and create a diminished stock of natural capital. Surplus food produced and disposed is rarely properly composted to recover and recycle its nutrients. Instead, waste is often diverted to landfills.

Water shortages are becoming increasingly common. The stock effects of global warming beyond Nature's tipping point has changed patterns of rainfall and drought. Modern farming intensity also increases water demand. This demand is large, with needs tallied in acre-feet. This is a measure that quickly allows a farm to calculate its water demand based on the number of acres irrigated and the total number of feet of water used each year for various crops.

This demand for crop irrigation, acre-feet at a time, has created water shortages around the world. The once-mighty Colorado River now tapers to a trickle and steals water from the most downstream users in Mexico as the price of the conversion of upstream deserts into agricultural oases. Various U.S. states have formulated water treaties that are now outdated as the increasing pace of demand for farming and urbanization has outstripped declining supply caused by drought and global warming. The *Hoover Dam*, built in the 1930s to protect and partition the Colorado River downstream supply, holds back the Mead Reservoir. Designed for a capacity of almost 30 million acre-feet of water, the reservoir had fallen to a quarter of its capacity by 2022.[2]

The intensive use of land for crops that are subsequently consumed by humans or harvested to produce fuels such as ethanol contributes to the accumulation of carbon dioxide into the atmosphere. In this sense, by converting forest land to farmland, the carbon dioxide cycle moves from centuries-long sequestration to constant reinjection of annual carbon dioxide removal through photosynthesis. While many of the chronic effects of modern agriculture can be considered flow discharges that can be remedied within a generation or generations, the conversion of more land to support a larger population is inducing global warming effects that most closely resemble stock pollutants, which are more fully described the next chapter.

SOLID WASTE AND RECYCLING

For the span of organic life on the Earth, species have lived and died, and have consumed and excreted organic material, all within Nature's balance and ability to absorb and process waste. Only since the transition away from nomadic life to settled agriculturally based economies has humankind and Nature been challenged with the need to deal with solid waste.

Humans now create more than two billion metric tons of municipal solid waste alone every year, with up to a third inadequately treated.[3] A small but increasing share of this waste is recycled. However, only in the last few decades has it become commonly understood that the exponentially growing scale of solid waste arising from human activity and prosperity now well exceeds any ability of Nature to process it. The rapid acceleration of our accumulation of solid waste exceeds the much more modest rate of natural decay of waste.

The most technologically advanced effluent treatment plants can be highly effective in the removal of biologically dangerous materials. Technologies even allow some communities to rely on reprocessed wastewater as a source of safe drinking water. In addition, a state-of-the-art sewage treatment plant can remove and recycle industrial compounds and elements such as phosphorous and heavy metals, and increasingly concentrate and sell the methane that occurs as biological materials are broken down within the plant. The remaining sludge is often nitrogen-rich and can be recycled as a natural fertilizer.

Such technologically sophisticated plants are expensive and hence are concentrated in high income jurisdictions. This economic reality gives rise to issues of environmental justice in which wealthier communities are afforded better protections and more efficient recycling than poorer communities and countries. The poorest communities and LDCs tend to be more rural and hence do not concentrate and treat human waste as do large urban areas. While LDCs may also have many of the world's most rapidly growing urban sectors and tend to produce the most unprocessed waste, the world's wealthiest nations are able to more effectively reprocess and recycle waste.

One of the most problematic and dangerous of all industrial waste arises from the process of nuclear fission. Fission from the collision of neutrons, atoms, and ions with heavy radioactive elements produces copious amounts of heat and a number of additional radioactive elements. Fission represents the only process Nature and humankind employ that can produce new elements. However, humans have developed few uses for these radioactive byproducts, some radioactive medical technologies and the use of spent uranium for very high strength military armor aside.

Until Generation IV fission reactors are perfected that can reprocess these potentially dangerous actinides that otherwise have half-lives measured in upwards of tens of thousands of years, these materials remain concentrated at operating and shuttered nuclear power plants across the United States and the world. Such fixed stock resources must be recycled someday both to preserve resources for future generations while they keep safe the ecosystem that can be harmed by radiation.

OPPORTUNITIES FROM IMPROVED FLOW RESOURCE POLICIES

The overarching philosophy in active management of flow resources is the need to internalize the intragenerational externalities they create. Recall that intergenerational dynamic efficiency required as a precedent that each generation must use their resources without waste or uninternalized externalities.

Such intragenerational externalities can be internalized through a variety of mechanisms describe above.

The Coase Theorem demonstrated that such internalization of externalities as pollution can be remedied through private contracting so long as resource ownership is well-defined. In practice, the Coase Theorem is difficult to administer, given asymmetries between affected parties. Under such a privatized regime, the value of these flow resources may be concentrated privately as well, unless the property rights are effectively auctioned off from the public trust.

Instead, public policy makers and regulators must typically play an essential role in internalizing flow pollutant externalities and collecting the resource rents that accrue. To the extent that these rents are renewed each year, the flow should be replenished within the generation and the wealth concentrated and employed to remedy the social costs across generations as well. Such fees should be aggregated into permanent funds to the extent that various pollutants may persist. In practice, pollution and effluent fees are often diverted to general revenue funds of national governments to offset taxes and stimulate fiscal consumption. Such diversion of wealth to support intragenerational income and consumption violates the provisions of intergenerational wealth optimization through the maintenance of a permanent fund in proportion to the stock of natural capital.

Recycling can also be used to greater effect. Humankind has the technology to almost completely recycle all emissions, be they airborne, waterborne, or solid waste. The two barriers to almost complete reuse of precious resources through recycling are collection and processing costs, and the energy burden of effective resource recovery and carbon sequestration; Public infrastructure for collection and processing is in place in some wealthier jurisdictions; We can collect human and industrial waste and recover critical natural capital; We know how to purify and reuse water, and we can remove effluents before they enter the air and carbon dioxide that can then be reinjected into the ground or converted to useful compounds. We must still overcome energy needs to make these methods economical. Ideally, we should impose these costs upfront in the purchases of goods, services, and factors of production to internalize these externalities.

We also know how to manage renewable resources, and what we must set aside as we extract and recover depletable resources. The greatest remaining challenges then rest with abundant, sustainable, and affordable energy for processing and recycling, and in dealing with the most significant threat to intergenerational sustainability—the degradation of our atmospheric capital. The balance of the book describes first this latter challenge and then what the public sector and corporations can do about it.

Summary

Policies that manage and limit the erosion of natural capital depend critically on whether the degradation of natural capital is temporary or permanent. The next chapter advances a perspective that helps delineate the present from the future. We can typically categorize natural resources and their degradation based on the persistence of the effects, the ability of nature to reestablish a resource, and the power of the Earth to cleanse itself of pollution and other harms. For short-term consequences, we can employ active management and internalization of externalities confined to a period of time or perhaps a generation. But, when our ability to harm the natural equilibrium is sufficiently profound that we must treat a natural resource as a stock rather than a flow, we must instead employ intergenerational tools rather than their shorter term intragenerational counterparts.

We must then deem some forms of degradation of natural capital to be effectively permanent if damages persist for many generations, centuries, or millennia. This necessarily raises issues of economic justice, not solely on behalf of those in the present, but of our regard for those who come after us and must face the cost of decisions we make today. In the next chapter, such intergenerational effects of the erosion of our stock of natural capital are explored.

ESG Toolkit

Are there resources a corporation uses for which they access through a Tragedy of the Commons?

Can the corporation offset this enrichment through other socially desired practices?

What are your intuition's policies and practices with respect to flow pollutants? Do you have an inventory of stock and flow pollutants in your organization and its supply chain?

Notes

1. Christy, F. T. (1973). Fisherman Quotas: A Tentative Suggestion for Domestic Management. *Occasional Paper no. 19*. Law of the Sea Institute, University of Rhode Island, Kingston, Rhode Island, USA.
2. https://en.wikipedia.org/wiki/Lake_Mead, accessed November 8, 2022.
3. https://datatopics.worldbank.org/what-a-waste/trends_in_solid_waste_manage ment.html#:~:text=The%20world%20generates%202.01%20billion,from%200. 11%20to%204.54%20kilograms, accessed November 8, 2022.

CHAPTER 19

An Intergenerational Stock Externality—Global Warming

The economic tools available for sustainability management can be adapted to differentiate between depletable resources for which the stock cannot be replenished, and renewable resources that, if managed well, can last indefinitely. First, externalities are internalized within each generation and resource rents have been collected by auction and reinvested into a permanent fund to benefit generations in perpetuity. Then, policy makers must ensure that non-market natural capital is likewise managed according to intergenerational efficiency.

Management of fixed resources in limited supply must contemplate the eventual transition to backstop technologies that replace them. Effective recycling may become economical if abundant and inexpensive sustainable energy is available. These sustainable returns can then benefit every generation equally. If we can successfully navigate these intergenerational challenges, there remains one existential problem that will continue to vex humankind for centuries. We know that a renewable resource can be abused to the point that recovery is difficult or impossible. Even renewable resources can become depletable, especially if the resource is prone to the tragedy of the commons arising from the absence of property rights.

Oxford References defines a depletable resource as[1]:

A resource the stock of which decreases whenever the resource is being used and **does not increase over the timescale relevant for economic decision-making**. Examples include deposits of coal, oil, or minerals. The adjustment speed of depletable resources is so slow that they can be modelled as made available once and only once by nature. (Emphasis added)

C. Read, *Understanding Sustainability Principles and ESG Policies*, https://doi.org/10.1007/978-3-031-34483-1_19

Delineation between a renewable and depletable resource, or a flow and stock pollutant, then depends on the effective planning horizon of economic decision-makers. If we find that the recovery time from the abuse of a natural resource is so far beyond the traditional scale of human decision-making, it is incumbent on us to treat such a resource as depletable.

Let us begin by pondering how mortals may delineate the present from the future. We each have an individualized regard for the future, many of us even for future generations beyond our individual existence on Earth. Let us define the future as beyond some threshold T in time beyond which our valuation is equally relevant to us as are all the years up to that point.

I begin by labeling the more immediate period of time the present and the latter the future. For example, depending on my individual discount rate, I may regard the next twenty years as of equal significance to me as the eternity beyond that threshold. We can then calculate the threshold T that separates the present from the future.

Such a threshold of equal regard for the present and the future depends on one's personal discount rate and various demographic, cultural, and other factors such as our age, faith, and expected longevity. Consider a constant value v that accrues evenly each year and a discount rate r. From these parameters, we can calculate a threshold T. We do so by comparing the sum of discounted values to that point in time T and the sum of discounted value beyond T. The value $V(T)$ of the present to time T is given as:

$$V = \int_0^T ve^{-rt}dt = v\frac{\left(1 - e^{-rT}\right)}{r},$$

while the future value, beyond time T to eternity, discounted to the present, is given as:

$$V = e^{-rT}\int_T^\infty ve^{-rt}dt = \frac{ve^{-rT}}{r}.$$

Equating these two expressions, we can solve to show that $rT = \ln(2)$, or $T = \ln(2)/r$. Graphing the relationship between our discount rate and the threshold that separates the present and the future, we see (Fig. 19.1).

Even if one disregards the potential irrelevance of all events beyond one's life on Earth, a mortal human with an exceptionally low discount rate of only 3% would find a flow of value up to 23 year threshold equivalent to the same flow from 23 years to eternity. This 3% discount rate is far lower than the weighted average cost of capital of most humans, corporations, and nations, especially the LDCs. Even at such a low discount rate, one would attribute now and the near future of perhaps one generation as equally significant as all future generations. John Maurice Clark, the economist son of the prominent turn of the twentieth-century economist John Bates Clark, once remarked that he believed humans were ill-adapted to regard a future beyond a decade or so. A one decade mortal planning horizon equates to an implicit

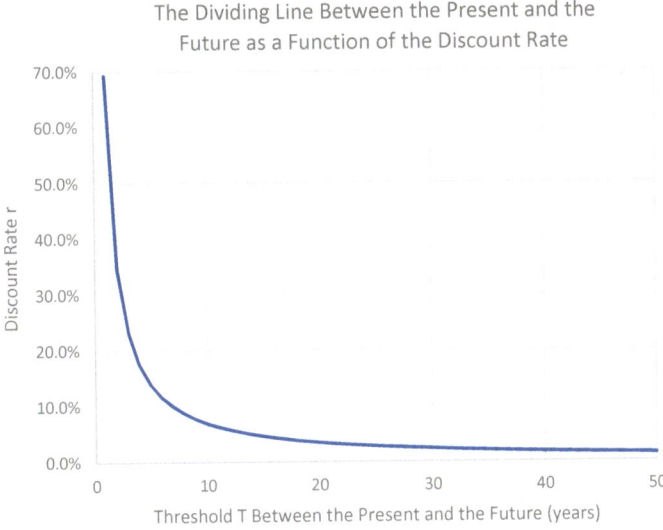

Fig. 19.1 The threshold between the present and the future based on personal discount rates

discount rate r of $\ln(2)/10 = 6.9\%$, which most economists would consider an underestimate of the discount rate of most individuals.

GLOBAL WARMING

Those who participate in democracies, govern our corporations, and lead our nations have the power to influence decisions that must balance our preferences today with repercussions that may persist for generations. In the case of persistent global warming emissions such as carbon dioxide, consequences can span centuries and millennia. These durations are well beyond the timescale relevant for economic decision-making and place the erosion of our atmospheric capital arising from global warming in the category of an intergenerational depletion of an essential asset in nature's bank. Not only has humankind failed to impose the proper fees to internalize global warming externalities intragenerationally, but we have in turn created perhaps the greatest intergenerational challenge in this Anthropocene Epoch. This challenge to sustainability arises from our failure to attend to the stock externalities that arise as the natural capital of our atmosphere is depleted or so degraded that its function is diminished for centuries or millennia.

The existential resource question is whether anthropogenic actions since the onset of the Industrial Revolution have squandered the effectiveness of an atmosphere that has evolved over a period spanning hundreds of millennia. All species and the entire ecosystem depend on our atmospheric natural capital. Given our shared use of the atmosphere and its inability to repair itself at a rate sufficient to overcome recent human environmental degradation, our atmospheric natural resource should be considered to have characteristics of a stock resource for the foreseeable future. To see that, let us delve into the nature of our atmosphere and depreciation over the Anthropocene Epoch.

TIPPING POINTS

The reason why the atmosphere shares characteristics more alike a fixed rather than a renewable natural resource is its inherent complexity and its propensity to be pushed past tipping points that induce millennial-scale changes in functions within our ecosystem. The complexities and non-linearities arising from such tipping points can result in exceedingly long-term implications, and occur as a result of decidedly short-term or myopic decisions, mostly over a handful of decades.

A theoretical model of an ecosystem component pushed beyond its crash threshold was described in Chapter 2. The atmosphere can be considered such a resource, subject to a crash or a sudden non-linear response to linear changes in its composition. It faces a tipping point beyond which even minimal human abuse will not prevent further depletion of the resource.

In a 2022 study published in the journal Science entitled *Exceeding 1.5 °C global warming could trigger multiple climate tipping points*.[2] McKay et al. described ways for which a global temperature increase above the accepted maximum threshold of 1.5° Celsius can induce positive feedback effects that increase global temperatures still further, even should anthropogenic emissions be curtailed.

The authors identified sixteen tipping points. These thresholds represent Nature's responses to global warming in key geosystems across the planet. The five most critical feedback processes that have already been triggered at the current level of global warming include (Fig. 19.2):

- The Greenland ice sheet
- The West Antarctic ice sheet
- Abrupt permafrost thaw
- Collapse of natural water flow by convection in the Labrador Sea and the Atlantic Ocean
- Massive coral reef die-offs in the tropics.

The *Intergovernmental Panel on Climate Change* (IPCC) has also emphasized that climate tipping point feedback loops become significantly more likely should global warming temperature increases approach the 2° Celsius threshold. These risks become very high when temperatures rise in the range of 2.5 to 4° Celsius. They note the following phenomena (Fig. 19.3).

The Inter-Governmental Panel on Climate Change (IPCC) subsequently reported in 2022 that there is only a 50% chance that the global temperature

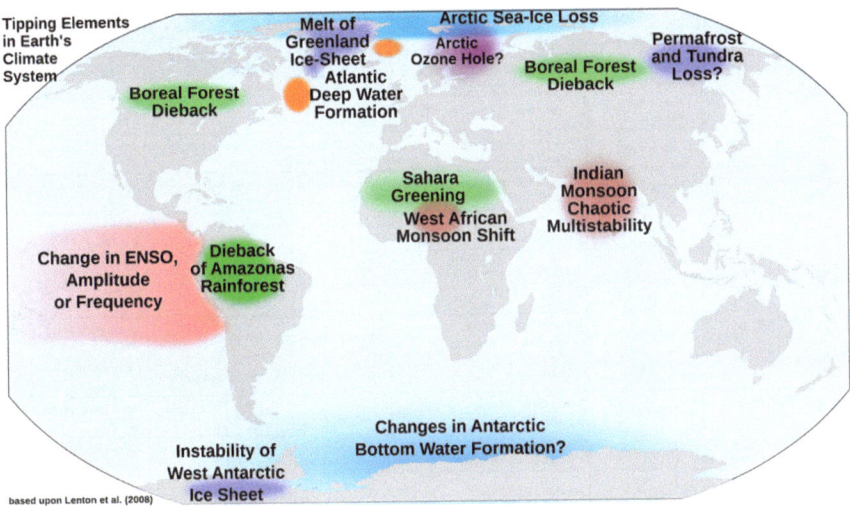

Fig. 19.2 Regions of various global atmospheric tipping points (Courtesy of—CodeOne [blank map], DeWikiMan [additional elements], CC BY-SA 4.0. https://creativecommons.org/licenses/by-sa/4.0, via Wikimedia Commons)

increase can be limited to the lower 1.5° Celsius threshold, even if greenhouse gas emissions could be decreased by 50% by the year 2030, followed by achievement of net-zero carbon emissions by 2050.

Timothy Lenton, a co-author of the 2022 *Science* article, and director of the Global Systems Institute at the University of Exeter wrote:

> Since I first assessed climate tipping points in 2008 the list has grown and our assessment of the risk they pose has increased dramatically. Our new work provides compelling evidence that the world must radically accelerate decarbonising the economy to limit the risk of crossing climate tipping points…To achieve that we now need to trigger positive social tipping points that accelerate the transformation to a clean energy future. We may also have to adapt to cope with climate tipping points that we fail to avoid, and support those who could suffer uninsurable losses and damages.

The bases of these conclusions that challenge sustainability and places the status of our atmosphere and oceans in the category of a natural capital stock rather than a flow resource are a series of scientific analyses and modeling show that, once a tipping point is crossed, changes persist over periods ranging from generations to millennia. These transitions into unstable atmospheric regimes can occur rapidly and result in sea levels rises that can exceed three or more meters over the next couple of centuries.

The ecosystem effects of these tipping elements can be organized into a number of geo- and bio-ecosystems, including:

Earth System Component/Tipping Element	Potential Abrupt Climate Change?	Irreversibility if Forcing Reversed (Time Scales Indicated)	Projected 21st Century Change Under Continued Warming	Change in Assessment
Global Monsoon (4.5.1.5; 8.6)	Yes, under AMOC collapse, medium confidence	Reversible within years to decades, medium confidence	Medium confidence in global monsoon increase; medium confidence in Asian-African strengthening and North American weakening	More lines of evidence than AR5
Tropical Forest (5.4.8; 8.6.2)	Yes, low confidence	Irreversible for multi-decades, medium confidence	Medium confidence of increasing vegetation carbon storage depending on human disturbance	More confident rates than AR5
Boreal Forest (5.4.8)	Yes, low confidence	Irreversible for multi-decades, medium confidence	Medium confidence in offsetting lower latitude dieback and poleward extension depending on human disturbance	More confident rates than AR5
Permafrost Carbon (5.4.8)	Yes, high confidence	Irreversible for centuries, high confidence	Virtually certain decline in frozen carbon; low confidence in net carbon change	More confident rates than SROCC
Arctic Summer Sea Ice (4.3.2; 4.6.2.1; 9.3.1)	No, high confidence	Reversible within years to decades, high confidence	Likely complete loss	More specificity than SROCC
Arctic Winter Sea Ice (4.3.2; 9.3.1)	Yes, high confidence	Reversible within years to decades, high confidence	High confidence in moderate winter declines	More specificity than SROCC
Antarctic Sea Ice (9.3.2)	Yes, low confidence	Unknown, low confidence	Low confidence in moderate winter and summer declines	Improved CMIP6 simulation
Greenland Ice Sheet (9.4.1)	No, high confidence	Irreversible for millennia, high confidence	Virtually certain mass loss under all scenarios	More lines of evidence than SROCC
West Antarctic Ice Sheet and Shelves (9.4.2; Box 9.4)	Yes, high confidence	Irreversible for decades to millennia, high confidence	Likely mass loss under all scenarios; deep uncertainty in projections for above 3°C	Added deep uncertainty at GWL >3°C
Global Ocean Heat Content (4.5.2.1; 4.6.2.1; 9.2.2; CCBox 7.1)	No, high confidence	Irreversible for centuries, very high confidence	Very high confidence oceans will continue to warm	Better consistency with ECS/TCR
Global Sea-Level Rise (4.6.2.1; 4.6.3.2; 9.6.3.5; Box 9.4)	Yes, high confidence	Irreversible for centuries, very high confidence	Very high confidence in continued rise; deep uncertainty in projections above 3°C	Added deep uncertainty at GWL >3°C
AMOC (4.6.3.2; 8.6.1; 9.2.3.1)	Yes, medium confidence	Reversible within centuries, high confidence	Very likely decline; medium confidence of no collapse	More lines of evidence than SROCC
Southern MOC (9.2.3.2)	Yes, medium confidence	Reversible within decades to centuries, low confidence	Medium confidence in decrease in strength	More lines of evidence than SROCC
Ocean Acidification (4.3.2.5; 5.4.2; 5.4.4)	Yes, high confidence	Reversible at surface; irreversible for centuries to millennia at depth, very high confidence	Virtually certain to continue with increasing CO_2; likely polar aragonite undersaturation	More lines of evidence than SROCC
Ocean Deoxygenation (5.3.3.2)	Yes, high confidence	Reversible at surface; irreversible for centuries to millennia at depth, medium confidence	Medium confidence in deoxygenation rates and increased hypoxia	Improved CMIP6 simulation

Fig. 19.3 Various climate tipping points (Courtesy the Intergovernmental Panel on Climate Change, AR6 WGI)

- The continent of Antarctica
- The Amazon Rainforest
- Western Africa monsoon patterns
- Equatorial coral reefs
- Labrador Sea convections
- East Africa subglacial basins
- Boreal Forests.

While these systems have been destabilized themselves, some of these systems are interrelated so that the collapse of stability in one system can trigger additional instability in another. Two additional driving forces have been deemed to require further research have also been identified:

- Arctic sea ice
- The El Nino Southern Oscillation.

The pathway toward these two other tipping points is nonetheless relevant. Arctic sea ice decline is believed to accelerate global warming because a vast white ice-covered polar region reflects significantly more energy back into space than a deep blue Arctic Sea that absorbs light and heat. This absorption of heat energy translates into higher Arctic Ocean temperatures. We have seen in this century an acceleration of a feedback loop that delays Arctic Ocean freezing into late fall and accelerates ice breakup in early spring. The declining scope and duration of reflective sea ice then further accelerates this process, labeled the *albedo effect*.

The albedo effect arising as dark blue ocean replaces reflective ice covers also occurs over snow- and glacier-covered land masses. Traditionally ice and snow covered areas are concentrated at the poles, which results in significantly greater warming from sea ice loss at the poles as global warming progresses. These areas have sequestered decayed masses of organic material in frozen peat bogs or covered by permafrost. With rising temperatures, the Arctic has begun to release methane as decay is accelerated. This effect further enhances global warming.

In addition, just as changing ocean parameters result in shifts in ocean currents, arctic temperatures that rise three to four times faster than average global increases elsewhere induce changes in temperature and pressure differentials which drive jet streams that determine prevailing weather patterns. This increased tendency for the Arctic Ocean to warm has led researchers to estimate that the Arctic Ocean will be completely ice-free for some period every summer by the year 2050. An ice-free Arctic Ocean has not occurred for more than 6000 years.[3]

This warming also induces changes in oceanic currents, much as changes in atmospheric temperatures induce pressure changes that influence global wind patterns. Shifting ocean current patterns have a profound effect. Should prevailing currents shift, decline, or reverse, the shuttling of atmospheric and ocean energy is altered, with destabilizing effects on climate and global temperatures.

Depletion of the Ozone Layer and the First Steps Toward Global Recognition of Climate Impacts

One of the features of the atmosphere that keeps the Earth habitable is a layer of ionized oxygen molecules called ozone, of the chemical form O_3^+. While its typical concentration in the upper atmosphere is only about ten parts per million within a layer in the stratosphere of an altitude ranging from 15 to 35 kilometers, this *ozone layer* is able to effectively and almost completely block light with a wavelength between 200 and 355 nanometers. This so-called UV-C range of the light spectrum is beyond the ability of our eyes to see, in the ultraviolet region just short of the X-ray range (Fig. 19.4).

Higher wavelength (lower frequency) light exhibiting blue, yellow, and red colors are not absorbed to the same degree in our upper atmosphere as is

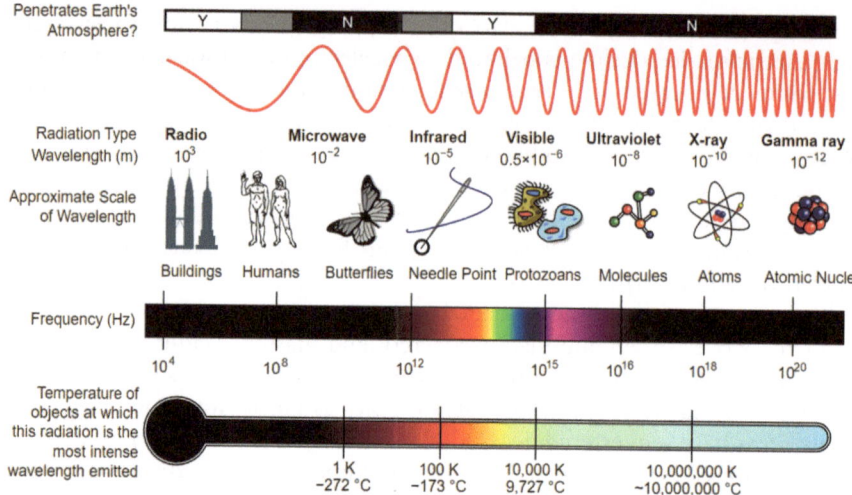

Fig. 19.4 The electromagnetic spectrum (Courtesy NASA—http://mynasadata.larc. nasa.gov/images/EM_Spectrum3-new.jpg)

higher energy ultraviolet light. These lower frequencies scatter to different degrees when light rays glance off molecules in the atmosphere. Blue light is scattered most, and hence fills the entire daytime sky, scattered from all directions. Red light is scattered little and is only noticed when we can observe the sun near dawn and dusk when scattered blue light dims. We also sense the heat of infrared light when the sun is most directly upon us.

Atmospheric ozone is uniquely capable of filtering out the very high energy ultraviolet light. High energy UV-C rays are used to sterilize medical equipment and etch circuits onto silicon wafers to produce the microprocessors that run our electronic world. In the absence of protection from such high energy light, the constant bombardment of near-X-ray UV-C light would destroy much of life on land.

In 1913, French physicists Charles Fabry and Henri Buisson discovered the unique ability of our atmosphere to filter out these dangerous ultraviolet rays. A British meteorologist named G. M. B. Dobson was motivated by their discovery to design a simple ***Dobsonmeter***. Dobson organized the establishment of a worldwide network of these sensors to monitor the effectiveness of the ozone layer to absorb dangerous UV light that protects biological life on terrestrial Earth.

Scientists began to observe through this network that the ozone layer was becoming less effective in filtering harmful ultraviolet light over the course of the twentieth century. In the 1960s and 1970s, scientists verified that this change was a consequence of ozone reaction with atmospheric ***chlorofluoro-carbons (CFCs)*** that were commonly used as a refrigerant in air conditioners and heat pumps and as a propellent in aerosol sprays. CFCs and their bromide counterparts, *bromofluorocarbons*, are chemically stable unless they are broken down into chlorine and bromine ions upon exposure to UV light. Each such radical can neutralize the UV absorption of more than 100,000 ozone molecules.

Constant reactions between molecular oxygen and high energy ultraviolet light splits ozone into oxygen molecules and atoms that soon reform into ozone ions O_3^+. In the process, the UV-C energy is absorbed and emitted in the form of much more benign infrared heat with each such ozone decay and reformation.

This constant shuttling reaction between oxygen and ozone constitutes a layer of protection which, if concentrated at sea level, would be only three millimeters thick.[4] Yet, even a modest ozone concentration has a profound effect of filtering out almost 98% of damaging UV rays that can kill organisms. Even so, the low rate of UV-C light that reaches the surface of the Earth kills thousands of people per year through skin cancer.

The greatest incidence of insolation containing potentially harmful ultraviolet light is at the equator. While this ozone is produced primarily in the tropics, prevailing stratospheric winds disperse ozone-depleting contaminants toward the poles. Ozone at the poles drops into the troposphere that contains our weather. The concentration of CFCs in the polar troposphere then induces dramatic ozone layer depletion in these polar extremes. Such *ozone holes* were the first telltale indicators of ozone depletion arising from CFC releases into the atmosphere.

Global awareness of this breakdown of ozone spurred groups such as the U.S. Environmental Protection Agency under the auspices of the Clean Air Act to include CFCs and BFCs in the act the *National Ambient Air Quality Standards*. The United States, Canada, and Norway, three nations with significant polar geographies, joined to ban CFCs in 1978. These nations issued a statement that evolved into the *Montreal Protocol*. The Protocol pledged as a first step to cap CFC production to 1986 levels and then, with subsequent negotiations of the protocol, ban CFC production altogether. The Montreal Protocol signed in Montreal, Canada on September 16, 1987 agreed to ban CFC production by January 1, 1989 if ratified by twenty nations. Eventually, all United Nations members and associates, totaling 198 states, agreed to a global ban, effective January 1, 1996. The United Nations also declared September 16 as the *International Day for the Preservation of the Ozone Layer*.

By 2003, the *American Geophysical Union* confirmed the effectiveness of the ban. While CFCs persist for decades in the upper atmosphere, it is expected that Nature's effectiveness in removing dangerous UV rays will return the ozone layer to 1980 levels sometime between the years 2050 and 2070. This reestablishment of ozone layer effectiveness will come almost a century after scientists first took notice of the ecosystem danger arising from ozone depletion, and about a lifetime after the world banned the gases that led to destruction of the ozone layer. This time it takes to reestablish adequate ozone protection, measured in a human lifespan, well exceeds the threshold that delineates now and the future. The timescale relevant for economic decision-making that delineates a depletable from a renewable atmospheric resource results in the classification of the ozone layer as a depletable rather than a renewable natural resource.

The Montreal Protocol is also unusual and significant because,

- for the first time, members of the United Nations successfully recognized a science- and evidence-driven process to protect the atmosphere,
- it was able to invoke the influence of the United Nations to secure the first global environmental agreement,
- it managed to resolve conflicting interests and the need to subsidize a transition within *Less Developed Countries* (LDCs) through funding by *More Developed Countries* (MDCs),
- it proceeded incredibly rapidly in integrating scientific evidence and consensus into global action, and
- it recognized that wealthier nations have a financial responsibility to prevent excess burdens of such policies on LDCs under the *Polluter Pays Principle (PPP)*.

The Montreal Protocol was the first example of a globally ratified environmental treaty in United Nations history. The groundbreaking and precedent setting protocol demonstrated countries could engage in a science- and evidence-based formulation of public policy, instill global cooperation, and establish the polluter pays principle. These innovations have far-reaching implications on the potential for climate change policy-making at the international level. This success offers some optimism that the same scenario-driven process and urgency could likewise be turned toward an even more dire and expensive erosion of our natural capital.

Soon after the Montreal Protocol was signed, the United Nations directed its attention toward global warming and climate change with the first *Conference of Parties of the United Nations Climate Change Accord* in 1995. However, it has not been until the recent Sharm El-Sheikh, Egypt *Conference of Parties (COP) 27* convention to the United Nations Climate Change Conference held in November of 2022 that nations have realistically discussed the need to burden-share on behalf of LDCs in transition toward an anthropogenic zero carbon policies. Should this convention succeed where others have failed since the Montreal Protocol, a pathway may be created to once again combine science and economics, this time to create a unified response to global warming.

GLOBAL WARMING

The Earth's atmosphere and the ecosystem it maintains function in a delicate balance. Earth's earliest atmosphere was little different than the gases, primarily hydrogen, that constituted the early cosmos and still linger in the atmosphere of Jupiter today. Methane, ammonia, and other simple combinations of hydrogen present on such gas giant planets of sufficient gravity to hold lighter gases in their atmospheres could not harbor life as we know it. Meanwhile, the greenhouse gas effect of high carbon dioxide concentrations on Venus today induces in the blisteringly high temperatures approaching 500 degrees Celsius.

The Earth's atmosphere evolved in a different way compared to our planetary neighbors because of one very fortunate circumstance. The second era of the Earth's atmospheric evolution occurred as asteroids that struck the Earth brought new elements and water held in ice, while volcanoes erupted and released yet more elements and compounds into the atmosphere. To then, the Earth had been nitrogen and carbon dioxide rich, much like Venus remains today. Nitrogen is a relatively inert gas, while the abundance of carbon dioxide became dissolved in newly formed oceans. Some oxygen arrived too, and, as the Earth cooled, oxygen and carbon dioxide-rich water created the medium for the Earth's first life forms.

One of the earliest forms of life on Earth, the *cyanobacteria*, absorbed carbon dioxide and sunlight to create oxygen. It initiated oxygenation and a carbon cycle that is the basis for organic life today. This carbon-based life created through photosynthesis resulted in sufficient oxygen production to give rise to the *great oxygenation event*. This increased atmospheric oxygen transformed the Earth's caustic *reducing atmosphere* to an *oxidizing atmosphere*. With this shift came a plethora of new life forms, some derived from the first cyanobacteria that depended on carbon dioxide, while others developed and thrived from the newfound oxygen in the atmosphere.

The balance between organisms that consumed carbon dioxide and produced oxygen and those that consumed oxygen and released carbon dioxide depended upon the relative proportion of oxygen and carbon dioxide molecules in the atmosphere. The stock of air we now breath represents this precarious balance that favors oxygen breathing animals. The human species can survive only within a narrow range of atmospheric oxygen. Should its component in the air drop from 21% to below 19.5%, humankind could no longer thrive. A proportion of 23.5% is also excessive for human life. The range of oxygen and carbon dioxide in the atmosphere has favored different life forms over the Earth's eras and still dictates the proportion of life forms today.

The evolution of this oxygenation process took place over billions of years to reach the point that favors human life. Free oxygen did not build up and establish a new equilibrium between atmospheric gases until oxygen was produced at a sufficient rate to exceed the rate of oxidation of metals on the

Earth's surface in what is known as the *great rusting*. Once minerals had most completely reacted with oxygen, the amount of oxygen in the atmosphere began to rise, and reached a peak of about 30% of the atmosphere a quarter billion years ago.

Today animals such as humans that breathe oxygen and plants that *photo respirate* to absorb oxygen are balanced by photosynthesizing plants that create oxygen. This equilibrium established an atmospheric composition of about 78% nitrogen, 21% oxygen, and 0.9% argon, among other trace amounts of neon, helium, methane, and krypton, and a carbon dioxide concentration that in the past 40,000 years has ranged from about 180 to 280 parts per million (ppm). Today's atmospheric carbon dioxide load now exceeds 420 ppm. In addition, the atmosphere can contain upward of 4%, or 40,000 parts per million of water vapor.

The level of oxygen has fluctuated to a high of about 30% of the atmosphere about a quarter billion years ago, and plunged to a low of 15% during the **Precambrian Era** that ended half a billion years ago. The relative richness of carbon dioxide in the atmosphere has created cycles of prolific plant production and decay that has maintained an atmospheric carbon dioxide and oxygen equilibrium that has been consistent for hundreds of millennia.

The atmosphere around us is a fluid that interacts profoundly as a part of the carbon cycle. It is both an oxidizer of inorganic and organic materials, and the molecular fuel that dictates the balance between the plant and the animal world. The relative concentration of gases in the atmosphere, especially water vapor, carbon dioxide, and methane also act to trap the heat that the Earth tries to reradiate as it absorbs light from the Sun.

The constituents of the atmosphere affect the relative balance of life on Earth, but they also determine the delicate balance of the atmosphere's temperature and pressure. In turn, the temperature of the atmosphere has a profound effect on the amount of water that is held in vapor, liquid, and solid form. The variation of atmospheric temperatures and pressures, the amount of water vapor the atmosphere holds, and the shifting balance between water held as vapor, liquid, and ice determines the climate as we know it. These various components of the atmosphere collectively increase the average temperature at the surface of the Earth by 33° Celsius compared to the temperature if the Earth did not benefit from the safe harbor our atmosphere and ecosystems provide and enjoy.

THE ATMOSPHERIC SCIENCE OF CLIMATE CHANGE

Heat from an Earth's surface that has been warmed by solar energy is radiated back toward space at a rate proportional to the temperature at the surface. This so-called *blackbody radiated heat* is trapped near the Earth's surface by greenhouse gases that act as a blanket that prevents this infrared energy from reradiating into space.

While the average temperature of the land mass is 14° Celsius, the average ocean temperature is 20° Celsius. Because 71% of the Earth's surface is ocean, the combination of a large surface area and a higher ocean surface temperature contributes to upwards of 90% of the Earth's warmth. Oceans are the most significant drivers of atmospheric energy accumulation and of our climate and weather.

Oceans in turn respond to this heat. While melting ocean ice does not contribute to sea level rise just as melting ice cubes do not raise the water level in a glass of water, higher atmospheric temperatures induce melting of ice and snow covering land masses. This runoff of fresh water flows into oceans and causes the sea level to rise and its salt content to change. In addition, water above 4° Celsius has a positive *coefficient of expansion* which causes ocean volume to increase at an accelerating rate with increased warming. We shall see that increased atmospheric temperatures also increases its water vapor contents and lowers atmospheric pressure. This trend allows for further expansion of oceans. The combination of land-based ice-to-water runoff and ocean expansion results in a sea level that is sensitive to oceanic and atmospheric temperatures (Fig. 19.5).

Scientists have employed a number of techniques to determine past atmospheric temperatures and sea levels dating back millions of years. These methods rely on measurement and composition of ocean sediment from core samples, observations of gases in bubbles contained in glacial ice cores, and the width and composition of tree rings from long-lived and fossilized wood.

These various techniques combine to describe sea levels, greenhouse gas compositions, and temperatures in the atmosphere dating back tens of millions of years. Scientists have discovered that the Earth has gone through periods in

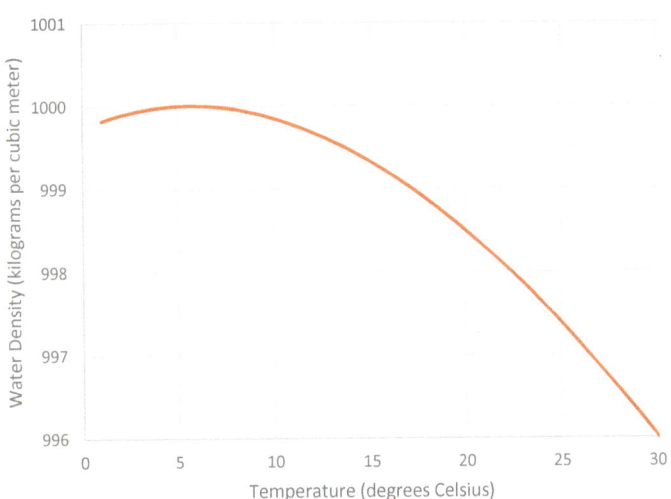

Fig. 19.5 Effect of ocean temperature on water density

which there was no permanent year-round ice to periods with significant ice covering much of the Northern Hemisphere. The cycling of natural processes has followed a consistent pattern spanning a hundred thousand years, with patterns in the runup and cooling down of the planet spanning tens and hundreds of millennia as natural processes are invoked (Fig. 19.6).

The most concerning aspect of the evolution and oscillations of this natural cycle of global surface temperatures is that its rate of change has accelerated in an unprecedented and alarming rate since humankind began burning fossil fuels to power the Industrial Revolution. Beyond pollution, the most obvious global effect of the Anthropocene Epoch has been increasing global surface temperatures (Fig. 19.7).

These changes in greenhouse gas levels, atmospheric temperatures, and sea levels are remarkably consistent not only in their repeated trends over time

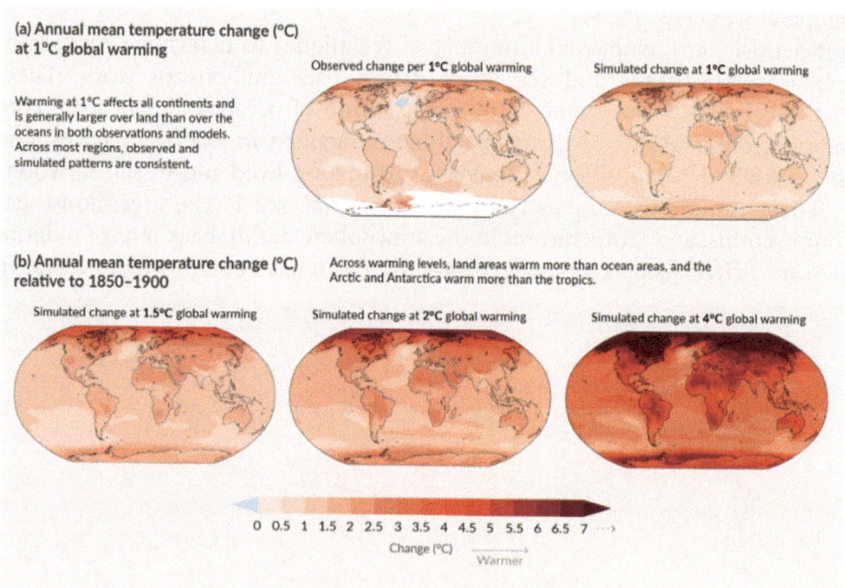

With every increment of global warming, changes get larger in regional mean temperature, precipitation and soil moisture

Fig. 19.6 Non-linear earth response to global warming as tipping points are exceeded (From Figure SPM.5 in IPCC, 2021: Summary for policymakers. In: *Climate Change 2021: The Physical Science Basis. Contribution of Working Group I to the Sixth Assessment Report of the Intergovernmental Panel on Climate Change* [Masson-Delmotte, V., P. Zhai, A. Pirani, S.L. Connors, C. Péan, S. Berger, N. Caud, Y. Chen, L. Goldfarb, M.I. Gomis, M. Huang, K. Leitzell, E. Lonnoy, J.B.R. Matthews, T.K. Maycock, T. Waterfield, O. Yelekçi, R. Yu, and B. Zhou (eds.)]. Cambridge University Press, Cambridge, UK and New York, NY, USA, pp. 3–32, https://doi.org/10.1017/9781009157896.001)

Cross-Chapter Box 9.1 | Global Energy Inventory and Sea Level Budget

Coordinators: Matthew D. Palmer (United Kingdom), Aimée B.A. Slangen (The Netherlands)

Contributors: Guðfinna Aðalgeirsdóttir (Iceland), Fábio Boeira Dias (Finland/Brazil), Catia M. Domingues (Australia, United Kingdom/Brazil), Gerhard Krinner (France/Germany, France), Johannes Quaas (Germany), Lucas Ruiz (Argentina)

Increased atmospheric greenhouse gas emissions since the 19th century have led to a net positive radiative forcing of Earth's climate (Sections 2.2 and 7.3) and a corresponding accumulation of energy in the Earth system. Quantification of this energy gain is essential to our understanding of observed climate change, and for estimates of climate sensitivity (Section 7.5). The global energy inventory is closely linked to our understanding of observed global sea level change, through the energy associated with loss of land-based ice and the effect of thermal expansion associated with ocean warming (Box 9.1, Sections 2.3.3.1 and 9.6.1; Table 9.5).

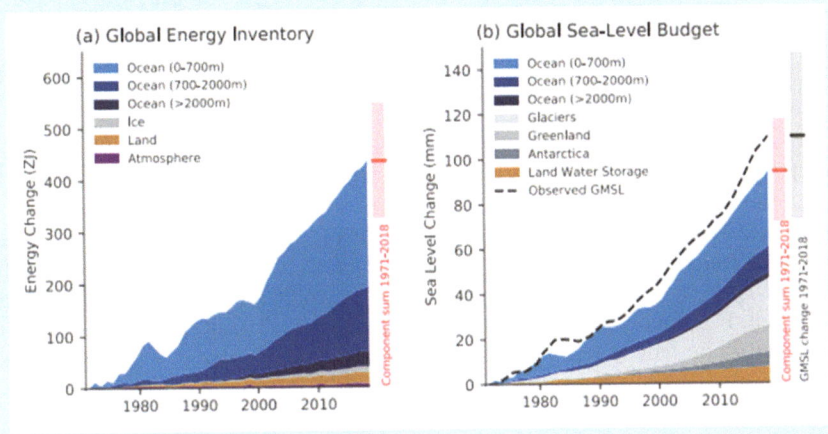

Cross-Chapter 9.1, Figure 1 | Global Energy Inventory and Sea Level Budget. (a) Observed changes in the global energy inventory for 1971–2018 (shaded time series) with component contributions as indicated in the figure legend. Earth System Heating for the whole period and associated uncertainty is indicated to the right of the plot (red bar = central estimate; shading = *very likely* range). Observed global mean sea level for 1971–2018 (shaded time series) as indicated in the figure legend. Observed global mean sea level change from tide gauge reconstructions (1971–1993) and satellite altimeter measurements (1993–2018) is shown for comparison (dashed line) as a three-year running mean to reduce sampling noise. Closure of the global sea level budget for the whole period is indicated to the right of the plot (red bar = component sum central estimate; red shading = *very likely* range; black bar = total sea level central estimate; grey shading = *very likely* range). Full details of the datasets and methods used are available in Annex I. Further details on energy and sea level components are reported in Table 7.1 and Table 9.5.

Fig. 19.7 Components of the earth's energy storage budget and sea level budget (Courtesy the Intergovernmental Panel on Climate Change, AR6 WGI)

but also in the relatively slow rate in which they had historically trended up—until now. For hundreds of millennia, gradual temperature trends are within a remarkably narrow temperature range averaging 10° Celsius across these peak-to-trough oscillations over a period of 50,000 years. That pace translates to about 0.2° Celsius per millennium, or about 1.2° Celsius rate of change of temperature for every 6,000 years. The estimate for global warming over the less than 200 years since 1850 is now approaching 1.2° Celsius, which is a pace about thirty times faster than the Earth typically encounters (Figs. 19.8, 19.9, and 19.10).

We now know that these temperature patterns over thousands and millions of years are related to the composition of our atmosphere and its ability to trap the heat that the Earth radiates. The amount of reradiated heat can be measured precisely. Of the insolation that reaches the Earth, some is reradiated

Changes in global surface temperature relative to 1850–1900

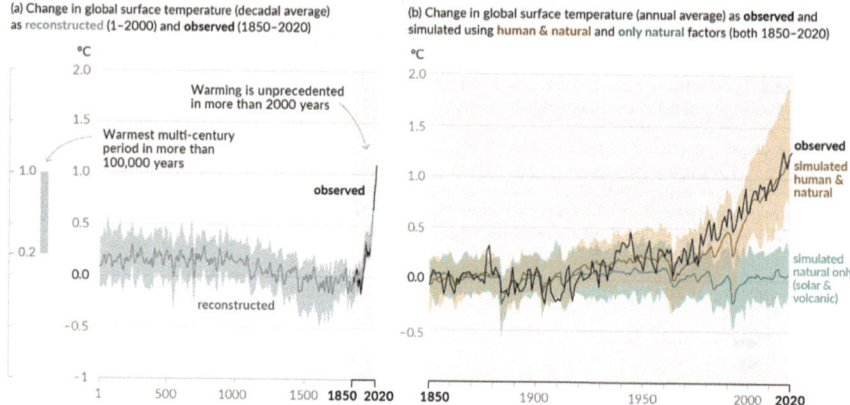

(a) Change in global surface temperature (decadal average) as reconstructed (1–2000) and observed (1850–2020)

(b) Change in global surface temperature (annual average) as observed and simulated using human & natural and only natural factors (both 1850–2020)

Fig. 19.8 Changes in global surface temperature (From Figure SPM.1 in IPCC, 2021: Summary for policymakers. In: *Climate Change 2021: The Physical Science Basis. Contribution of Working Group I to the Sixth Assessment Report of the Intergovernmental Panel on Climate Change* [Masson-Delmotte, V., P. Zhai, A. Pirani, S.L. Connors, C. Péan, S. Berger, N. Caud, Y. Chen, L. Goldfarb, M.I. Gomis, M. Huang, K. Leitzell, E. Lonnoy, J.B.R. Matthews, T.K. Maycock, T. Waterfield, O. Yelekçi, R. Yu, and B. Zhou (eds.)]. Cambridge University Press, Cambridge, UK and New York, NY, USA, pp. 3–32, https://doi.org/10.1017/9781009157896.001)

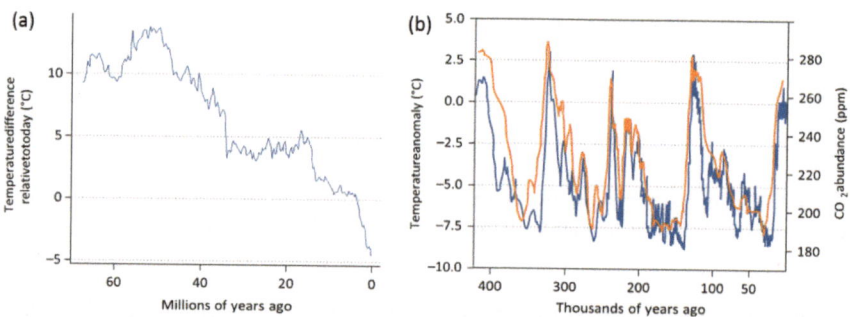

Fig. 19.9 Patterns of global temperatures over millions of years and millennia

and some is absorbed into the ecosystem. A net of about 240 watts/m² of insolation acts to warm the Earth and our atmosphere. Even a change of a few watts per square meter in net insolation can then have a significant effect of warming over time.

Scientists have asserted for two centuries that gases in the atmosphere serve a critical role in this equilibrium that results in the warming and reradiation rates. In 1824, the preeminent French mathematician and physicist Joseph Jean-Baptiste Fourier published "Remarques Générales Sur Les Températures Du Globe Terrestre Et Des Espaces Planétaires,"[5] in which he calculated the

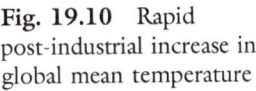

Fig. 19.10 Rapid post-industrial increase in global mean temperature

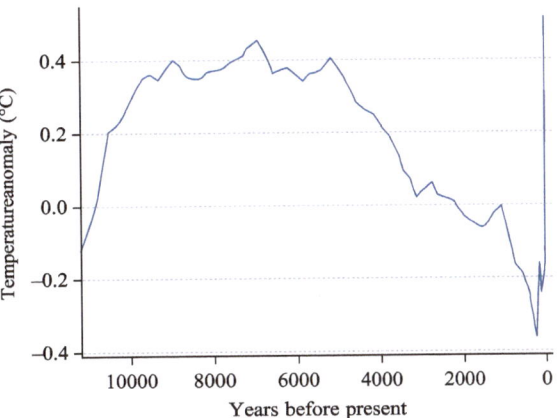

theoretical temperature of the Earth given its distance from the Sun. He postulated that the much higher temperature observed on Earth could be explained by an atmosphere that acts as an insulator to prevent reradiation of the solar energy the Earth receives.

In 1856, Eunice Newton Foote, a self-trained scientist in New York State, performed experiments of the temperature in glass tubes exposed to the Sun and containing various gases. She concluded that these gases indeed have an insulating effect, and postulated that the presence of both carbon dioxide and water vapor could lead to higher atmospheric temperatures. Separately, John Tyndall made the same observation later that century, while Svante Arrhenius in 1896 provided predictions of the effects of a doubling of atmospheric carbon dioxide. In 1901, Nils Gustaf Ekholm used the analogy of a greenhouse to describe this theory.

The nature of atmospheric insulation depends on specific qualities of each gas. The predominant gases nitrogen, oxygen, and argon, at 79, 19, and 0.5% respectively in our atmosphere, have negligible heat trapping qualities. Just as the temperature of the atmosphere is an equilibrium parameter that balances insolation and reradiation, other components in our atmosphere establish equilibria. For instance, new carbon dioxide is naturally emitted into the atmosphere, through processes such as volcanic eruptions, forest fires, exhalation from animals, and other sources, and is reabsorbed by natural absorption of oceans, plants, and other processes in an equal amount. The burning of fossil fuels has disrupted and overwhelmed this natural equilibrium.

The scientist Charles D. Keeling first began measuring the correlation between atmospheric temperatures and its carbon dioxide content, which is typically measured in mere parts per million rather than percentage, given its minute content in our atmosphere. Scientists have been observing the upward trend the Keeling Curve ever since. As emissions increase beyond traditional levels since 1850 through human activity, primarily in the form of combustion of fossil fuels, about half are reabsorbed through natural processes called *carbon sinks*, while the remainder raises the atmospheric concentration of carbon dioxide.

The Various Greenhouse Gases

While carbon dioxide emissions are the single most significant causes of anthropogenic global warming, carbon dioxide itself is only one of the greenhouse gases, and is not even the largest contributor to global warming, despite the oft-cited need to decarbonize the atmosphere. The largest contributor to global warming is a very familiar gas to most people: water vapor.

Water Vapor

The Earth has experienced a shift in the relative share of global warming gases held in equilibrium in the atmosphere since the onset of the Industrial Revolution. The most significant component in the determination of climate and atmospheric temperature is the level of water vapor in the atmosphere. Because a molecule of water, made of two parts hydrogen and one part oxygen, is less dense than molecules of nitrogen, oxygen, and carbon dioxide, atmospheric water vapor lowers atmospheric density and pressure and in turn drives the Earth's weather.

Water vapor represents up to 4% of the atmosphere. It is in equilibrium, with evaporation primarily over oceans or large lakes as a *source* of the atmosphere's vapor content, while precipitation acting as a *sink* to lower atmospheric vapor. The ability of the atmosphere to hold water depends globally on the atmosphere's average temperature and more locally on temperature changes from air currents that move water molecules to different levels of the atmosphere. Changes in atmospheric conditions affect the atmosphere's water vapor content significantly, while atmospheric conditions have insignificant effects on its ability to hold the other gases.

Water vapor content is proportional to temperature. A higher atmospheric temperature results in a greater ability for the atmosphere to hold water vapor. But, while a higher temperature allows the atmosphere to hold more vapor, and hence increase the greenhouse effect, the components of water absorbed from evaporation and precipitated are in equilibrium at a given time. Hence, higher ocean and atmospheric temperatures significantly increase the atmosphere's water vapor content, the evaporation rate, and the rate of precipitation globally. While precipitation is also dependent on local conditions, more intense precipitation on average is associated with global warming, even while some regions may become drier with changing weather patterns.

The capacity of the atmosphere to hold water vapor creates a destabilizing positive feedback loop. As the temperature of the atmosphere increases, so does the amount of this greenhouse gas held in the atmosphere, which further increases the greenhouse effect arising from water vapor.

A higher atmospheric temperature exponentially increases its maximum water vapor pressure, which then results in an accelerated greenhouse gas effect, still greater atmospheric vapor pressure and more global warming, etc. Water in the form of vapor and as ice crystals in clouds represent 50% and 25%, respectively, of global warming atmospheric components. In addition, for a similar atmospheric column weight that generates approximately one atmosphere of pressure (the equivalent of 29.92 inches of mercury or 101.325 kilopascals, or kPa) at the Earth's surface, the *ideal gas law* shows that, for a given quantity of atmosphere, the pressure of water vapor in the atmosphere is given by the *August-Roche-Magnus* equation:

$$\boldsymbol{Vapor\ Pressure(kPa)} = 0.61094\ e^{\frac{17.625}{1+243.04/T}},$$

where T is measured in °Celsius. Figure 19.11 shows the exponential effect of an increase in temperature on water vapor content.

In 1850, commonly considered the beginning of the period of global warming with the onset of fossil fuel use in the Industrial Revolution, the average global temperature was 13.6° Celsius. At that temperature, the average vapor pressure of water, as a share of average air pressure was 1.55%. A 1.6° Celsius rise in the average global temperature increases the maximum water vapor content by 0.15 percentage points to 1.7%, which correspond to a 10.2% increase in potential water vapor content and a commensurate increase in the ability of the atmosphere's most significant greenhouse gas to contribute to global warming.

At 30° Celsius, the amount of water vapor the atmosphere can hold is 4.2% of the atmosphere's pressure, and rises by 0.4 percentage points to 4.6% with a further 1.5° Celsius rise in air temperature. This concentration is almost

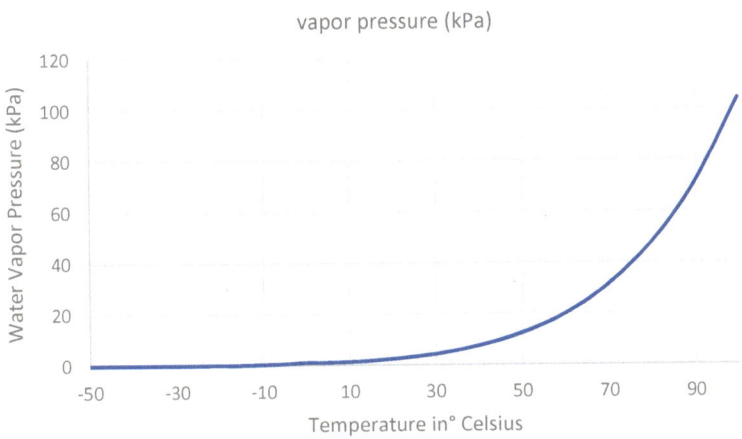

Fig. 19.11 Relationship between temperature and potential atmospheric water vapor pressure

three times the increase in moisture content that occurs on average in the atmosphere. We see that this positive feedback loop is non-linear. Increases in the temperature of the atmosphere results in an accelerating ability of the air to hold moisture, and hence exponentially accelerates global warming.

This greater moisture content also permits the atmosphere to retain more heat energy. This lower air pressure and greater ability for the atmosphere to hold moisture and latent heat makes for more powerful storms and hurricanes that have an increased propensity for intense and more widespread precipitation. Higher water vapor concentrations also induce lower atmospheric pressures and more pronounced regional pressure differences. This greater capacity of the atmosphere to hold water vapor and heat energy is most pronounced in the tropics that drive the Earth's cyclones.

Higher moisture content can also create more significant cloud cover. A cloud is water vapor that has precipitated in the atmosphere into suspended liquid or ice under conditions of sufficient updraft and lower temperatures at higher altitudes. Clouds can act to trap warm air from below and prevent it from reradiating. These driving forces represent 25% of global warming. They are also positively correlated with the moisture content of air and hence atmospheric pressure. Both water vapor, and the clouds they promote, are positive feedback loops that exacerbate global warming and become increasingly significant as the atmospheric temperature rises.

Carbon Dioxide

Atmospheric carbon dioxide has risen from 285 ppm in 1880 to an estimated 423 ppm in 2023.[6] While its effect is not as profound as that of water vapor in the atmosphere, carbon dioxide is still an important driver of global temperatures because its rising atmospheric concentration is extremely rapid and almost entirely human-induced. This 48% increase in the content of carbon dioxide since the onset of the Industrial Revolution is the single most potent anthropogenic driver of global warming (Fig. 19.12).

In more recent years, the dramatic contribution to global warming from carbon dioxide had been accelerating. This trend shows some fluctuation throughout the year, especially with the seasonal growth of flora in the Northern hemisphere each summer and its intendent ability to absorb carbon dioxide throughout the growing season (Fig. 19.13).

Until the last few generations, carbon dioxide in the atmosphere tended toward equilibrium. Carbon dioxide increases arise from *carbon sources* such as the creation of carbon dioxide from respiring animals, ocean outgassing, biomass decomposition, volcanic eruptions, forest fires, and the belching and flatulence of ruminant animal belching and flatulence as a source of carbon dioxide and methane emissions. Meanwhile, the ocean and the Earth's flora represent a *carbon sink*.

The ocean alone removes about 30% of carbon dioxide, first by dissolving it at the surface and then cycling the carbon to the ocean's bottom where it can

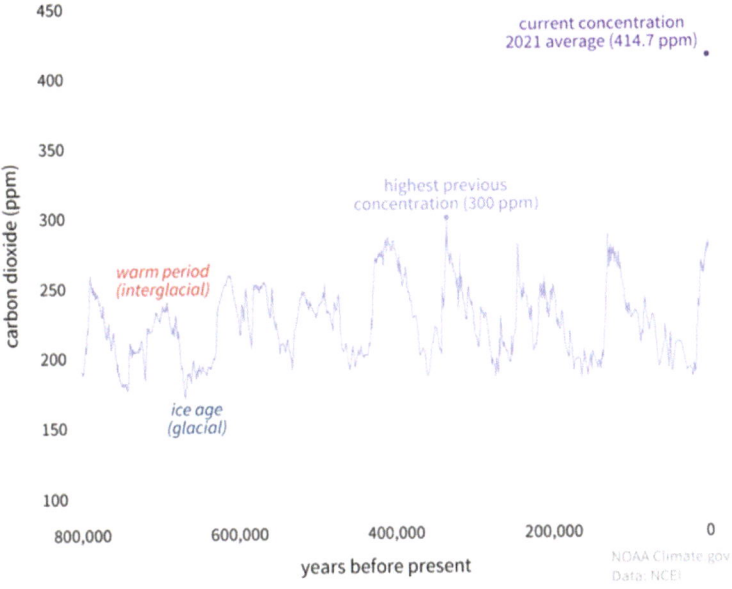

Fig. 19.12 Carbon dioxide patterns over the past 800 millennia (Courtesy of the National Oceanic and Atmospheric Administration, www.climate.gov)

precipitate and is sequestered in the form of carbonates and limestone. Forests also act as a natural carbon sink, through photosynthesis and soil fixation, which constitutes about 29% of natural carbon removed annually.

METHANE

Natural gas emissions, constituted primarily of methane, represent a smaller volume of total emissions, but are much more potent per molecule in global warming. Methane represents about 1.7 parts per million of the atmosphere compared to carbon dioxide's 423 parts per million. While methane represents only 0.4% of the number of molecules in the atmosphere relative to carbon dioxide, it contributes an equivalent warming capacity of 18% that of carbon dioxide. On a molecule-to-molecule basis, methane is 45 times more potent.

Beside leaks from poor or aging natural gas distribution infrastructure or incomplete combustion, methane is released from manure decomposition and from livestock, rice cultivation, coal and other mining, and solid waste decomposition in landfills and sewage treatment plants, all of which have accelerated since the onset of the Industrial Revolution.

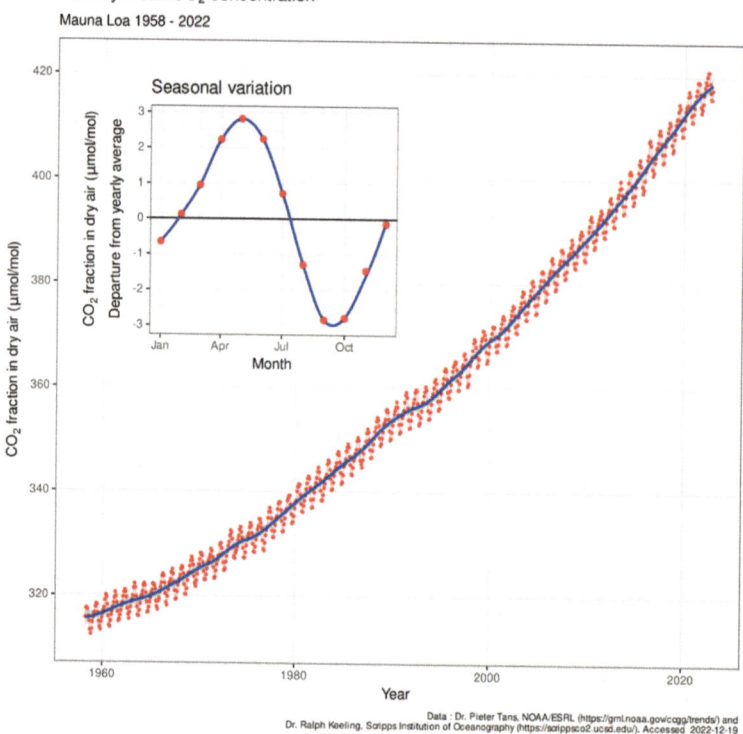

Fig. 19.13 Trend in atmospheric carbon dioxide levels (Courtesy Delorme, CC BY-SA 4.0 and https://creativecommons.org/licenses/by-sa/4.0> via Wikimedia Commons, https://commons.wikimedia.org/wiki/File:Mauna_Loa_CO2_monthly_mean_concentration.svg)

Natural geological emissions of methane amount to about 1–2 million metric tons per year while industry emits about 175 million metric tons annually.[7] These quantities are only a small fraction of the 40.8 gigatonnes of carbon dioxide equivalent emissions annually in 2021. While methane decomposes more quickly in the atmosphere than carbon dioxide, it can persist upwards of two or three generations, at which time these molecules decompose into carbon dioxide and water vapor. For these reasons, methane is both a highly potent and dangerous greenhouse gas and yet is also one of the simplest emissions to avoid through diligent improvements in methane containment and capture.

OTHER GREENHOUSE GASES

Nitrous oxides, hydrocarbons, and chloro- bromo- and various other fluorocarbons, also called ozone-depleting gases, further contribute to global warming, while aerosols tend to reduce global warming. The ozone gases combine with oxygen and water vapor to produce molecules that can reduce the level of the ionized oxygen molecule called ozone O_3^+ which otherwise protects organic matter from the harmful effects of the most energetic ultraviolet rays that radiate from the Sun.

On the other hand, aerosols help mitigate global warming by reflecting back into space insolation before it reaches the Earth's surface. The production and release of human-made aerosols has been proposed to geoengineer the atmosphere and mitigate global warming. The emission and suspension of soot and other materials may also have similar global warming moderation effects. These can arise from industrial and agricultural processes, mining, cement manufacturing, and natural volcanic eruptions. Thick daytime clouds can also reflect insolation back into space, but thinner clouds and evening clouds help to prevent the Earth from reradiating heat.

Various fossil fuels produce different amounts of greenhouse gases for an equivalent amount of heat, typically measured per million British thermal unit (MMBtu) of energy. Fossil fuels mostly produce carbon dioxide, so their emissions are measured in equivalent kilograms of carbon dioxide, while other emissions are measured in grams (Fig. 19.14).

The combination of effects from the growing quantity of these greenhouse gases in the atmosphere has increased net solar heating by about 4 watts per square meter, while aerosol suspension has decreased net insolation by less than 1 W/m^2. The net effect is an increase in average insolation of about 2.7 W/m^2, which is more than a 1% increase in the net atmospheric warming rate.

There are a variety of natural processes that also induce fluctuations in global warming, but they are relatively minor. These include variations in the insolation rate because of the relative location and strength of the Sun. Solar radiation changes in predictable ways but these changes in actual solar radiation are measured over billions of years. Small periodic shifts in the Earth's

Fossil Fuel Emissions:	kg CO2/MMBtu	g CH4/MMBtu	g N2O/MMBtu
Coal	1002	11	1.6
Motor gasoline	70	3	0.60
Propane	63	3	0.60
Natural gas	53	1	0.10

Fig. 19.14 Greenhouse gas emissions for various fossil fuels (Courtesy of US EPA emission hub. https://www.epa.gov/system/files/documents/2022-04/ghg_emission_factors_hub.pdf; https://www.epa.gov/climateleadership/ghg-emission-factors-hub)

orbit and tilt also occur, with the Earth moving up to 1% closer to the Sun, or tilting on its axis by upwards of a degree or so. However, these effects too occur over 100,000 years. These astronomical effects occur incredibly slowly to the degree that their effects are outswamped by natural and now human-produced variations on the planet itself.

The benchmark measure of the potency of greenhouse gases is their equivalent carbon dioxide equivalency CO_2E (Fig. 19.15).

Natural fluctuations in the atmospheric equilibrium have been outswamped by the 100-fold increase in emissions arising from human commerce, and primarily from the combustion of the fossil fuels that had acted as an immense carbon sink for a hundred million years. While an increased atmospheric carbon dioxide content does not induce the same direct positive feedback effects as does water vapor, the general increase in temperature induces a positive feedback loop from other sources.

These tipping points, once activated, accelerate global warming still further, and are expected to soon contribute to the costs of global warming to a degree that out-swamps the damage caused by human-induced carbon emissions. The main sources of accelerated natural temperature-induced greenhouse gas emissions arise primarily by the release of naturally sequestered gases. For instance, the uncovering of peat by wetlands erosion and permafrost melting accelerates the decay of biomass that has been protected for thousands or hundreds of thousands of years (Fig. 19.16).

Anthropogenic carbon dioxide represents approximately 75% of carbon dioxide equivalent emissions.[8] An additional 18% of human greenhouse gas-induced global warming arises from methane emissions, while 4% arose from nitrous oxide emissions and 2% from fluorinated gases. The carbon dioxide component arises primarily from the burning of fossil fuels, but forest fires accelerate global warming, while deforestation reduces Nature's ability to

Gas	Persistence	Relative Warming Potential	Increase	Percent
Carbon dioxide	500 years	1	130 ppm	56%
Methane	12.4 years	28	1.1 ppm	15%
Nitrous oxide	121 years	265	75 ppb	5%
Halocarbons	Years to millennia	100s to 1000s	A few ppb	11%
Ozone	Weeks to months	N/A		12%

Fig. 19.15 Greenhouse gas persistence and post-industrial revolution contributions (Courtesy Metrics of major greenhouse gases. Increases in abundance and fraction of heating for the years 1750–2010. GWPs are calculated assuming a 100-year time horizon (IPCC, 2013: Climate Change 2013: The Physical Science Basis. Contribution of Working Group I to the Fifth Assessment Report of the Intergovernmental Panel on Climate Change. Cambridge University Press, Cambridge, United Kingdom and New York, NY, USA, 1535 pp.)

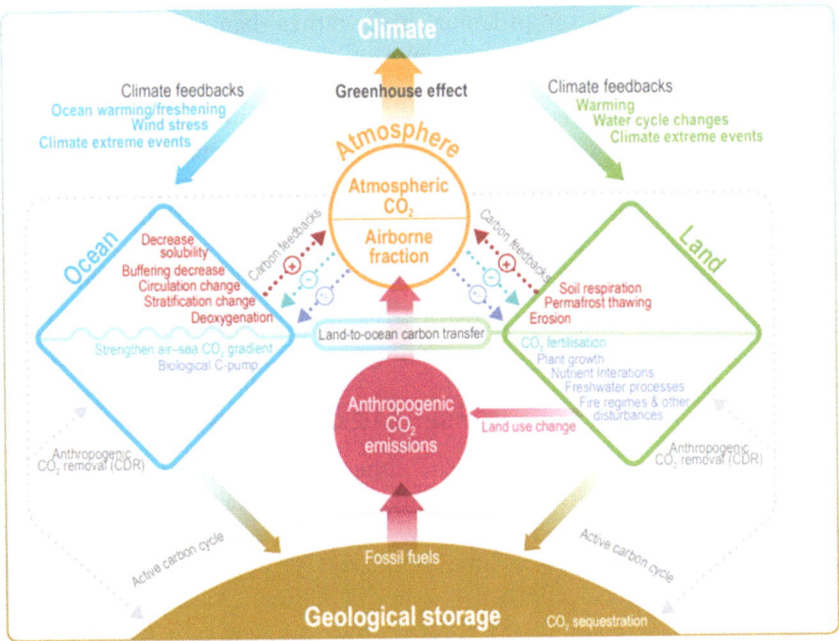

Figure 5.2 | Key compartments, processes and pathways that govern historical and future CO₂ concentrations and carbon–climate feedbacks through the coupled Earth system. The anthropogenic CO₂ emissions, including land-use change, are partitioned via negative feedbacks (turquoise dotted arrows) between the ocean (23%), the land (31%) and the airborne fraction (46%) of anthropogenic CO₂ that sets the changing CO₂ concentration in the atmosphere (2010–2019; Table 5.1). This regulates most of the radiative forcing that drives the heat imbalance that drives the climate feedbacks to the ocean (blue) and land (green). Positive feedbacks (red arrows) result from processes in the ocean and on land (red text). Positive feedbacks are influenced by both carbon-concentration and carbon–climate feedbacks simultaneously. Additional biosphere processes have been included, but these have an as-yet-uncertain feedback impact (blue-dotted arrows). CO₂ removal from the atmosphere into the ocean, land and geological reservoirs, necessary for negative emissions, has been included (grey arrows). Although this schematic is built around CO₂ (the dominant greenhouse gas), some of the same processes also influence the fluxes of CH₄ and N₂O and the strength of the positive feedbacks from the terrestrial and ocean systems.

Fig. 19.16 Determinants of carbon dioxide atmospheric equilibrium (Courtesy the Intergovernmental Panel on Climate Change, AR6 WGI, Chapter 5, Fig. 5.2, page 682, https://www.ipcc.ch/report/ar6/wg1/downloads/report/IPCC_AR6_WGI_Chapter05.pdf)

act as a carbon sink. Chemical and industrial processes further contribute to carbon dioxide emissions, with cement production the single largest manufacturing process that accelerates global warming.

DEFORESTATION

About 34% of land globally is devoted to agriculture while 26% is forested. Deforestation to create more farmland and feed a growing population is the most significant way that land use has been accelerating global warming over the past century. Widespread deforestation releases carbon that may have been sequestered for generations before a forest is burned or otherwise allowed to decay naturally. Once deforested and subsequently converted to grassland, crops, or grazing, accumulate biomass that is returned to the carbon cycle much more quickly. This conversion that accelerates carbon dioxide and methane emissions on an annual basis, rather than over many decades, is the second largest anthropogenic cause of global warming, after fossil fuel combustion.

If sequestered carbon is considered a stock that preserves the atmospheric status quo, the conversion of forest to farmland depreciates our atmospheric natural capital and initiates processes that contribute to further annual stock depreciation. Only about 24% of deforestation results in sequestered forest products such as lumber. Most deforestation is primarily from forest fires and conversion to agricultural land, especially for ranching and production of cooking oils that quickly return sequestered carbon to the atmosphere which took generations to remove.

Negative Feedback Effects

Most of the processes described create either direct positive feedback effects that accelerate global warming, such as increased water vapor greenhouse gases with higher temperatures, or indirectly, through the release of sequestered natural carbon arising from melting permafrost or eroding wetlands. There are some minor feedback effects that slow down these processes.

For instance, thin clouds act to trap heat but extensive daytime cloud cover can reflect insolation away from the Earth. Such heat reflection can then act as a negative feedback loop that mitigates positive feedback arising from a greater amount of atmospheric water vapor as global temperatures rise. The dark surfaces of a forest may also absorb more heat than straw colored ranchland, some greenhouse gas creation effects of deforestation are mitigated (Fig. 19.17).

In addition, higher carbon dioxide levels are partly mitigated by nature. While elevated carbon dioxide emissions may destroy ocean coral, they may also accelerate plant metabolization and hence the absorption of carbon dioxide. Higher temperatures may also extend growing seasons and hence remove more carbon dioxide through photosynthesis. However, these mitigating negative feedback loops are outswamped by the consumption and decomposition of crops and their release of carbon dioxide. Many positive feedback loops are initiated as the climate exceeds various tipping points.

Climate Change Modeling

The science of global warming is exceedingly complex given the large number of variables involved, the geographical diversity of greenhouse gas emissions, the various tipping points that cause otherwise linear scientific relationships to become non-linear, and the unpredictable ways humans affect these relationships. Even if models can successfully predict the effects in the aggregate on the average global temperature, an understanding of climate change necessarily requires an understanding of often interrelated systems.

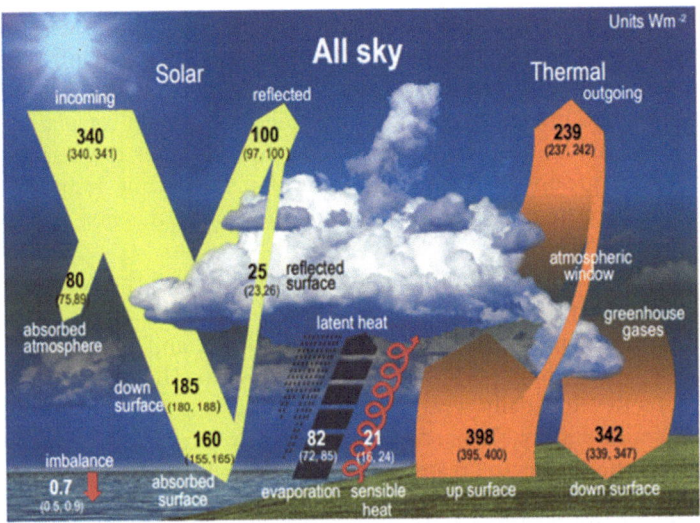

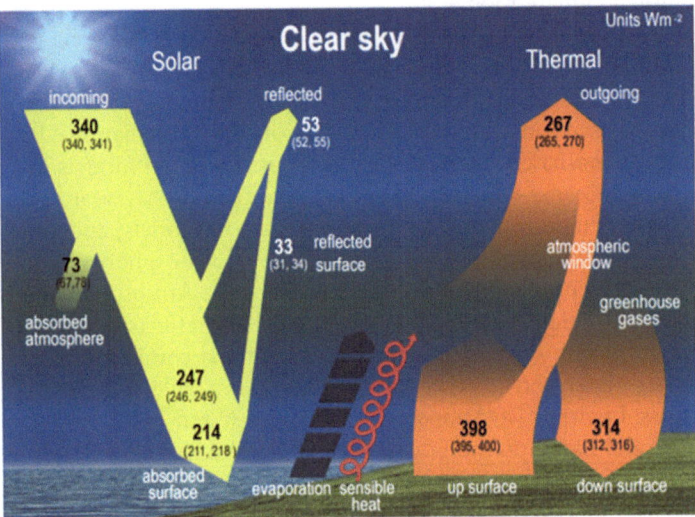

Figure 7.2 | Schematic representation of the global mean energy budget of the Earth (upper panel), and its equivalent without considerations of cloud effects (lower panel). Numbers indicate best estimates for the magnitudes of the globally averaged energy balance components in W m⁻² together with their uncertainty ranges in parentheses (5–95% confidence range), representing climate conditions at the beginning of the 21st century. Note that the cloud-free energy budget shown in the lower panel is not the one that Earth would achieve in equilibrium when no clouds could form. It rather represents the global mean fluxes as determined solely by removing the clouds but otherwise retaining the entire atmospheric structure. This enables the quantification of the effects of clouds on the Earth energy budget and corresponds to the way clear-sky fluxes are calculated in climate models. Thus, the cloud-free energy budget is not closed and therefore the sensible and latent heat fluxes are not quantified in the lower panel. Figure adapted from Wild et al. (2015, 2019).

Fig. 19.17 Insolation and reradiation budget and the effect of cloud cover (Courtesy the Intergovernmental Panel on Climate Change, AR6 WGI, Chapter 7, Fig. 7.2, page 934, https://www.ipcc.ch/report/ar6/wg1/downloads/report/IPCC_AR6_WGI_Chapter07.pdf)

The first computational global change models began in the wake of World War II. John von Neumann was a Jewish-Hungarian polymath who immigrated to the United States and assisted in the Manhattan Project. Neumann proposed that the Eniac Computer he designed to assist in calculations necessary to model the first atomic bombs be repurposed after the war to forecast the weather. A group was formed by him at his academic home, Princeton University, where he and Jule G. Charney developed the first computerized weather forecast models.

As with climate modeling today, their research divided the two-dimensional surface of the Earth into grids. They modeled air, water vapor, and temperature movement within each grid across the globe, and then connected the flow between grids. Based on their work, by 1955 an IBM 701 computer was employed to develop daily weather forecasts. By the end of the decade, the computer could reasonably predict weather a few days in advance. Neumann also directed the group to create three-dimensional circulation models of the atmosphere and to extend their analysis to the global atmosphere. These were the first global climate models.

Toward the latter half of the 1950s, a Russian climatologist named Mikhail Budyko authored The Heat Balance of the Earth's Surface,[9] which used an early heat energy balance system of equations to predict the Earth's average temperature based on the rate of insolation and reradiation. By 1967, a *global fluid dynamics* expert named Kirk Bryan extended models by including the effects of oceans. Syukuro Manabe collaborated that year with Richard Wetherald to include the effects of water vapor on models, and were the first to calculate the effects of changes in the level of atmospheric carbon dioxide on climate. Their paper Thermal Equilibrium of the Atmosphere with a Given Distribution of Relative Humidity[10] described the effects of greenhouse gas emissions on global warming within a well-specified computational model.

Two years later, on March 4, 1969, William Sellers, of the University of Arizona's Institute of Atmospheric Physics, stated in his paper A Global Climatic Model Based on the Energy Balance of the Earth-Atmosphere System that

> The major conclusions of the analysis are that removing the Arctic ice cap would increase annual average polar temperatures by no more than 7° Celsius, that a decrease in the solar constant by 2-5% might be sufficient to initiate another ice age, and that man's increasing industrial activities may eventually lead to a global climate much warmer than today.

These profound words remain consistent with current estimates and marked both the beginning of climate change modeling and concern within the scientific community of the anthropogenic drivers of climate change.

By 1970, as one group at MIT was preparing The Limits to Growth, another group released The Study of Critical Environmental Problems (SCEP).[11] Just as the concerns of the Club of Rome inspired the United

Nations Brundtland Commission on Sustainability, the meetings of scientists facilitated by MIT inspired a subsequent meeting in Stockholm in the following year, which produced Inadvertent Climate Modification: Report of the Study of Man's Impact on Climate.[12] These two reports influenced the 1972 *United Nations Conference on the Human Environment* to found their *UN Environment Programme* and permanently placed climate change on the table of political leaders worldwide.

By 1975, Manabe and Wetherald published their The Effects of Doubling the CO_2 Concentration on the climate of a General Circulation Model[13] in which they estimate that a doubling of the carbon dioxide level would increase global temperatures by 2.9° Celsius. This number was recently affirmed as the reasonable case estimate of the global temperature increase that can be expected by the mid-twenty-first century should policymakers fail to return to their pledges toward carbon neutrality.

Over the decades, models have become increasingly advanced and complex, and have teased interactions and results that could not have been gleaned from less sophisticated models. The increasing sophistication of scientists' understanding of complex interactions has described global climate change in increased detail and accelerated alarm. Each iteration creates greater certainty and, typically, generates results of profound climate change that are increasingly immediate.

On December 6, 1988, the *World Meteorological Organization* and the *United Nations Environmental Programme* created the *Intergovernmental Panel on Climate Change* (IPCC) based on UN General Assembly Resolution 48/58. The IPCC's mission is to accumulate and disseminate the best scientific analyses and policy strategies related to the scientific, social, and economic aspects of climate change. It has now produced six assessments, beginning in 1990, with the most recent published in 2022.

The Sixth Assessment Report of 2022 contained the following highlights:

- Anthropogenic greenhouse gas emissions continue to rise, with the greatest decade to decade rise occurring in the last decade, an increase of an estimated 59 Gigatonnes of Carbon Dioxide equivalent ($GTCO_2$) over the decade from 2010 to 2019.
- The largest growth components are carbon dioxide and methane emissions. 58% of accumulated anthropogenic greenhouse gas emissions occurred in the 140 years between 1850 and 1989, and an additional 42% in the next thirty years. The decade from 2010 to 2019 alone represented 17% of the total accumulated emissions.
- To limit global warming to a median warming of 1.5° Celsius would require that total additions to accumulated greenhouse gases be limited to 500 $GTCO_2E$, which is only 22% more than was emitted in the decade from 2010 to 2019. Emissions of 1150 $GTCO_2E$ would result in 2.0° Celsius warming.

- Households in the top decile of incomes contribute a disproportionate 34–45% share of emissions, while the bottom half of households contribute 13–15% of emissions.
- Limiting global warming to between 1.5° Celsius and 2.0° Celsius would require "deep, rapid, and sustained emissions reductions."
- The risk of triggering significant tipping points is high should global warming rise to 2.0° Celsius and are very high should 2.5–4.0° Celsius be exceeded.

Another article in 2022 in the journal Science by David Armstrong[14] noted that the Earth has already exceeded the 1.0° Celsius threshold that may induce five of sixteen trigger points. These include the accelerated breakdown of Greenland and West Antarctic ice sheets, massive tropical coral reef die-offs, a collapse of the Labrador Sea convection current, and widespread and abrupt thawing of boreal permafrost.

Four of these five tipping points are highly likely to be triggered once the temperature rise reaches the 1.5° Celsius threshold set by nations at the Paris Agreement in 2015, and an additional five more triggers become possible. These scientists also note that, even if UN Paris Agreement emission reduction thresholds are met to limit warming to 2.0° Celsius, dangerous climate change can no longer be avoided. To avoid even 50% odds of tripping significant tipping points, global warming must be limited to 1.5° Celsius, and global greenhouse gas emissions must be cut in half by 2030 while countries must reach net-zero emissions by 2050.

Meanwhile, the **United Nations Framework Convention on Climate Change** noted in October of 2022 that current emissions trajectories place the planet on a path toward a 2.1–2.9° Celsius temperature increase by the end of the century.

While the UN IPCC outlines causes and effects to the year 2100, human activity since 1850 and responses within the next generation will affect Nature for centuries and millennia to come. Once the implications of exceeding tipping points accumulate, longer term tipping points, such as the breakdown of the **Atlantic Meridional Overturning Circulation** *(AMOC)* can invoke very long-term effects on the Earth. Ocean warming and cause sea level to rise, while glacial and floating ice melt can induce changes measured somewhere between a millennium (a thousand years) and an epoch (millions of years).

Even should anthropogenic emissions cease and humankind reach net zero carbon emissions by 2050, the Earth must go through a centuries-long cycle to reequilibrate. In this case, it will endure a far greater shock to the carbon dioxide level than has been measured over recent millennia and epochs.

As a consequence of cascading complexities, scientists have estimated a long run sea level rise of 2.3 meters per degree Celsius after 2000 years,[15] the extension of our current interglacial period by 100,000 years,[16] and increased ocean acidification for hundreds to thousands of years.[17] Scientific analyses conclude that the size and the unprecedented rate of climate change since 1850 are correlated with the accumulation of excesses in greenhouse gases in proportion to the level of emissions arising from fossil fuel combustion by humans.

Based on the science developed in an understanding of the effects of anthropogenic releases of greenhouse gases, various scenarios can be developed. An international consortium of scientists has assembled to estimate upcoming global warming based on past observations and correlations and a variety of future emission and mitigation scenarios. These projections by the Assessment Modeling Consortium have been developed based on various *shared socioeconomic pathways (SSPs)*, denoted SSP1 to SSP5 (Fig. 19.18).

The first pathway, labelled SSP1, is the most realistic ideal response by national economies to the challenges of global warming. This SSP assumes fossil fuel-related emissions fall from current levels, with net human emissions that reach net zero by 2075. This pathway is sufficient to hold the global atmospheric temperature to 2° Celsius by the end of the century.

The second pathway assumes that the current trend to alternative energy sources continues but not with the ambition of sustainability goals set in SSP1. Under this pathway, temperatures continue to increase beyond the 2° Celsius threshold, but are limited to a 3° Celsius increase by the end of the century.

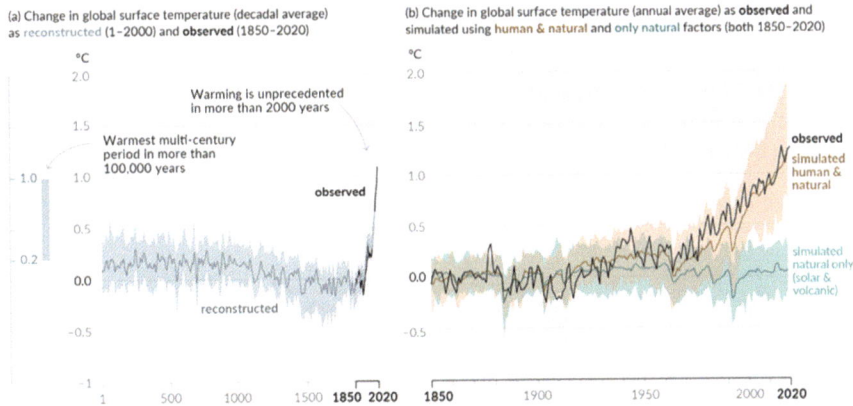

Fig. 19.18 Global surface temperatures from 1850– (Courtesy of Figure SPM.1 in IPCC, 2021: Summary for Policymakers. In: Climate Change 2021: The Physical Science Basis. Contribution of Working Group I to the Sixth Assessment Report of the Intergovernmental Panel on Climate Change [Masson-Delmotte, V., P. Zhai, A. Pirani, S.L. Connors, C. Péan, S. Berger, N. Caud, Y. Chen, L. Goldfarb, M.I. Gomis, M. Huang, K. Leitzell, E. Lonnoy, J.B.R. Matthews, T.K. Maycock, T. Waterfield, O. Yelekçi, R. Yu, and B. Zhou (eds.)]. Cambridge University Press, Cambridge, UK and New York, NY, USA, pp. 3–32, https://doi.org/10.1017/9781009157896.001)

SSP3 results if MDCs are unable to reach an agreement with LDCs to ensure these LDCs invest in sustainable energy sources as their energy demands increase with economic development. Because of a failure to invest in sustainable energy alternatives, these developing nations repeat the fossil fuel mistakes of past development in the most developed nations, which result in sufficient greenhouse gas emissions to raise global temperature to 4.5° Celsius on average.

The fourth pathway, SSP4, assumes that there is little cooperation between most and least developed nations, and LDCs continue to develop in ways that draw down their extensive fossil fuel resources to permit their rapid development. Because there is little difference between SSP3 and SSP4 assumptions, analyses generally do not include SSP4.

The fifth pathway, SSP5, assumes that each economy develops based on local interests to maximize development with no regard for the resulting implications on global warming. This *free-rider scenario* assumes rapid continued economic growth with little regard for the emissions arising from continued and accelerating fossil fuel use.

In this scenario, the global temperature is expected to increase by 5.5° Celsius or more. This total temperature increase over the span between the apex and nadir of the cycle between planetary warming and ice ages approaches 8–10.0° Celsius. Rather than spanning 100,000 years, a similar scale of temperature increases may occur over just two centuries.

The IPCC estimates the implications of each of these human response scenarios on various planetary parameters. The scenarios are denoted by their SPS scenario and the degree of additional net solar forcing, in watts per square meter, separated by a hyphen. For instance, the most optimistic scenario is SSP1-1.9 to describe the results from an expected 1.9 W/m² increase in solar forcing (Fig. 19.19).

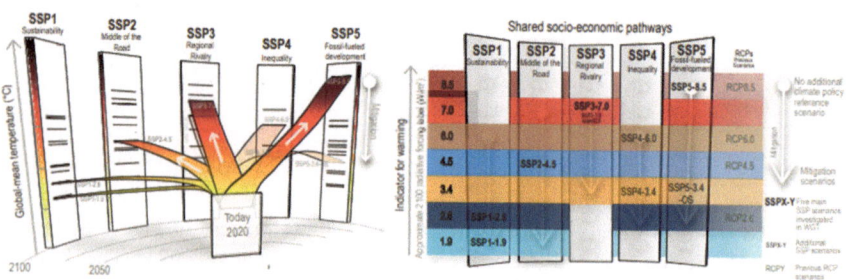

Fig. 19.19 Various socioeconomic pathways as determined by the IPCC

Cost and Consequences of Global Warming

Global warming is occurring at an unprecedented pace and at a rate that far exceeds the ability of species to adapt and to migrate to optimal habitats. Higher temperatures, new regions of perennial drought, more intense heat waves, and increasing water temperatures have already driven migration of species toward the poles. Scientists expect these changes to profoundly accelerate species loss.[18] Less is known about potential destabilization of the ecosystem given the complex relationships between plant and animal species. Marine life is expected to suffer the most, while those species that depend on wetlands have already been displaced by a 50% loss of wetlands due to human activity and global warming.[19]

The human species is also expected to suffer. The **World Health Organization (WHO)** has labelled global warming its greatest threat.[20] Infectious diseases will propagate more rapidly, and children especially become more vulnerable to disease and drought. Between 2030 and 2050 the WHO estimates that an additional quarter million people will die annually as a consequence of global warming, while by the end of the century, between half and three quarters of the world's population will face life threatening conditions such as extreme heat and humidity.[21]

Food security is also affected, especially as staple products such as wheat, corn, and soybeans grown in tropical latitudes are unable to thrive under more extreme temperatures. These costs are not evenly distributed, though, with agriculture in temperate latitudes likely benefiting from global warming. However, most of the world's population live in regions that will most profoundly suffer extreme heat, humidity, and drought. These populations have also historically been the most likely victims of a lack of food security.

Such factors of economic injustice are expected to accelerate. As early as 2030, the World Bank predicts that more than a hundred million additional people will be pushed into poverty.[22] These factors will even further accelerate migration by economic refugees, challenge immigration policies of wealthier nations, and may even threaten to worsen armed conflicts globally (Fig. 19.20).

Scientists and social scientists predict a variety of regional effects arising from global warming. Some nations, such as the Maldives and Tuvalu, will likely lose almost all their landmass. The most extreme climate scenarios predict that upwards of a third of the world's population may be forced to live in inhospitable conditions similar to the Sahara Dessert,[23] while the cost to global economies tallies into the trillions of dollars per year.

An estimate by Callahan and Mankin (2022) published on October 28, 2022 in Science Advances estimates that the global economy has already lost between $5 trillion US and $29 trillion US between 1992 and 2013 because of global warming, with low income nations experiencing upwards of a 6.7% drop in income, while high income nations suffered more modest 1.5% drop.[24] Global warming has become a driver of *economic injustice*. Meanwhile, the

Human activities affect all the major climate system components, with some responding over decades and others over centuries

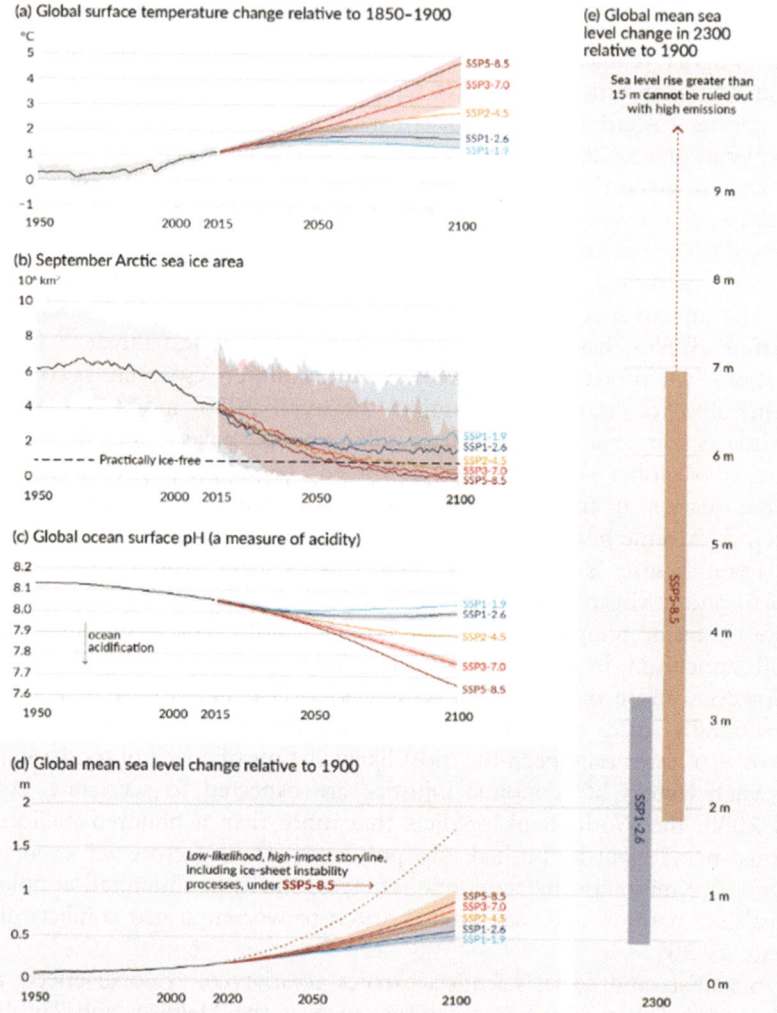

Fig. 19.20 Implications of global warming based on various IPCC scenarios (From IPCC 2021 Figure SPM.8: Summary for Policymakers. In: Climate Change 2021: The Physical Science Basis. Contribution of Working Group I to the Sixth Assessment Report of the Intergovernmental Panel on Climate Change [Masson-Delmotte, V., P. Zhai, A. Pirani, S.L. Connors, C. Péan, S. Berger, N. Caud, Y. Chen, L. Goldfarb, M.I. Gomis, M. Huang, K. Leitzell, E. Lonnoy, J.B.R. Matthews, T.K. Maycock, T. Waterfield, O. Yelekçi, R. Yu)

consultancy Deloitte estimates that unchecked climate change may cost the global economy $178 trillion US from 2021 to 2070 in net present value terms, a sum that exceeds $3.6 trillion US annually, in 2021 dollars.

These losses include declines in water stocks, such as the drought in the Western United States that has reduced the headwaters of the Colorado River by 75%. Large areas of irrigated land will be abandoned, and major coastal cities and coastal buffer zones may be lost, especially in regions such as Bangladesh, Florida, the Gulf Coast, the Maldives, and Tuvalu. Meanwhile, storms and tropical cyclones will become increasingly intense with much greater rainfall and flooding, and the seasons conducive to cyclone formation will be extended.

These losses will induce pressures on nations and governments to protect their own economies and resources at the expense of others. Borders between nations may become increasingly politicized and militarized, and the economic gulf between MDCs and LDCs shall expand and will further challenge economic justice. Conflicts will arise over water resources that shall make oil resource wars pale in comparison. The United Nations will be particularly challenged to help moderate these conflicts, especially given the right to veto given to some of the world's wealthiest nations on the Security Council.

While these consensus scenarios are dire, a series of Conference of Parties meetings, including the 2022 COP 27 meeting in Egypt, harbor hope to defuse potential future conflicts by fomenting policy responses and investments today. These Conference of Parties meetings have made a great deal of progress in theory, and have set the agenda for a series of major climate change meetings, with the Paris Agreement the most recent example in 2015.

However, the pledges made at these conferences and protocol meetings, especially by wealthy nations, while often ambitious, have almost universally failed to produce the promised results. The reasons are a study in political posturing, the dichotomy between national leaders and domestic leadership, and the difficulty of world leaders to fully integrate intergenerational benefits against intragenerational costs.

THE PARIS AGREEMENT

The meeting of the 196 countries that participate in the Conference of Parties to the United Nations Climate Accord set a standard to reduce greenhouse gas emissions by 50% over 2005 levels by 2030, and become carbon–neutral by 2050. The benchmark from 2005 was 44,153 million tons of carbon dioxide equivalent (44.153 Gtonnes CO_2E). This goal would bring global emissions annually to 22.076 Gt CO_2E) by 2030 (Fig. 19.21).

Global emissions in 2021 reached 40.8 Gt of CO_2 equivalent, primarily due to a substantial rise in energy-related carbon dioxide emissions. Yet, by 2020, 80% of global energy consumption still remained fossil fueled. In 2019, BP estimated that there remains approximately 53 years of known oil reserves at current consumption levels, and more than a century of reserves at 2030 levels

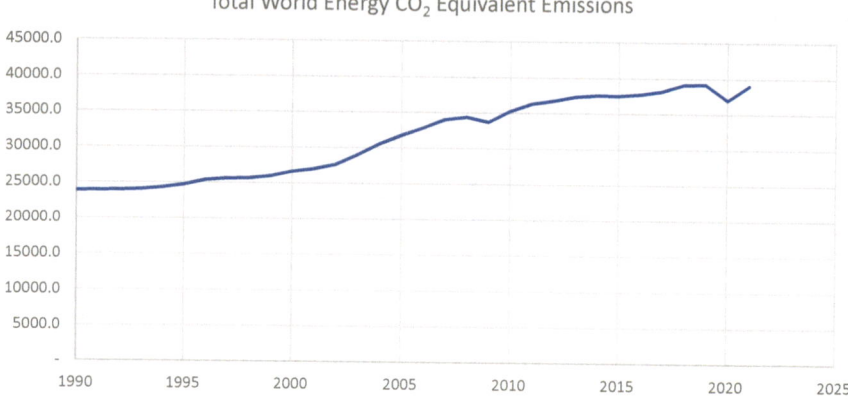

Fig. 19.21   Total world carbon dioxide equivalent emissions

of consumption necessary to limit global warming to the level agreed upon at the Paris Agreement.

But, while there remain known reserves lasting well into the next century based on limits negotiated at the Paris Agreement, fossil fuel corporations continue to expand their exploration budgets. Oil refining capacity, coal, and natural gas exploration and processing investments are estimated to rise to over $650 billion US in 2022 after two years of increases above 2020 levels. While the pace of clean energy development and capacity spending is also accelerating to twice the level of fossil fuel investments, it is evident that the fossil fuel industry is not quickly ratcheting down exploration and their intended expansion of reserves.

This is despite adequate reserves lasting almost four times longer than the point that scientists and the Conference of Parties agree is necessary to reach carbon neutrality and avoid exceeding the 1.5–2.0° Celsius tipping point.

Recall the SSP5 scenario assumed each nation acts as a *free-rider* by pursuing domestic economic development goals without regard to its effects on other nations. Under this path, the correlation between global Gross Domestic Product, energy usage, and atmospheric depletion is expected to continue under the *business-as-usual* scenario (Fig. 19.22).

Instead, the *business-as-usual* upward trend in emissions shows a growing gap between the agreement nations forged in the Paris Agreement and expected emissions given the current path. The energy budget necessary to attain pledges made in the Paris agreement results in substantially reduced fossil fuel extraction compared to known reserves totaling more than 3.1 trillion barrels. Figure 19.23 shows that the total need for barrels of oil equivalent (coal, oil, natural gas) from 2025 to 2035 is 1.207 trillion barrels, while known oil reserves already account for 1.763 trillion barrels, with peak oil estimated to not occur before 2025. If we relied on oil alone, we already have more

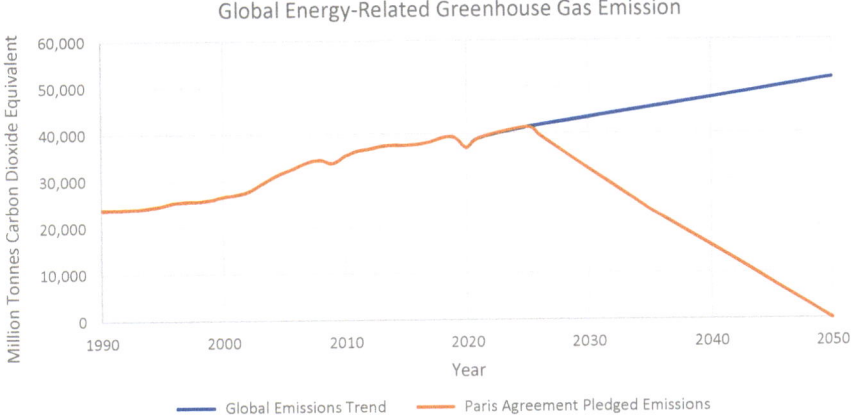

Fig. 19.22 Greenhouse gas emission trends and Paris Agreement pathway

	Total Known Petroleum Reserves	Million Barrels of Oil Equivalent	Value of Fossil Fuel Reserves (trillion dollars US) Based on $90/barrel Oil
known oil reserves tonnes	236,294,751,172	1,762,759	$159
known natural gas reserves m³	188,073,776,245,117	1,106,316	$100
known coal reserves tonnes	1,074,108,000,000	237,635	$21
	Total:	3,106,710	$280
price of oil:	$90	size of global economy 2020 Trillions $US	$85
Maximum Consumable BOE to Taper to Carbon Neutrality by 2050		1,207,715	$109
Jacobson et al Sustainability Investment Estimate (trillions)	$61.5	Necessary Carbon Surtax per BOE or Equivalent	$51

Fig. 19.23 Known petroleum reserves and allowable fossil fuel usage to meet climate goals

than half a trillion surplus barrels of oil than we need to meet climate target pledged in the *Paris Agreement*.

Meanwhile, the total value of fossil fuel reserves, imputed at a price of oil of $90 per barrel, remains a staggering $280 trillion, a sum that oil producing countries and companies are loathe to sacrifice. Figure 19.24 illustrates some of the sources of greenhouse gas emissions in our fossil-fuel-based economies.

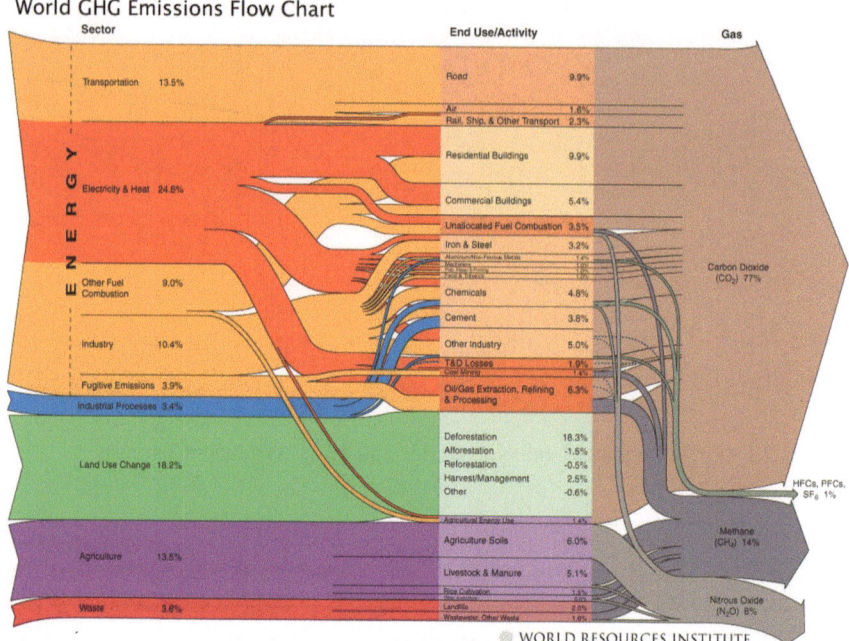

Fig. 19.24 Greenhouse gas emissions by sector and type (Courtesy World Resources Institute, https://www.wri.org/data/world-greenhouse-gas-emissions-2005, accessed January 22, 2023)

The economies especially in the MDCs are highly dependent on fossil fuels. We see that the level of fossil fuel consumption and carbon dioxide emissions very accurately tracks the level of GDP growth globally (Fig. 19.25).

It is apparent that, for the most part, countries and corporations still tie economic growth to increased fossil fuel consumption and hence greenhouse gas emissions. Until a new paradigm is adopted that demonstrates to governments and the private sector that economic growth can occur while greenhouse gas emissions and fossil fuel combustion decline, it appears that the more pessimistic SSP5 business-as-usual scenario may prevail.

Summary

While the tools of economics give us insights into what must be done to satisfy intergenerational natural capital externalities, the single most significant shared natural resource upon which all humankind and our ecosystem relies continues to be depleted despite the pleadings of scientists and social scientists alike. The gulf between good policy and politics seems more intractable than the challenges we have partially overcome in internalizing intragenerational externalities. The next section describes the types of intergovernmental agreements

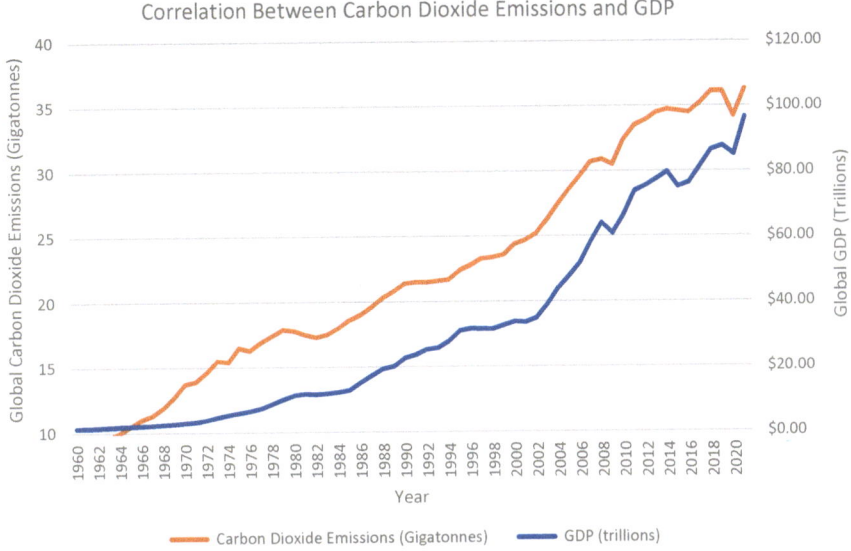

Fig. 19.25 Relationship between world GDP and carbon dioxide emissions

and corporate strategies that must be forged to avoid the greatest consequences of global warming. In addition, the elements that must be established to promote human and natural sustainability are discussed, and the barriers to implementation of these elements are explored. Tools are outlined that distribute the responsibility and awareness of these issues across both public and private sectors.

ESG Toolkit
The major economies of the world have departed from the trajectory agreed upon to stem the worst of global warming.
    Does this departure absolve corporations of their responsibility?
    What can corporations do collectively to meet sustainability goals?

NOTES

1. https://www.oxfordreference.com/display/10.1093/oi/authority.201108030 95711482;jsessionid=E66DEBCEBDF56AF410C4C2B021B55F02, accessed February 19, 2023.
2. Armstrong, David L., Arie Staal, Jesse F. Abrams, Ricarda Winkelmann, Boris Sakschewski, Sina Loriani, Ingo Fetzer, Sarah E. Cornell, Johan Rockstrom, and Timothe M. Lento (2022). "Exceeding 1.5 °C Global Warming Could Trigger Multiple Climate Tipping Points", Science. **377**(6611).

3. https://www.sciencedaily.com/releases/2008/10/081020095850.htm#:~:
   text=7000%20years%20ago.-,The%20Arctic%20Ocean%20may%20have%20b
   een%20periodically%20ice%20free.,about%206000%2D7000%20years%20ago,
   accessed November 9, 2022.
4. https://en.wikipedia.org/wiki/Ozone_layer, accessed November 9, 2022.
5. Fourier, Joseph Jean-Baptiste. (1824). "Remarques Générales Sur Les Températures Du Globe Terrestre Et Des Espaces Planétaires", Annales de Chimie et
   de Physique. 27: 136–167.
6. https://www.metoffice.gov.uk/research/climate/seasonal-to-decadal/long-
   range/forecasts/co2-forecast#:~:text=This%20is%20a%20good%20guide,of%
   20419.2%20ppm%20in%202023, accessed February 17, 2023.
7. https://www.nature.com/articles/s41586-020-1991-8, accessed November 9,
   2022.
8. https://en.wikipedia.org/wiki/Climate_change, accessed November 9, 2022.
9. Budyko, M. I. (1961). "The Heat Balance of the Earth's Surface", Soviet
   Geography. 2(4): 1961.
10. Manabe, Syukuro, and Richard T. Wetherald. (1967). "Thermal Equilibrium
    of the Atmosphere with a Given Distribution of Relative Humidity", American
    Meteorological Society, May 1, 1967: 241–259.
11. The Study of Critical Environmental Problems, MIT Press, 1970.
12. Study of Man's Impact on Climate, editors. (1971). Inadvertent Climate
    Modification: Report of the Study of Man's Impact on Climate, MIT Press.
13. Manabe, Syukuro, and Richard T. Wetherald "The Effects of Doubling the
    CO_2 Concentration on the Climate of a General Circulation Model", American
    Meteorological Society, January 1, 1975.
14. Armstrong, David, G. McKay, Aris Stahl, Jesse F. Abrams, Ricarda Winkelman,
    Boris Sakschewski, Sina Loriana, Ingo Fetzer, Sarah E. Cornell, Johan Rock-
    strom, and Timothy M. Lenton. (2022). "Exceeding 1.5 °C Global Warming
    Could Trigger Multiple Climate Tipping Points", Science. 377(6611).
15. Smith, Joel B., Stephen H. Schneider, Michael Oppenheimer, Gary W. Yohe,
    et al. (2009). "Assessing Dangerous Climate Change Through an Update of the
    Intergovernmental Panel on Climate Change (IPCC) 'Reasons for Concern'",
    Proceedings of the National Academy of Sciences. 106 (11): 4133–4137.
    Bibcode:2009PNAS.106.4133S. https://doi.org/10.1073/pnas.0812355106.
    PMC 2648893. PMID 19251662.
16. Levermann, Anders, Peter U. Clark, Ben Marzeion, Glenn A. Milne,
    et al. (2013). "The Multimillennial Sea-Level Commitment of Global
    Warming", Proceedings of the National Academy of Sciences. 110 (34):
    13745–13750. Bibcode:2013PNAS.11013745L. https://doi.org/10.1073/
    pnas.1219414110. ISSN 0027-8424. PMC 3752235. PMID 23858443.
17. Collins, M., R. Knutti, J. M. Arblaster, J.-L. Dufresne, et al.
    (2013). "Chapter 12: Long-Term Climate Change: Projections, Commitments
    and Irreversibility" (PDF). IPCC AR5 WG1 2013, pp. 1029–1136.
18. IPCC. (2021). "Summary for Policymakers" (PDF). IPCC AR6 WG1 2021.
19. Bindoff, N. L., W. W. L. Cheung, J. G. Kairo, J. Arístegui, et al.
    (2019). "Chapter 5: Changing Ocean, Marine Ecosystems, and Dependent
    Communities" (PDF). IPCC SROCC 2019, pp. 447–587.
20. Watts, Nick, W. Neil Adger, Paolo Agnolucci, Jason Blackstock, et al.
    (2015). "Health and Climate Change: Policy Responses to Protect Public

Health". The Lancet. **386** (10006): 1861–1914. https://doi.org/10.1016/S0140-6736(15)60854-6.

21. IPCC. (2022). Pörtner, H.-O., D.C. Roberts, M. Tignor, E. S. Poloczanska, K. Mintenbeck, A. Alegría, M. Craig, S. Langsdorf, S. Löschke, V. Möller, A. Okem, B. Rama, et al. editors. Climate Change 2022: Impacts, Adaptation and Vulnerability. Contribution of Working Group II to the Sixth Assessment Report of the Intergovernmental Panel on Climate Change, Cambridge University Press.

22. Hallegatte, Stephane, Mook Bangalore, Laura Bonzanigo, Marianne Fay, et al. (2016). Shock Waves: Managing the Impacts of Climate Change on Poverty. Climate Change and Development (PDF). Washington, DC: World Bank. https://doi.org/10.1596/978-1-4648-0673-5. hdl:10986/22787. ISBN 978-1-4648-0674-2.

23. Balsari, S., C. Dresser, J. Leaning. (2020). "Climate Change, Migration, and Civil Strife", Current Environmental Health Reports. 7 (4): 404–414. https://doi.org/10.1007/s40572-020-00291-4. PMC 7550406. PMID 33048318.

24. Callahan, C. W., and J. S. Mankin. (2022). Science Advances. **8**, eadd3726.

PART V

Attainment of ESG Goals—A Public and Private Partnership

The *sustainability ethic* is based on the premise that one generation should not have the privilege by the order of our birth to deprive a subsequent generation of the same capacity to provide for its needs. The principle asserts that each generation should ensure human sustainability by providing the same opportunity to ensure prosperity and human rights across all generations. The ethic requires preservation of the ability of the natural system to be maintained so that the ecosystem remains healthy. It also argues for a fair and efficient economic distribution system that can be sustained to provide the infrastructure that mediates human needs.

Efficient and sustainable natural resource usage is generally the benchmark for anthropocentric sustainability. The Club of Rome concerns about population growth that outstrips resource availability revived consideration of the dismal Malthusian prophecy from more than a century and a half earlier. The Brundtland Commission concern that sustainability must be ensured, but not at the expense of the prevention of less developed nations to aspire for their own development placed even greater challenges on the preservation of fixed resources across human generations.

A further challenge was the realization that sustainability depends crucially upon the decisions of corporations that remain the core of the private sector in free-market-oriented economies. Government can help guide but does not typically substitute for the production and resource usage decisions of the private sector. If corporations do not recognize their *corporate social responsibility (CSR)* in the mix of principles that preserve resources necessary to support human consumption, then the onus of sustainability is placed entirely on governance.

Sufficiently far-sighted and effective governance across all nations interconnected by our dependence on and influence over the environment is difficult to coordinate. An *Environmental, Social, and Governance (ESG)* perspective was first defined by the 2005 United Nations Global Compact. It was subsequently included in the *United Nations Principles for Responsible Investment* a year later. Nations increasingly appreciate the need to integrate sustainability goals from the Brundtland Commission, the Paris Agreementand COP 27. These agreements hinge on the ability to also marshal private sector capacities and objectives.

These ESG principles act as a benchmark for the measurement and communication of corporate responses to sustainability challenges. They are not confined solely to environmental preservation. Indeed, while such issues as corporate carbon footprints provide avenues for corporations to contribute to energy sustainability, sustainable development also requires sustainable social practices that include worker health and safety, diversity, equity and inclusion, and protection of human rights within the corporate veil. In addition, governance practices must be established that align and communicate these corporate values to consumers, suppliers, employees, and markets.

The United Nations and leading corporations and investor groups now recognize the need for a partnership if humans wish to succeed in established sustainability goals. The UN has increasingly tied sustainability development goals into a broader set of principles called the *Millenium Development Goals (MDG)* that established eight criteria for sustainable human development in the year 2000 on the cusp of the new millennia. These include the need to:

1. Eradicate extreme poverty and hunger
2. Achieve universal primary education
3. Promote gender equality and empower women
4. Reduce child mortality
5. Improve maternal health
6. Combat HIV/AIDS, malaria, and other diseases
7. Ensure environmental sustainability
8. Develop a global partnership for development

Figure VI-0-1—United Nations Millennium Development Goals
(Courtesy of the United Nations - https://www.un.org/millenniumgoals/)

While the United Nations had originally aspired to reach these goals by the year 2015, failures to attain them led to a rephrasing of these and other goals that became the basis of the *Agenda for Sustainable Development by 2030*. It articulated the restatement to "build on the Millennium Development Goals and complete what they did not achieve." They also added *Peace* to the *People, Planet, and Prosperity* (3P) mantra that was described in Chapter 1. While progress to success remains elusive, these well-articulated goals in the Agenda nonetheless remain influential and have been involved in many subsequent discussions.

The eight *Millenium Development Goals* have grown to seventeen specific *Sustainable Development Goals*, with 169 specific targets defined to help realize these broader goals. These new goals span the social, economic, natural, and environmental aspects of sustainability. They include traditional goals such as the alleviation of poverty and hunger and the provision of clean water in goals 1, 2, and 6, environmental goals such as affordable and clean energy, preservation of the ecosystem in SDGs 7, 14, and 15, human economic infrastructure through economic growth innovation, reduced workplace inequality and sustainable communities and cities in goals 8, 9, 10, 11, and 12, climate change in 13, and social goals for reduced inequality, good health, consistent and high-quality education, gender equality, and peace in goals 10, 3, 4, 5, and 16, respectively. They state in goal 17 that entities shall attain these goals through the necessary partnerships, including among nations and with corporations.

Figure 2.2 **SUSTAINABLE DEVELOPMENT GOALS.** *Reprinted with permission of the United Nations Sustainable Development Goals [https://www.un.org/sustainabledevelopment/]. The content of this publication has not been approved by the United Nations and does not reflect the views of the United Nations or its officials or Member States.*

Figure VI-0-2—The United Nations Sustainable Development Goals (Courtesy of the United Nations)

Much of the success then depends not only on the ability to construct sound and sustainable public policy, but also on the ability of scientists and technocrats to assist in the conversion of principles to effective regulations. Success must also be mindful of the degree to which stakeholders and corporations in the private sector will embrace these frameworks.

These targets allow various stakeholders to determine and measure their engagement and success. Targets for various types of stakeholders are tailored to the extent of their footprints and roles and the potential for their practices to affect sustainability. Some sectors such as energy or resource extraction interact much more profoundly with climate or resource objectives. All sectors span concerns for worker rights and diversity, equity, and inclusion. Likewise, all sectors depend on effective governance to establish standards and enforce regulations designed to promote success in the attainment of the SDGs.

Corporate participation is critical because corporations extract and purchase the vast bulk of natural resources and produce the goods and services that contribute to the modern market economy. Commerce also determines the degree to which corporations devote their financial capital to purchase natural capital rather than human-made capital. A corporate farm can use more fertilizer derived from the mining of the fixed factor phosphorous, or it can instead employ more labor and machine-intensive techniques in managing the land. A utility must decide whether to purchase natural gas to continue to fuel a gas-powered peaker electric generation plant rather than use a combination of solar power and battery storage. A steelmaker determines whether to use coal or hydrogen to operate their ovens or use scrubbers to cleanse emissions rather than purchase pollution allotments that enable their smokestacks to continue to operate.

In conducting their affairs, corporations make decisions that affect the stock of abiotic natural capital in the form of the resources in the ground, the oceans, and our atmosphere. These physical natural resources are augmented by our biotic resources, such as fisheries, forests, and farms that can be managed to provide a sustainable flow of biotic resources. Each resource has an optimal sustainability path, and each path is managed primarily in the private sector, with influence exercised by enlightened governance.

Recent research has identified a pathway that permits nations to succeed in arresting and slowly reversing global warming without destroying economic growth. A *Green New Deal* is designed to limit any further depletion of our atmospheric stock that protects the ecosystem. If successful, over ensuing millennia, the atmosphere shall be repaired and the ecosystem will be rebuilt, albeit in a new form. Upon the completion of the necessary infrastructure buildout of its alternative energybackstop, fossil fuels never need to be extracted. Any fixed resources necessary in the creation of alternative and sustainable energy are directed to an infrastructure that is not consumed over time and can be recycled at the end of its service life.

A Green New Deal may then end the consumption of certain natural resources such as the combustion of fossil fuels that emit greenhouse gases. The resulting depletion of the atmosphere should be considered depletion of a fixed stock resource because it reduces natural capital for centuries and millennia. This unabated global warming induces atmospheric depreciation that suffers an extremely long repair time necessary for the natural regeneration of the resource.

A Green New Deal is consistent with *strong sustainability*, at least in energy production, because it provides an ongoing sustainable energy resource without incremental natural resource depletion. It also promises the necessary energy to ensure more effective recycling of other resources to move toward ongoing sustainability in other sectors as well. The next section describes the respective roles of the public and private sectors in the creation of sustainability and social responsibility toward our natural capital.

CHAPTER 20

Responses to Sustainability and Climate Change

The millennium-scale degradation of our atmospheric resources compromises our ability to maintain the Earth's ecosystem. This ecosystem degradation is the existential sustainability issue of the Anthropocene Epoch. Beyond the obvious need to cease the release of greenhouse gases, there are other challenges and opportunities humankind can explore to mitigate harm to future generations.

Ecosystem Implications

The implications of global warming on human and other species differ considerably. Among human populations, peoples who reside in semi-temperate and polar regions such as Canada and Russia, the two largest countries by land mass in the world, will discover that great swaths of previously challenging human habitats become hospitable and more natural resources become accessible. Large regions that could not support a sufficiently long growing season become agriculturally productive.

Temperate regions become warmer while their coastal regions become prone to flooding and more severe storms. The bulk of human populations in the MDCs who live in such regions have the capacity to adapt to and mitigate global warming consequences. Meanwhile, hotter regions in the tropics shall suffer temperatures difficult to bear and for extended periods every year to the degree to which economies are hampered.

These differential effects create conflicts between those who benefit, those who can adapt, and the most vulnerable who suffer the greatest human damage due to heat, flooding and drought, disease and an inability to grow crops.

C. Read, *Understanding Sustainability Principles and ESG Policies*, https://doi.org/10.1007/978-3-031-34483-1_20

Ocean changes are most dramatic because our oceans absorb the vast bulk of the increasingly intense net solar power arising from increases in greenhouse gases. For each degree Celsius increase, scientists predict a one meter sea level rise. Rising seas will erode the important ecosystem component that is the buffer between land and water. Loss of coastal grasses and mangrove stands that harbor myriad species no longer provide their buffers, nor their protection from coastal storms.

At the same time as the sea level rises substantially, the ocean's chemical composition changes rapidly. Increased runoff of cold fresh water in glaciated areas causes a change in ocean temperature and salinity. Underwater currents such as the Gulf Stream may cease entirely as differences in temperatures and salinities that drive oceanic currents are disrupted.

So too are nutrient balances disrupted. Species optimized to thrive at certain temperatures are forced to migrate toward the poles but nutrients may not be able to support displaced species populations. Oceans also become more acidic as higher carbon dioxide atmospheric concentrations induce a greater amount of carbonic acid to form in water. These lower ocean pH values especially harm species for which their calcium-based shells dissolve with increased oceanic acidity. Such effects accelerate rapidly because of the compounding positive feedback that occurs when various tipping points are triggered.

These various tipping points erode resiliency buffers that protect against physical or ecosystem collapse. Tripping of tipping point thresholds makes meteorological and ecological systems more vulnerable to regular perturbations that impinge randomly on any system. The result is an acceleration of the degradation of a wide variety of phenomena in a nonlinear and sometimes even exponential rate. These tipping points include:

- melting of permafrost that exposes stored methane in the form of peat bogs and methane hydrate crystals, which increases methane emissions and further accelerated global warming.
- the cutoff of waterflow from the tropics to the temperate region via the Gulf Stream, which causes significant changes to North Atlantic temperatures and climate.
- rapid melting of grounded ice sheets in Greenland and the West Antarctic, which can raise sea levels by upwards of ten feet by the end of the century.
- a shift in seasonal monsoons in India, Pakistan, Bangladesh, and Indonesia that changes patterns of flooding and drought and modified agricultural productivity in habitats that support more than two billion residents.

Scientific models of the climate and the geophysical system are increasingly sophisticated and able to discern subtle interactions between effects, differential regional impacts, and the risks arising as resiliency buffers are reduced.

Humankind has put in motion shifts of unprecedented rate and scale. Necessary responses are massive in scale, expensive, and uncertain given the lack of experience of humankind in any such past circumstances.

We shall see in Chapter 22 that solutions with significant global ramifications require potentially confounding diplomacy to be effective. Given that nations have consistently failed to mount and fund much more modest methods to mitigate rather than reverse global warming, international cooperation and sufficient redistribution of resources from MDCs to LDCs are likely elusive.

GEOENGINEERING

The range of potential remedies to humankind and the ecosystem requires atmospheric greenhouse gas concentration reversals through geoengineering or mitigation of damages imposed and adaptation to a new steady state. A variety of prospective responses to geophysical conditions never before contemplated by humankind have been proposed. Most of these responses involve a reduction in insolation impinging on the Earth. Such *geoengineering* mimics the natural ability of volcanic eruptions to induce periods of global cooling that can persist for years. With a volcanic eruption comes emissions of sulfur that causes water vapor to condense into droplets and clouds that can reflect solar radiation back into space before it reaches and warms the surface of the Earth.

Other proposals include the injection of large numbers of satellites into low Earth orbit that can unfurl very large reflective blankets of mylar or other lightweight materials to reflect the Sun's rays before they penetrate the atmosphere. Such solutions remain fraught with uncertainties and unintended consequences, and are likely extremely expensive and resource intensive.

Reversal of geophenomena has never been tried on the massive scale necessary to remediate the effects of global warming. It has taken generations of scientists to refine our understanding of global warming to the current state-of-the-science, and will likely take decades to fully understand the consequences and potential successes of geoengineering. Much research must be done to determine consequences of an increasingly aerosoled atmosphere engineered to reflect solar energy, or suspensions of particulates that are also known to harm plants and animals.

Carbon sequestration can potentially allow society to avoid some of the pitfalls of more massive and more consequential atmospheric geoengineering. Sequestration can involve the removal of carbon dioxide from the atmosphere by injecting the gas into underground cavities that can trap it indefinitely. Depleted oil and natural gas wells and mining caverns can act as permanent sinks for carbon dioxide, but are highly geographically dependent. The volume of gaseous carbon dioxide, even if it can be compressed and remain compressed underground, requires multiple volumes relative to the oil and natural gas extracted that generated the carbon dioxide.

Chemical and organic processes can also be used to mimic or amplify the natural process of photosynthesis or other chemical processes that harness insolation. Abundant alternative energy can transform carbon dioxide into more stable or usable materials or reengineered into sustainable synthetic fuels. The bonds that form to create wood or carbohydrates can produce foods, fertilizers, filler materials, or building materials.

These chemical compounds can sequester carbon by placing them into storage or by recycling these compounds in economically beneficial ways. While the carbon-to-double-oxygen is quite robust, conversion to carbon monoxide is not prohibitively energy intensive. The resulting carbon monoxide can be used as a precursor for a variety of industrial chemicals, from fuels to plastics.

Direct removal of carbon dioxide from the air and its reinjection into the ground can also be engineered to mimic nature's conversion of carbonates and calcium deposits to geologically stable calcium carbonate. A company called Climeworks has produced a trial plant in Iceland that can convert 4000 tons of carbon dioxide into a carbonate called calcite. By dissolving carbon dioxide into water, the resulting bubbly soda water reacts with specific geologies to create stone. Basalt rocks from volcanic deposits that are magnesium-, calcium-, and iron-rich readily react with the soda water. Such carbon-based products can even be used to reduce the carbon dioxide generation and fortify cement production, an activity that generates about 8% of anthropogenic carbon dioxide emissions.

The volcanic island of Iceland can tap its rich endowments of hydro-electricity to create an excellent opportunity for permanent sequestration of carbon dioxide sequestered in soda water. However, the cost of scaling plants capable of sequestering thousands of tons of carbon dioxide each year pales compared to the tens of billions of metric tons emitted each year. Ten million equivalent facilities, with massive energy costs associated with them, would be necessary to completely mitigate anthropogenic carbon dioxide emissions.

These engineered solutions tend to be expensive and do not reduce the original cause of global warming since they are designed to counteract anthropogenic effects rather than remove the root causes of global warming. Sufficient scaling to remove more than 40 billion tons of carbon dioxide equivalents annually is prohibitively expensive. More research and development is warranted to determine the viability of geoengineering in the adaptation to the consequences of global warming.

MITIGATION

By addressing the root cause of global warming, mitigation reduces the damage caused by consequences such as rising sea level or changing ocean currents. Reductions in fossil fuel combustion and their associated emissions, elimination of methane releases arising from natural gas extraction, distribution, and incomplete combustion, and reduced release of nitrous oxides and ozone-depleting gases all constitute mitigation. Unlike geoengineering, these solutions offer permanent solutions to the consequences of global warming.

Until relatively recently, mitigation of global warming has been in the substitution away from the generation of heat from the worst greenhouse gas-emitting forms of fossil fuel combustion toward more efficient fuels as measured in the energy generated per ton of carbon dioxide produced. Cleaner coal substituted for less efficient and more polluting coal, oil-based fossil fuels replaced coal, and then natural gas replaced forms of liquid petroleum. Each of these innovations reduced carbon dioxide emissions, often by a factor of two or more, but has only slowed the increase in accumulated emissions of carbon dioxide that approaches two trillion tons since 1850.

Jacobson (2022)[1] describe the scale of investment necessary to mitigate fossil fuel emissions by substituting sustainable energy sources for fossil fuel combustion. The scope of their recommendations includes a number of initiatives that will require an investment by humankind approaching $61.5 trillion US. The researchers note that there is not a one-size-fits-all solution to the provision of alternative forms of energy to substitute for the use of fossil fuels. Rather, mitigation solutions must be carefully tailored to local conditions and opportunities.

On the demand side, Jacobson recommends that alternative technologies must be made available to replace the fossil fuels used for heat and industrial processes. Abundant electric capacity can permit the large-scale production of such alternatives as ammonia and hydrogen for combustion, although there remain infrastructure challenges to distribute these fuels given their requirements for high pressure and or low storage temperature.

Every region of the world has unique energy characteristics and varying potential to fill local demand with unique local supply opportunities. Iceland, for instance, has significant endowments of geothermal energy and the potential for additional hydroelectricity, but smaller island nations may not have the available land for onshore wind generation or the land and necessary spacing for photovoltaic panel farms. Instead, nations may instead need to depend on offshore wind generation and floating solar power at higher costs.

Mitigation must be carefully tailored to geographic differences in natural capital. For each region, Jacobson was able to determine the most efficient mix of land-based, floating, and rooftop solar power, on and offshore wind, hydroelectricity, geothermal energy, centralized steam production, wave and tidal power, and the necessary expansion of distribution and energy storage networks necessary to support optimal national energy portfolios and energy needs. Their analyzes were developed to optimize regional solutions across 145 countries that represent almost all of the world's population and land mass.

They determine that land and water area intensity to support a *Green New Deal* is large but not overwhelmingly so. About 0.17% of available land must be devoted to alternative energy production and distribution. Another 0.36% of land must be shared with other uses while still supporting necessary spacing, for instance to optimize the orientation of the Sun's incidence throughout the day or the spacing between wind turbines.

Significant resources must also be diverted to manufacturing. The massive investment costs of such new energy infrastructure can be recouped in a matter of years rather than decades. This expansion of physical and resource capacity is significant but not overwhelming. In addition, a large amount of human resources must also be employed in energy production. Jacobson estimates that 55.6 million long-term jobs will be created in a Green New Deal, while 27.2 million jobs will be lost from such industries as fossil fuel exploration, extraction, and distribution, and energy production.

Not all economies share equally in the employment benefits and costs of a Green New Deal, though. Net jobs will be lost in a few countries that are highly energy production dependent, such as Canada, Russia, and some nations in Africa. However, most nations will benefit from new and well-paying direct jobs in alternative energy production, indirect jobs in related industries, and induced jobs from those who benefit from the spending of income generated in these sectors.

The physical footprints of alternative energy vary considerably across each energy source. Hydroelectric power is very land-intensive, with a requirement of 502,380 m^2/MW, while wind generation needs only an average of 3.22 m^2/MW. There can also be synergies between these land uses. Hydroelectric reservoirs are already freshwater, accessible, and less vulnerable to wave action, and large-scale power distribution is nearby. This power source can also be throttled relatively quickly to adjust to compensate for the loss of capacity for solar power when the Sun is not shining.

Reservoirs are also ideal for floating solar power installations that convert insolation to electricity at upwards of 30% efficiency. Evaporation of water is reduced and solar panels can run cooler and more efficiently when located above a water source. Reservoirs collocated with hydroelectric dams also have existing access to the electric grid to move solar electricity to end users.

The authors determine that a total power capacity of 41.742 Terawatts of power will be provided by the following sources derived from data from their Fig. 20.1:

ADAPTATION

Finally, more local and idiosyncratic solutions involve *adaptation* to the consequences induced by global warming, such as construction of seawalls and planting of mangrove stands in coastal areas, better insulation of homes so they may withstand greater outside temperatures, and even the reconstruction of villages as has proven necessary in northern coastal communities to protect against ocean encroachment.

Under any scenario, humankind and the ecosystem will necessarily require a combination of adaptation and other measures given the little time available

Onshore wind	32.1%
Offshore wind	12.9%
Residential Rooftop PV	5.7%
Commercial, Public Rooftop PV	9.9%
Utility PV	30.0%
Central Steam Production	2.7%
Geothermal Electricity	0.7%
Hydroelectricity	4.9%
Wave electricity	0.08%
Tidal electricity	0.04%
Solar thermal heat	0.42%
Geothermal heat	0.49%
Total	100%

Fig. 20.1 Mix of energy resources in a sustainable economy according to Jacobson et al.

to reduce greenhouse gases and, ideally, remove these gases from the atmosphere so it may return it to a state that predated the Industrial Revolution. More than 63 trillion tons of carbon dioxide alone has been emitted into the atmosphere since the onset of the Anthropocene Epoch. About half of this additional carbon dioxide released into the atmosphere has been removed by natural processes that have at the same time changed for centuries the level of acidification of the Earth's oceans and the relative and differential abundance of the ecosystem's flora.

Adaptation typically invokes techniques that have been employed by humans and by insurers for centuries to mitigate risks. Homes built in an increasingly flood-prone area are moved to higher ground. Berms and dunes are reestablished to replace those lost in storms. Drought-prone or salt tolerant species may be replanted to ensure continued agricultural production. Trees and grasses are planted to prevent further erosion, and rivers are dammed or diverted to control increased precipitation. Even entire cities such as Amsterdam and Venice are preserved through giant movable tidal barriers. Following such interventions to adapt to shifts or misunderstanding of natural processes, humans may impose planning and zoning changes to ensure past mistakes are not repeated.

These techniques have been employed by economies that are sufficiently advanced and wealthy to overcome engineering challenges and financial burdens. The scale of global warming is so large that it will challenge even the most affluent MDCs with efficient markets and forms of government that can redefine various property rights and standards. Nations less endowed with the luxury of strong and deep financial markets, sufficient expertise and technology transfer, insurance markets, and appropriate government institutions must instead endure cycles of building, destruction, and rebuilding in ways that may not adequately indemnify them from future climate calamities. Differences among peoples in their ability to adapt have profound environmental justice consequences.

While there is barely a century and a half between the onset of the Anthropocene Epoch and a pledge of nations to attain carbon neutrality by 2050, solutions will require vigilance for scores of generations of humankind to come so they may guide the recovery of the atmosphere. These are the long-time spans of geological epochs, and well exceed the ability of humankind to plan for effective remediation. Hence, adaptation remains an existential challenge in our human response to global warming.

SUMMARY

The challenge arising from the depletion of our atmospheric natural capital is significant, but there are solutions. They tend to be expensive, and require geoengineering, adaptation, and mitigation. But, while the required investments challenge the ability of the global investment community, sovereign funds, and the capacities of the World Bank and the International Monetary Fund, the returns from such investments are equally profound. The planet is afforded an opportunity for unlimited sustainable energy. The next section outlines what the public and private sectors can do to bring the planet closer to sustainability.

ESG Inquiries

To what degree does your institution mitigate its carbon footprint within your organization?

To what degree has your institution adapted to sustainability challenges?

Does your organization's strategy include elements with regard to its depletion of natural capital?

NOTE

1. Jacobson, Mark, Anna-Katharina von Krauland, Stephen J. Coughlin, Emily Dukas, Alexander J. H. Nelson, Frances C. Palmer and Kylie R. Rasmussen. (2022). "Low-Cost Solutions to Global Warming, Air Pollution, and Energy Insecurity for 145 Countries", Energy and Environment Science. **15**, 3343, https://doi.org/10.1039/d2ee00722c.

CHAPTER 21

The Public Sector and ESG—Environmental Policy

Challenges to sustainability, of our environment, among our peoples, and within and across our institutions may appear overwhelming. A systematic approach to the various facets that demand our attention will provide the blueprint for humankind to construct strategies to attain sustainability. The ESG paradigm can be used to categorize and organize these sustainability challenges. Each of the three pillars is invoked to describe the process in the resolution of interests often in conflict when faced with significant challenges to sustainability. They are described in turn:

THE ENVIRONMENTAL PILLAR

In the environmental category, the Environment, Social, and Governance paradigm identifies:

- Air, Water, and Solids Pollution, including plastics chemicals and microplastics
- The intergenerational use of fixed resources
- Sustainable use of resources
- Renewable Energy
- Environmental justice
- Climate Change and Atmospheric Pollution
- The ecosystem and markets
- Lifecycle stewardship
- Green bonds and carbon markets.

C. Read, *Understanding Sustainability Principles and ESG Policies*, https://doi.org/10.1007/978-3-031-34483-1_21

The economics of pollution, the importance of recycling, the use of fixed and sustainable resources, and the development of renewable energy, are quite well understood. The issue of optimal extraction of these resources requires a combination of Pigouvian taxes to internalize externalities in the case of such regional harms as pollution, the imposition of quotas or the establishment of property rights for renewable resources that can be used sustainably, and the creation of permanent funds to spread the benefits of the extraction of a depletable resource over time to benefit all generations.

Regulators and markets within a nation can establish principles by which these harms can be mitigated and future generations protected. The greater challenge is when decisions made today by one nation harm other nations or generations. This challenge is complicated for issues such as climate change when damage to an ecosystem spans generations and political boundaries, and when a resource defies easy establishment of property rights. Issues of economic justice, market failures, and the frustration of environmental justice can only be partially ameliorated through traditional market mechanisms such as sustainability bonds and carbon markets.

The greatest challenge to the application of sustainability theory in the management of issues such as global warming is often the lack of jurisdiction and well-established property rights. Negotiation of global treaties in the presence of border-spanning externalities is difficult. Externalities arise because the abuse by one generation or nation of natural capital may deprive other nations or generations of the sustainable use of the resource.

Global warming encompasses all these challenges. The ecosystem and peoples rely in myriad ways on the sustainability of our atmosphere that is being eroded at an unprecedented rate. The ESG paradigm can be invoked to address the pressing existential sustainability issue of our era.

While market mechanisms function reasonably well for a people that is able to establish appropriate property rights, markets also introduce environmental justice concerns. Any market solution allocates resources proportional to one's wealth and ability to pay. Nations or individuals of higher income command a greater amount of resources in a market economy. Hence, if the avoidance of environmental harm is considered a market privilege rather than a human right, those of lower income will be outbid by wealthier entities. Likewise, health or aesthetic benefits of a better environment will flow to those who can pay more to preserve their health or enjoy environmental amenities. These factors create economic justice conflicts between MDCs and LDCs.

The nature of market allocations based on income and ability to pay also challenges the principle of Benthamite utility that provides for the greatest good for the greatest number of people. To better understand the ways in which markets allocate items valued by humans, it is instructive to recall how economics models human decision-making.

Let us begin by acknowledging that humans are the only species able to dominate and sculpt our environment based on our preferences. Humans shall maintain a healthy and resilient ecosystem only if we incorporate broader ecological values into our human values and reflect these values in markets. Ecological sustainability must then be cultivated or expressed as a human value.

ECONOMY-WIDE EFFICIENCY

Let us assume humans are or can be instilled with a preference for the preservation of the ecosystem or the prevention of environmental harm. How might these preferences be expressed?

Economics models how humans make choices among scarce alternatives. Let us consider a simple example of two values described in Chapter 16. One is the consumption of some sort of human-produced good or service that has no negative environmental consequences. The second value is some sort of a composite environmental quality labeled B, such as our appreciation of a sustainable and resilient ecosystem, separate from the resources that may provide for the human-produced good labeled C.

Let us further assume that these two values compete only for human-made resources such as labor and capital. In other words, we can use our available human resources to consume or to preserve and enjoy our environment. This construct avoids for the moment Herman Daly's observation that many forms of human production necessarily require the consumption of natural resources in finite supply.

We can express the interplay between our preferences for consumption and environmental amenities using our familiar indifference curve approach (Fig. 21.1):

This diagram assumes that there is a tradeoff between consumption goods C and an environmental amenity good B that changes as income rises and affords wealthier individuals greater opportunity to attain satisfaction. A wealthier nation with a more extensive capacity to produce or enhance both goods may favor the maintenance and improvement of environmental amenities once basic consumption needs are met. Hence, an MDC may be more willing to trade off some consumption for an improved environmental or natural capacity (point 2) compared to an LDC (point 1). Economists call goods for which demand increases quickly with rising income as luxury goods. An improved environment is a luxury an MDC can more easily afford.

ECONOMY-WIDE EQUITY

Even within one nation, demographic and income differences delineate categories of consumers, each with their own preferred tradeoff between consumer and environmental preferences. As an example, those of lower income may have insufficient access to health care that can ameliorate various consequences arising from pollution. When evaluating the costs of pollution over shortened lifespans and a lower quality of life to the immediacy of food and shelter, a lower-income household or nation may be willing to sacrifice an

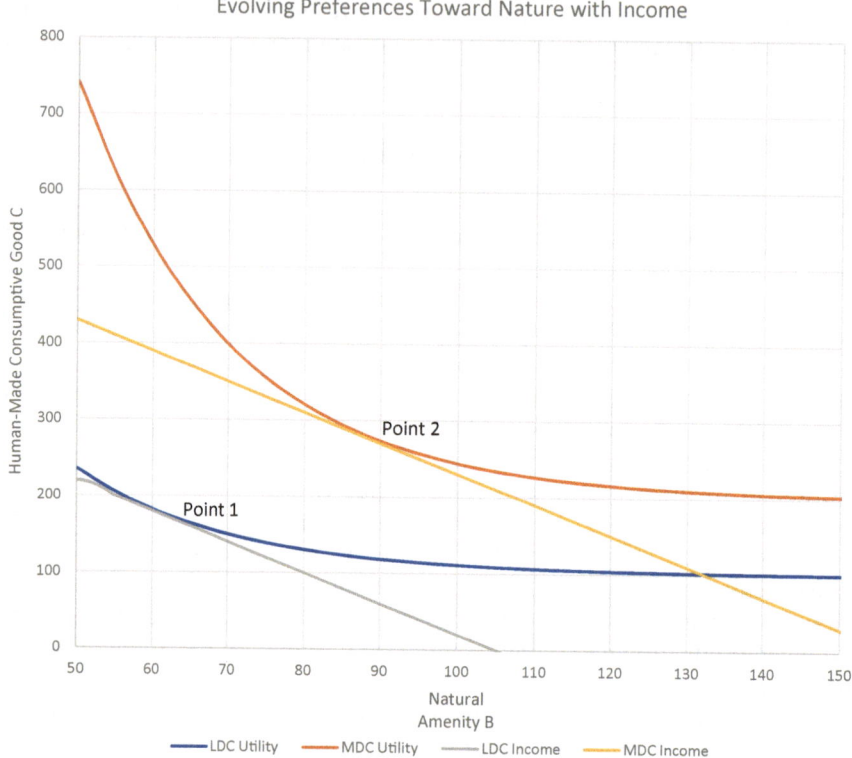

Fig. 21.1 Evolving preferences toward nature and the environment with increased income

improved environment for greater consumption of immediate importance. Alternately, a healthier and more sustainable environment may be preferred by higher-income individuals whose consumption needs have been easily met.

An economy as a whole is an aggregation of various individuals who may value the environment differentially because of either difference in their planning horizon or in their ability to pay for environmental preservation. In the aggregate, resources in a market economy are allocated based on the values of those willing to bid the highest prices. A wealthy economy with degraded natural capacity may then be willing or required to pay some share of their wealth to enhance environmental resources, which will afford those of lesser means an opportunity to enjoy environmental quality beyond what they could generally afford.

These variations in income, preferences, demographics, and capacity have implications on the ability to coalesce socially or politically within a nation to avoid depletion of our natural capital. The difference in the mean preferences versus those at the extremes represents a social cost and a measure of social resiliency. While those who wish to purchase greater environmental preservation can exercise their market power by subsidizing improved environmental quality, market dynamics may leave some citizens unable to afford the costs of environmental preservation. As a matter of economic justice, action to preserve the environment may warrant subsidization for those who can least afford additional environmental amenities.

Within an economy, there are avenues to align such preferences. For instance, federal governments have offered tax credits or subsidies for those of low income to purchase electric cars. Subsidies to low-income households can improve home insulation to minimize energy loss. Revenues from an appropriately high carbon tax to accelerate the movement off of fossil fuel consumption can help fund these initiatives and the construction of sustainable energy sources. A rebate of part of those revenues can also be transferred to lower-income households to afford a measure of economic justice and relieve these households with the income burden of the tax without reducing their incentive to seek cleaner energy sources and avoid the tax themselves. The costs to low-income households arising from higher transitional sustainable energy costs can be offset through government programs funded by carbon taxes. Since energy consumption is correlated with income, the burden of such transfers would then fall primarily on higher income and higher energy consuming households. Such solutions are more elusive since they would require wealth transfers from MDCs to lower income LDCs.

GLOBAL EFFICIENCY

The tools of economic efficiency demonstrate mechanisms to optimize environmental stewardship within a nation. We next explore whether these tools can also ensure greater environmental justice between MDCs and LDCs. Let us postpone for the moment the issue of compensation for damage already caused by the erosion of natural capital that will continue to generate harm for many generations to come. In the meantime, is it possible to better leverage the technologies available to nations globally so that a more efficient mix between consumption and environmental protection can be collectively enjoyed by MDCs and LDCs alike?

If the capacity to produce either the consumption or the environmental good exhibits diminishing returns, the tradeoff between the environment and consumption is not linear. Protection against greater environmental degradation comes at increasingly large sacrifices in consumption. Figure 21.2 shows the case of two nations, one with a lower ability to produce human capital because of a lack of past investment in production capacity, but also with a greater remaining endowment of environmental natural capital.

MDC production relies on its greater availability of technology at equilibrium point MDC-a. The production possibility curve for the LDC relies on a more substantial endowment of natural capital that has yet to be degraded through intensive economic development and yields utility at the equilibrium point LDC-a.

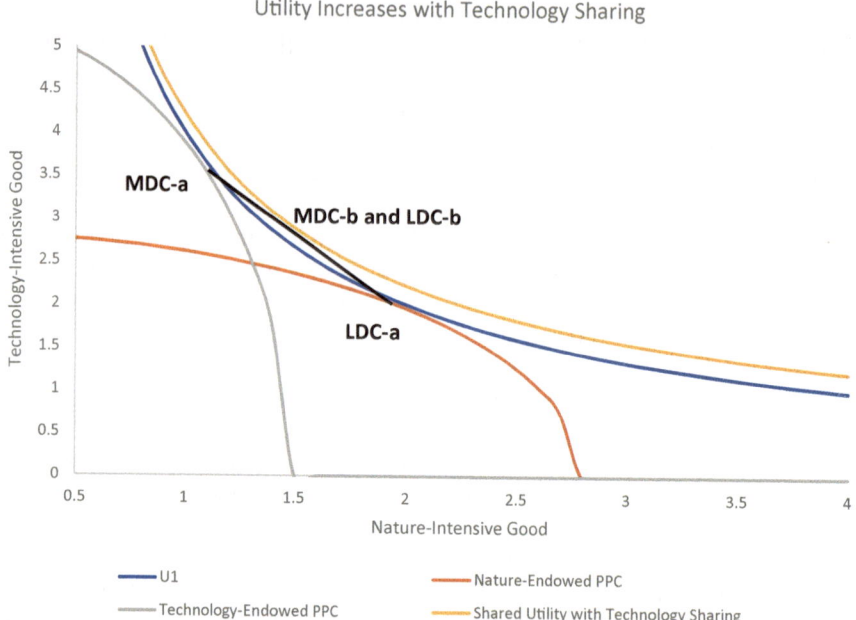

Fig. 21.2 The efficiency gains arising from technology transfer

Even if citizens of two nations have the same preferences toward consumption of technology-intensive goods and of nature-intensive goods, along the common lower indifference curve, an MDC will tend to choose a point that relies on their relatively greater technology endowment, while the LDC will take better advantage of their greater natural endowment. But, by sharing technology and natural capital between the nations, these nations can move along the line that joins both their equilibria to attain an even higher indifference curve at MDC-b and LDC-b.

Here we see the value of global cooperation. Even if nations share identical preferences and indifference curves, cooperation allows the more consumptive economy to trade some consumption for greater environmental stewardship with the less consumptive society. Mutually advantageous trade allows both nations to generate a greater level of overall utility.

The model illustrates the value of global cooperation on a shared environment. We can view the environment as a tradeable good for which the most developed nations can sacrifice some of its consumption to more easily purchase greater environmental stewardship from nations yet to fully develop in the same environmentally consumptive manner.

GLOBAL EQUITY

Such an equitable and efficient approach often requires significant transfer payments between participating nations, in the form of consumption, income, and technology transfers, to secure such an equilibrium. It is the lack of willingness of MDCs to regard the natural capital of the atmosphere as a tradeable good worthy of preserving that has stymied global climate negotiations.

The model demonstrates that there can be another valuable role in capacity transfer. Investments can be made in environment-preserving productive infrastructure in nations yet to fully develop their consumptive infrastructure as have MDCs. Through such technology transfers, LDCs can develop sustainability as the Brundtland Commission had prescribed, but without repetition of the environmental degradation MDCs suffered.

For instance, MDC nations can invest heavily in clean energy infrastructure in LDCs to afford these nations a pathway to develop and even export energy, but without the environmental degradation that MDCs caused through their history of intensive fossil fuel usage. This realization and promise to explore such opportunities, combined with recognition of *loss and damage* funding for LDCs that suffer from climate change, offered a breakthrough at the 2022 COP27 Conference of Parties to the United Nations Climate Accord meeting.

THE FEASIBILITY OF SUSTAINABLE ENERGY

Jacobson (2022) demonstrated that sustainable energy is feasible and even profitable. The study assumes a transition to 100% sustainable energy by 2050 at the latest, but preferably by 2035, with 80% of the transition by 2030 consistent with the recommendations of the Paris Agreement. They note that, along the current path of emissions, the additional amount of carbon dioxide equivalent emissions will exceed the preferred path of a maximum of 1.5° Celsius warming stated by *Paris Agreement* signatories by 2032. It is estimated that, if current emissions levels are maintained to 2041, the 2.0° Celsius threshold will be breached.

They motivate their discussion not only based on the threats to sustainability and ecosystems arising from global warming, but also on the cost of pollution on human health and the risks imposed by energy insecurity. Their primary goal is to measure the return on investment in sustainable energy, capitalized in savings in energy costs over time.

On the demand side, these researchers analyze the current and future estimated needs for energy by considering a baseline *business-as-usual* (BAU) scenario in which energy demand for traditional fuels (coal, oil, natural gas, solar and geothermal sources, wood, and waste heat) is modeled in the residential, commercial, industrial, military, food and building product harvesting, and transportation sectors. They group 145 countries and their energy needs and sources across 24 regions worldwide that represent 99.7% of the world's population.

The study analyzes expected energy demand in the 24 regions across six user groups and seven sectors and estimates the most efficient mix of potential *wind-water-solar* (WWS) and geothermal resources that can augment and replace existing fossil fuels. They combine modeling of global climate patterns over the 2018–2050 period to optimize the employment of natural wind, water, and solar resources.

This transition requires not only investments in energy replacement but also energy storage and improvements in the electricity grid to ensure economies can efficiently transport electricity derived from sustainable sources while they meet the needs of peoples both day and night. Also modeled are investments in how and where sustainable energy is produced and how it must be stored and transported to meet the needs of existing population patterns and their energy use throughout the day and night. The study determines that some changes in energy use technologies are necessary. These include:

- Heat from combustion of fossil fuels will instead be provided by air and ground heat pumps, residential natural gas stoves and dryers by induction stoves and heat pump dryers, and heat for industrial processes be replaced with induction, resistance, and electric arc furnaces and dielectric and electron beam heaters for high temperature sources, and with central steam plants and heat pumps for low temperature needs.
- Fossil-fueled vehicles will, for the most part, be replaced with battery-electric vehicles, with hydrogen fuel-celled or combustion vehicles used for very long distance and weight sensitive transport.
- Existing hydroelectric power will be employed for generation but also for load shifting as a lower cost alternative than battery storage, augmented by pumped hydro.
- Current energy distribution mechanisms will be optimized and improved, which results in further reductions in energy losses.

Under these assumptions, total (fossil fuel and electrical) energy consumption can be reduced by 56.4% in the process. Of the 56.4% improvement in energy consumption, 38.4 percentage points arise through the greater efficiency of electricity for combustion, including heating and internal combustion engines used in transportation and industry. This improvement includes 20.5 percentage points arising from greater transportation reliance on wind-water-solar power, 4.3 points from efficiency gains in industrial heat, and 13.6 points from optimized use of heat pumps rather than coal, oil, and natural gas furnaces.

Because fossil fuel exploration, extraction, and distribution are energy intensive, an additional 11.3 percentage points arise if fossil fuels must no longer be extracted. Energy efficiency improvements, including better insulation of homes in hot and cold climates, produce another 6.64 percentage point improvement.

To meet the greater reliance almost solely on electricity, total electricity consumption would rise by 85% relative to the *business-as-usual* scenario. Of this greater need, 7.56% of onshore wind, 0.8% of offshore wind, 9.13% of photovoltaic generation, 14.4% of geothermal needs, 0.001% of tidal requirements, 2.76% of tidal power needs, and 100% of hydroelectricity are already in place.

While hydroelectric power is already in place, it can be used more effectively to remedy unique aspects of other power sources. Hydroelectric potential already constitutes great potential for the bulk of storage needs by utilizing hydro generation throttling as necessary to meet changing demand patterns over the 24-hour clock and backfill the intermittent qualities of solar and wind generation. However, the authors find that a significant amount of battery storage is still necessary to augment load shifting, especially in regions not endowed with hydroelectric power. In addition, charging of battery-electric vehicles should be shifted primarily to the day.

These various sustainability factors are optimized for each nation and region, and may well differ for some regions. For instance, Canada has an immense potential for both sustainable energy production and for a large reduction in fossil fuel extraction, but also faces different needs for energy distribution in a physically large country. On the other hand, island nations have excellent access to water and offshore wind resources very near their population centers. These regional differences are modeled.

The study finds some of these transition costs, for instance of appliance and vehicle updating, can be subsumed as existing consumer durables need replacement. Such replacements on a life cycle basis then need not add to the new investment total if replaced in a regular upgrade cycle.

Some nations already rely substantially on sustainable energy. For instance, the cost of full conversion Iceland must make is a relatively modest $2.8 billion investment, while the United States and Europe must invest $6.7 trillion and $5.9 trillion respectively to attain energy sustainability. Still, the present value of new net investment and grid stabilization is $61.5 trillion is 2020 U.S. dollars. Of this estimated $61.5 trillion cost for a *Green New Deal*, $45.7 trillion arises from investments in sustainable electricity and heat generation, including increased employment of central heat plants and geothermal heat. The rest of the $61.5 trillion total is for heat and cooling capacity and electricity storage, electrolysis and compression infrastructure for hydrogen production and distribution, and long-distance electricity transportation, including a greater reliance on high voltage direct current (HVDC) transmission.

These investments are sizable but they also generate significant savings even in the medium term. The Jacobson study shows that this total investment of $61.5 trillion will generate energy savings of $17.8 trillion annually by 2050, reduce health care annual costs by $33.6 trillion, and mitigate climate costs by an additional $31.8 trillion. In combination, social and private costs from the energy sectors are reduced by $6.6 trillion annually, which is a 92% reduction in ongoing energy costs. While the investment is large, the dividends in cost reductions are also very sizeable and will spur new economic growth.

In a new sustainable energy equilibrium, direct energy costs fall by 14.3% per unit of energy but also from the 56.4% decrease in global combined energy consumption. Energy cost savings alone result in a payback in the necessary investments over 5.5 years across all nations, while the social cost payback, including energy, health care, and climate mitigation, is recovered in less than a year. Individual nations recover their investments over a range of 0.9–21.9 years from energy cost reductions, and 0.1–6.7 years for total social cost recovery.

The study notes that about seven million people die annually from outdoor and indoor (cooking and heating related) pollution each year. This constitutes the second largest global mortality factor, after heart disease. Health and other effects are one of their two measures of the benefits of carbon dioxide equivalent emission reductions. Reliance on fossil fuels also imposes energy insecurity costs arising from fossil fuel environmental damage and extraction instabilities, the overreliance on centralized rather than decentralized power, and conflicts fought over fossil fuels.

Health savings are concentrated in higher pollution countries such as China (20.6%) and India (27.3%). The authors note that carbon dioxide equivalent emission reductions in China alone represent total emissions of 128 of the 145 countries representing 99.7% of the world's population.

Finally, net job creation is also enhanced in the long term, beyond the initial infrastructure investment period. A total of 55.6 million new long-term full time equivalent jobs are created, while 27.2 million jobs will be lost as the fossil fuel industry is displaced. Net jobs are increased in all regions except Canada, Russia, and Africa. Not included in these totals are potential jobs arising from improvements in building energy efficiency and manufacturing of more energy sustainable appliances.

The Jacobson study describes necessary improvements on electrical grids, more efficient employment of district and centralized heating and cooling, greater effectiveness of load shifting across the 24-hour clock, and increased reliance on developments in the hydrogen-augmented economy. Their estimates are elaborations on previous studies by Jacobson et al. (2019),[1] Ram et al. (2019),[2] and Teske ed. (2019),[3] but draw similar conclusions as these previous studies even as it goes farther in its modeling of uncertainties in the pattern of demand and population.

However, despite the substantial benefits versus costs ratio and the attractive economics, a significant challenge remains. The greatest challenge may arise in the lack of collective political will, given the large investments necessary and the inevitable transfer of financial capital between nations.

Methods to Fund a Transition to Sustainable Energy

The issue remains. Estimates of the costs of a rapid transition to 100% renewable energy are high, even if they are matched with even greater savings and a short payback period on investments. Global funding and national burdens to create sustainable energy and reverse degradation of our atmospheric capital must still be determined.

We can ask the question whether a surtax on each barrel of oil equivalent would be sufficient to raise sufficient revenue for the transition. Carbon taxes per ton or metric ton of carbon dioxide equivalent have been proposed to address pollution from the emission of greenhouse gases.

Nations such as Canada have proposed a carbon tax that is scheduled to approach $170 per ton of carbon dioxide emitted by the year 2030. Let us explore a modest $51 per barrel of oil, equivalent to a surcharge of less than $1.50 per gallon of gasoline. These increases are of a scale that are not unfamiliar. For instance, the price of oil and gasoline rose by similar amounts in response to the invasion of Ukraine.

The total fossil fuel oil equivalents must be limited to about 1.2 trillion barrels by 2050 if the Earth is to limit global warming to 1.5° Celsius. The sale of this volume of oil equivalent would then yield sufficient carbon surtax revenue to fund the $61.5 trillion investment Jacobson et al. determine to substitute sustainable energy for fossil fuel usage and limit global warming to 1.5° Celsius (Fig. 21.3):

	Total Known Petroleum Reserves	Million Barrels of Oil Equivalent	Value of Fossil Fuel Reserves (trillion dollars US) Based on $90/barrel Oil
known oil reserves tonnes	236,294,751,172	1,762,759	$159
known natural gas reserves m^3	188,073,776,245,117	1,106,316	$100
known coal reserves tonnes	1,074,108,000,000	237,635	$21
	Total:	3,106,710	$280
price of oil:	$90	size of global economy 2020 Trillions $US	$85
Maximum Consumable BOE to Taper to Carbon Neutrality by 2050		1,207,715	$109
Jacobson et al Sustainability Investment Estimate (trillions)	$61.5	Necessary Carbon Surtax per BOE or Equivalent	$51

Fig. 21.3 Necessary carbon tax to fund transition to sustainable energy

Such a solution, if initiated early enough, ensures that global warming tapers to a level that does not induce runaway global warming as successive tipping points are exceeded. Accelerating damage is averted and nations must then cope with damages that occur annually at the 1.5° Celsius threshold.

Once accelerating damage is stemmed, efforts can then proceed to mitigate annually the new steady state damage caused by global warming. Who then pays for the *loss and damage* reparations and movement toward optimal sustainable development must still be determined.

Intergenerational Economic Justice

From an ecological perspective, the least risk sustainability solution is to return atmospheric greenhouse gas levels to their pre-industrialization concentrations. A less satisfactory but more politically pragmatic approach is to eliminate future growth of greenhouse gas concentrations and indemnify nations experiencing annual losses arising from past emissions.

Policy mechanisms and carbon taxes can be established to ensure that there is no further environmental or natural capacity degradation, including retreat from tipping points that may be tripped. Global partners must cooperate to reduce the effects of past atmospheric degradation on vulnerable nations that do not have the resilience to prevent environmental shocks. Such shocks can push LDCs beyond national economic tipping points and induce ecosystem and economic collapse.

Borrowing a concept from *weak and strong sustainability*, we may also frame the discussion as one of *weak or strong reparations*. Under strong reparations, the actual losses incurred as nations absorb the brunt of climate calamities, in increased droughts or flooding, the consequences of heat on the human and natural ecosystem, and related casualties of global warming are indemnified by offending nations in proportion to their emissions of greenhouse gases.

If agreement can be forged on the cost and remedy of climate catastrophes, nations must still negotiate who must pay. The loss and damage approach broached by LDCs at the 2022 COP27 gathering implies this responsibility must be shouldered by the MDCs who caused global warming rather than the LDCs who suffer from it.

Even if it was established that emitting nations did not understand the consequences of their actions at the time of their emissions, it does not resolve them of the responsibility under the Polluter Pays Principle. Offending nations must deal with their consumption of the atmosphere and the deprivation by MDCs of the possibility of LDCs to claim an equal amount of environmental degradation ignorance. Nations even if in ignorance degraded our shared natural atmospheric capital on a multigenerational basis benefited substantially in their economic development. If their resulting economic development was Pareto efficient, in the sense that the commercial value they created exceeded the value of atmospheric degradation, then the wealth MDCs enjoyed was sufficient to pay for the damage caused. Such strong reparations are hence feasible, even if they are perhaps unpalatable among those who benefitted from past emissions.

An argument often leveled against strong reparations is that LDCs arguably benefitted through their relationships with MDCs that generated wealth in the absence of proper Pigouvian taxes on past emissions. Even if an LDC grew because of their partnership with an MDC, this does not absolve the MDC of responsibility based on the Polluter Pays Principle.

As an example, a colony may benefit from the sale of commodities to a colonial power, but that does not absolve harm or damage committed by the colonial power. Civil and international law addresses compensation for harms imposed rather than retribution over motivations, without regard for motives or intent.

However, weak reparations is a more pragmatic approach that may not entirely indemnify the losses arising from past emissions, but instead provides those harmed with a compensating sustainable energy infrastructure investment to further their economic development and mitigate future emissions. Assuming first that LDCs are given the tools to develop sustainably, such *weak reparations* are a series of mitigations and adaptations that efficiently minimize future damages and also compensate LDCs for their losses arising from global warming.

This loss in productive capacity among LDCs willing to forego repetition of past development mistakes by MDCs that resulted in environmental damage can be compensated by a transfer of human-made capital from MDCs to LDCs. These transfers to LDCs can enhance their production and permits them to enjoy consumption that is more sustainable. If such technology infrastructure investment is less expensive than compensation for direct losses arising from environmental damage, and if this capital investment has the potential to grow an economy at an appropriate rate without environmental degradation, then this mitigation solution is more efficient than loss reparation.

A Possible Sustainability Breakthrough

At the 2022 Conference of Parties 27 (COP27) to the United Nations Climate Accord, participating nations successfully placed on the table for the first time an acceptance of some form of reparations. This proposal was accepted by COP27 delegates for ratification by the 197 signatories to the United Nations Climate Accord. While the level of such loss and damage compensation was not determined, delegates from LDCs further proposed a form of weak reparations. They suggested that compensation could come in the form of large investments by MDCs in forms of sustainable energy for LDCs such as the enhancement of wind and solar power and the means to store and export sustainable energy.

Such compensatory investments may restrict from access to some amounts of land for solar or onshore wind generation or water for offshore wind generation. However, these losses are compensated by the potential for sustainable energy that can spur other forms of economic development and the creation of new exportable energy resources.

Climate change solutions fall into two categories. One is the prevention of additional future damage of a scale pledged by the Paris Agreement and tabulated by the Jacobson study. The second is in repair of ongoing damage caused by past actions, through loss and damage payments.

From a political perspective, the large upfront investment costs, both in prevention and reparation, have caused sovereign participants at the various UN-sponsored conferences to balk and delay. A more pragmatic solution, from the political perspective, is the carbon tax described above. A carbon tax has both the effect of properly internalizing future externalities and hence preventing future damage, and may also be made sufficiently large to address loss and damage.

Greater reliance on carbon taxes would still require coordination of global market prices across myriad sovereign peoples. If such market intervention proves to be politically unpalatable, then it may be possible to capitalize on the significant rewards that accrue to the investment in sustainable energy that already yields the lowest cost of all new power sources, even if properly calculated and instituted carbon taxes are neglected.

This solution can be mobilized at the private level with some sovereign facilitation. **Venture capital funding** for sustainability is expected to rise to half a trillion dollars US per year by the middle of the 2020s. While this level of investment is inadequate by a factor of ten compared to the extent of investments necessary along the lines of the Jacobson study, the total investment cost of energy sustainability is but a fraction of the estimated $250 trillion of total investment funds in global capital markets or the losses from global warming between 2021 and 2070 of $178 trillion in 2021 dollars as tallied by Deloitte. A sufficient investment scale exists if only barriers to such massive international investment and intergenerational externalities can be solved.

However, such private investment is very sensitive to the cost of capital, rate of return, and investment risks. Estimates of the rate of return that can be earned in sustainability investments can be calculated based on current technologies and anticipated growth projections should less expensive and sustainable energy become available and the extract fossil fuels is reduced. The greatest uncertainties are in the cost of capital and inherent investment risk.

Sufficient capital can be raised either through borrowing or the issuance of new equity. Finance theory demonstrates that, under certain conditions, either method to raise investment capital is equivalent, and both are priced in proportion to the amount of risk inherent in these investments.

International investments are riskier than domestic investments because of the lack of protections afforded multilateral international transactions. Given the necessary scale of sustainable energy sector investments and the potential costs of loss and damage payments, traditional financial markets are inadequate to mobilize the large sums necessary and pool the inherent uncertainties sufficiently. This is especially true given that the most promising investment opportunities are in LDCs that may not have a long history of protecting financial investments.

Governments of sovereign MDC nations have sufficient capacity to pool and diversify inherent environmental investment risks. They also have the ability to subsidize or guarantee loans so that the risk premium inherent in the interest rate charged is removed. *Sustainability bonds*, described more fully in Chapter 25, may then be offered with real interest rates (adjusted for inflation) of just 2% or 3%, which dramatically reduces the debt service of investments amortized over the 30–50 year lifetime of a sustainable energy project, and dramatically improves the economics of sustainability investments.

For instance, the total interest paid over 40 years for a 2% bond at the cost of government capital compared to a private sector 5% bond priced at a more customary private sector cost of capital, discounted to the present, sums to $2.6 trillion in savings on a $6 trillion investment. These sizeable savings dramatically improve private investment returns and the prospects for private markets to fund necessary sustainability investments.

To ensure sufficient investor interest, MDCs can also provide a guaranteed purchase price for sustainable energy for the duration of the investment. Examples abound of such guarantees. For instance, to stimulate retail investment in solar power, Germany had offered a fixed purchase price to domestic producers of solar and wind power. MDCs can pledge to purchase any excess sustainable energy generated in LDCs at a fixed price, provide a guaranteed price to investors sufficient to compensate them for their risk. LDCs are afforded the opportunity to collect the difference in risk free revenue for their own development purposes and are permitted to access a share of the power for their own development needs. This solution also provided the technology transfer necessary to allow LDCs to develop along the UN Sustainable Development Goals.

An alternative to market solutions is direct sovereign investment. This approach may be more difficult to muster than the intervention of sovereign nations in issuing and guaranteeing investment bonds. In 2020, in the wake of record foreign aid to combat COVID-19, total global aid only reached $161.2 billion US.[4] The scale of investments for the stemming of future climate change damage alone is 300–400 times larger, and in addition to the legitimate needs for other forms of foreign aid.

The world's MDCs may instead offer grants and low-cost loans directly to LDCs with the most potential for positive returns over time. In doing so, these LDCs may be able to invest in future revenue streams and technologies that permit them to substantially increase their path of sustainable economic development.

To best mobilize and protect such investments, a UN- or World Bank-administered global sustainability investment fund, with returns subsidized by funding nations to reduce private sector risks for their investment, and the creation of mechanisms to protect large international investments, may offer a breakthrough in the loss and damage impasse.

SUMMARY

A missing dimension that makes global coordination among governments difficult is the difficulty to coordinate global efforts when there remain differences in social and governance capacities to cooperate on sustainability issues. The second dimension involves the ability of the corporate sector to mobilize action to support sustainability. While science and engineering offer guidance on what we need to do, we are still left to determine how. The ESG paradigm provides opportunities for the private sector as described beginning in Chapter 23.

The public sector plays a small role relative to markets in the employment and diminishment of natural capital. However, national governance has the tools to regulate the private sector and encourage the direction of private sector investments. The next chapter explores the global imperative for cooperation in remedying global warming.

ESG Toolkit

Corporations can find advantageous gains through cooperation with corporations elsewhere that can develop resources more sustainably.

Can a corporation with a challenging carbon footprint form a subsidiary in a nation capable of producing sustainable energy? In doing so, the carbon footprint per dollar of revenue is reduced.

How might synergistic partnerships be formed to diversity such carbon footprints?

NOTES

1. Jacobson, M. Z., M. A. Delucchi, M. A. Cameron, S. J. Coughlin, C. A. Hay, I. P. Manogaran, Y. Shu, and A. K. von Krauland. (2019). One Earth 1: 449–463.
2. Ram, M., D. Bogdanov, A. Aghahosseini, A. Gulagari, S. A., Oyewo, M. Child, U. Caldera, K. Sadovskaia, J. Farfan, L. S. N. Barbosa, M. Fasihi, S. Khalili, C. B. Dahlheimer, G. Gruber, T. Traber, F. De Caluwe, H.-J. Fell and C. Breyer. (2019). "Global Energy System Based on 100% Renewable Energy—Power, Heat, Transport and Desalination Sectors", Study by Lappeenranta University of Technology, and Energy Watch Group, Lappeenranta, Berlin.
3. Teske, S., editor. (2019). Achieving the Paris Climate Agreement Goals—Global and Regional 100% Renewable Energy Scenarios with Non-energy GHG Pathways for + 1.5 1C and +2 1C, Springer Nature, Switzerland.
4. https://www.weforum.org/agenda/2021/04/foreign-aid-2020-covid-19-oecd/, accessed December 1, 2022.

CHAPTER 22

The Public Sector and ESG: Social and Governance Pillars

The study of sustainability draws together a large number of elements from the sciences, social sciences, and philosophy. These various tools can be used to answer a wide variety of questions. To convert analysis into action requires a broadly accepted way to organize the strategies of decision-makers. The ESG paradigm provides that scaffolding so that nations and corporations can enhance economics well-being and protect the ecosystem.

The societal perspective identifies the relationship between citizens and their state, or between peoples across sovereign states. It considers the social contract that citizens and stakeholders have with their institutions and states. It also identifies ways to improve the social contract to empower employees, improve productivity, promote social and economic justice, and encourage intergenerational equity. In doing so, the ESG goal is to enhance and clearly delineate the policies and property rights each of us has that protect us and define our relationships within the institutions we belong.

In the social category, the Environment, Social, and Governance Paradigm identifies the following elements to enhance societal sustainability:

- Employment laws
- Public, corporate, and employee policies
- Employee and executive compensation and incentives
- Privacy and information governance
- Diversity, Equity, and Inclusion (DEI)
- Economic justice
- Private and public property
- Migration and economic refugees.

© The Author(s), under exclusive license to Springer Nature Switzerland AG 2023
C. Read, *Understanding Sustainability Principles and ESG Policies*,
https://doi.org/10.1007/978-3-031-34483-1_22

Societal sustainability within the ESG paradigm can be enhanced to resolve issues that frustrate markets or defy the assertion of property and human rights. To help resolve challenges to social sustainability, ESG must address rights and responsibilities with respect to the environment and climate change, values inherent in diversity, equity, and inclusion (DEI), economic justice, private, public, and common property rights, and migration between nations.

Peoples depend on the establishment and preservation of a modicum of human rights, especially in relationships between markets, citizens, and those who we elect to govern us. The first four categories of employment laws, laws and policies, compensation and incentives, and privacy all establish essential property rights to promote well-functioning societies, economies, and markets. The appropriate establishment of such human property rights have dual roles. One is to ensure that individuals can function efficiently within institutions and society in ways that allow us to pursue our interests and enhances productivity. The other goal is to ensure these individual pursuits do not violate the rights of others at worst, and further broaden collective interests at best.

Society sustains itself best when these rights are protected. The legal system, our religion and ethics, and the principles that govern our various behaviors within different cultures have all evolved to protect societal and individual interests. Some cultures emphasize societal interests over individual pursuits, while other cultures have determined that, when the ability of individuals to pursue happiness and prosperity is least impeded, society also generates the greatest value. All social systems strive to evolve to create social harmony that can be sustained over time.

These various systems are typically enshrined within sovereign borders, with a relatively smaller set of values considered universal and global, but with only weak mechanisms for their enforcement across borders. Recognition of a certain set of human rights is essential to meet the challenges of the establishment of global sustainability and the amelioration of conflicts between nations over resource usage. While attempts to protect universal human rights remain challenging and ineffective, sovereign nations have established conventions that govern and protect mutual economic interests.

An Economic Justice Ethic

An *international economic justice ethic* that enshrines the protection of property, protects the flow of economic and political refugees and victims of persecution, and celebrates diversity, equity, and inclusion across peoples ultimately conflicts with national notions of sovereignty.

Recall John Rawls' theory of economic justice within a nation, which we define as a set of peoples who reside within common borders. Rawls' principles have significant implications on the nature of mutual responsibilities to fellow citizens. Those who implicitly agree to be governed within their nation broadly consider their social contract to be fair and just.

The notion that rights are conferred upon a people within a nation was first imagined in 1215 in England with the Great Charter, or *Magna Carta*. That charter provided a set of rights on behalf of the church and rebel barons in protection from an unpopular king. It acted as a contract between the monarch and his subjects as a way to purchase peace and avoid rebellion.

While the Magna Carta may be viewed as a series of conditions that constitute a peace treaty and the establishment of basic human rights, it was nonetheless influential as the first recognized contract between a governing entity and its citizens. Charters also govern a corporation, a county, or a constitutional democracy. They form the legal foundation for both the existence and legitimacy of the governing body and those who agree to be governed.

Such an advocacy of the rights of individuals within a sovereign nation was advocated by Jean-Jacques Rousseau in his On the Social Contract; or, Principles of Political Right (French: Du contrat social; ou, Principes du droit politique) as an extension of his earlier Discourse on the Origin and Basis of Inequality Among Men (French: Discours sur l'origine et les fondements de l'inégalité parmi les hommes) in 1755. Rousseau wrote in an era of frustration among subjects of an increasingly despotic monarchy in pre-revolutionary eighteenth century France. Rousseau stated, "Let us then admit that force does not create right, and that we are obliged to obey only legitimate powers." He argued for a mutually beneficial social contract as a substitute for the coercion over people who defy submission. In his social contract, a state cannot enslave the conquered. Instead, a set of principles were set forth to define the relationship between government and the governed.

John Rawls elaborated on the principles that might govern such contracts among citizens to provide liberty and opportunity to the least advantaged. Rawls' principles were based on his notion of a veil of ignorance over one's original position. The Rawlsian provisions can be viewed as protection of a set of property rights which provides something of value to each citizen and hence acts as a measure of economic justice within a people. Rawls asserted that well-ordered peoples should have liberty to determine issues such as the distinction between public and private property, savings, population, and education according to each nation's sovereign values rather than international prescripts.

Rawls' theory of justice applies to a people who, in return for the granting of these property rights, accept the legitimacy of the governor. Within these sovereign people we label a nation are also collective rights regarding how resources under the sovereign will be owned, extracted, and distributed, how personal property shall be protected, and how the powers of the governing administration shall be limited and modified.

If all peoples considered themselves to be members of a global society rather than a nationality, Rawls principles at the state level can be appropriated to the global level. However, there is no analogy to a global state to which all citizens are subject. Rather, individuals typically only pledge their allegiance to a sovereign nation state, with all its cultural and societal implications and obligations.

Following A Theory of Justice,[1] Rawls produced an international analog, The Law of Peoples[2] (1999) to better describe how nations of peoples should interact with each other. While Rawls argues that within a set of peoples is a mutual obligation for fairness and for policies without regard for original position, between sets of peoples is only one mutual obligation—to ensure that peoples within sovereign nations are governed and can function within fair and well-ordered societies.

These principles include (from Rawls [1999], p. 37):

1. Peoples are free and independent, and their freedom and independence are to be respected by other peoples.
2. Peoples are to observe treaties and undertakings.
3. Peoples are equal and are parties to the agreements that bind them.
4. Peoples are to observe a duty of non-intervention.
5. Peoples have the right of self-defense but no right to instigate war for reasons other than self-defense.
6. Peoples are to honor human rights.
7. Peoples are to observe certain specified restrictions in the conduct of war.
8. Peoples have a duty to assist other peoples living under unfavorable conditions that prevent their having a just or decent political and social regime (Rawls 1999a, p. 37).

Rather than a set of rights granted those who are subjects of a sovereign people, the principles between sovereign nations do not constitute a social contract in the same way as those governed within a sovereign nation. Instead, nations agree to practices that define transactional relationships between sovereign equals. These principles enshrine the rights of sovereigns to protect their property and peoples while they create an obligation that sovereigns respect the sovereignty of other peoples. Such an approach creates harmony in international transactions and adherence to treaties between sovereigns. In adhering to these principles, sovereigns need not devote excessive resources to defend their own sovereignty among equals.

While nations of peoples conduct international arm's length trade, their trade is ultimately voluntary and transactional rather than constitutional. Since international trade is voluntary, it is by definition noncoercive and inherently mutually beneficial. While Rawls described the moral imperative to prevent economic injustice within a nation of peoples, there is no such imperative to ensure equal prosperity or the prevention of economic injustices between peoples. Just as one cannot judge the morality of mutually beneficial and voluntary market transactions, one cannot judge the actions of nations that do not violate Rawls' principles between peoples, beyond the need to ensure that peoples function within well-ordered societies.

Instead, Rawls argued for some international facilitation of voluntary trade among nations so they can operate in a well-ordered manner. But, the burden of mutual responsibility ends there. Within Rawls' global scheme, rules of arms-length transactions and fair play apply. Violators of basic principles that efficiently govern multilateral transactions only run the risk of ostracization and sanctions that may restrict or prevent their participation in future international transactions. Treaties can further define the rules of transactions, but sovereign states cannot otherwise be punished for breaches beyond sanctions that constrain subsequent trade.

There remains one dimension that is of paramount importance for the success of such a global compact that defines principles of multilateral trade and mutual respect of sovereign property. If nations agree to enshrine the protection of property owned within the borders of a sovereign nation, including a respect for whatever mechanism a well-ordered society deems appropriate, how should property be treated that spans or is not confined within borders or their extensions to *exclusive economic zones*?

As an example of such international governance principles, the *Convention of the High Seas*, signed by 63 nations in 1958, states that "no State may validly purport to subject any part of them (the high seas) to its sovereignty." Since the extension of exclusive economic zones to 200 nautical miles of recognized sovereign coastlines, for which nations have sovereign rights to the natural resources within these extended boundaries, *high seas* beyond the jurisdiction of sovereign nations represent about 50% of the Earth's surface. In 1994, the United Nations Convention of the Law of Seas was ratified to enshrine the principle of resource access within sovereign boundaries, including exclusive economic zones. It also reaffirmed the principle that nations do not have unrestricted rights beyond these sovereign limits. In such international waters, every nation has universal jurisdiction to prosecute any violations of the law.

To attain an orderly assignment of rights to mineral extraction, median lines are drawn halfway between adjoining sovereign nations. Each nation is then given qualified property rights on such fixed stocks of resources such as minerals as a way to avoid the Tragedy of the Commons. The *International Seabed Authority*, empowered by the UN Convention of the Law of the Sea, governs and arbitrates resources that represent the *common heritage of all mankind*. Under UNCLOS, Part XI, Section 2, Article 140[3]:

Benefit of mankind:

Activities in the Area shall, as specifically provided for in this Part, be carried out for the benefit of mankind as a whole, irrespective of the geographical location of States, whether coastal or land-locked, and taking into particular consideration the interests and needs of developing States and of peoples who have not attained full independence or other self-governing status recognized by the United Nations in accordance with General Assembly resolution 1514 (XV) and other relevant General Assembly resolutions.

The Authority shall provide for the equitable sharing of financial and other economic benefits derived from activities in the Area through any appropriate mechanism, on a non-discriminatory basis, in accordance with Article 160, paragraph 2(f)(i).

Article 160, 2(f) (i) states that the Authority can:

...consider and approve, upon the recommendation of the Council, the rules, regulations and procedures on the equitable sharing of financial and other economic benefits derived from activities in the Area and the payments and contributions made pursuant to article 82, taking into particular consideration the interests and needs of developing States and peoples who have not attained full independence or other self-governing status. If the Assembly does not approve the recommendations of the Council, the Assembly shall return them to the Council for reconsideration in the light of the views expressed by the Assembly.

While almost all nations have ratified the UNCLOS, sixteen sovereigns have not acceded to the convention: Andorra, Eritrea, Israel, Kazakhstan, Kyrgyzstan, Peru, San Marino, South Sudan, Syria, Tajikistan, Turkey, Turkmenistan, United States, Uzbekistan, Holy See, and Venezuela.

The provisions allow for a sharing of the benefits accruing from resource extraction in the shared region of the High Seas. However, mechanisms remain unformulated and untested. It was not until 2023, for instance, that the UCLOS clarified the protection and sharing of intellectual property, for instance the discovery of valuable genetic stocks, from the shared high seas.

The UNCLOS principle provides a framework to prevent the Tragedy of the Commons in shared resources that contribute the **common heritage of all mankind**. It potentially invokes global protections from the intergenerational nuisances that arise when fossil fuels are burned and atmospheric oxygen is consumed and replaced with carbon dioxide in the atmosphere that protects the ecosystem. If so, nations would arguably have responsibility for the consumption of our shared atmospheric resource and the global warming that results.

This obligation is currently the subject of negotiation toward a climate change treaty akin to the Montreal Protocol that banned ozone-depleting CFC emissions. The United Nation Climate Accord has provided a framework for negotiations, with the goal of ensuring no further environmental degradation beyond agreed limits and providing for transfers to indemnify nations that suffer from past atmospheric degradation.

If there exists such an implicit multilateral social contract between peoples that no nation has the right to consume a critical stock resource and hence harm all peoples, how must international and national governance processes formulate such a treaty?

The Governance Pillar

Underpinning functionality of well-ordered societies is the existence of an implicit contract or an explicit constitution that preserves certain rights and creates obligations. This quid pro quo depends critically on the definition of property. The working definition of property from a contractual or social perspective is a right or title to something of value that cannot be expropriated without due process or compensation. For instance, the rights elucidated in the United States Bill of Rights, Canada's Charter of Rights and Freedoms, and equivalent contracts between government and the governed provide a series of protections of interests.

National constitutions and common law typically assert that the taking of established rights which create value for people of a nation requires compensation as administered by criminal and civil penalties. For some forms of property, free markets permit the efficient exchange of formal or informal title to property through a simple and streamlined process. Such fluid free markets provide a well-understood framework for the transaction of implicit or explicit titles of ownership of goods and services. They may even permit the surrender of other rights such as the right to be free of bodily harm in exchange for a professional sports contract or the right to free speech through a non-disclosure agreement. On occasion, by creating market processes to exchange a broad range of property rights, what would be otherwise cumbersome and expensive exchanges become routine with low transactions costs.

People within a sovereign nation regularly exchange titles to the homes in which we live, shares of companies we own, the right to be repaid for money we have lent, or the sale of resource on property we own, in addition to the right to exchange cash to purchase a cup of coffee. Markets are efficient mechanisms to exchange these rights so that they may be allocated to those who most value them. This highest value allocation in a free-market system is based on a combination of the intrinsic utility such a transaction provides to its purchaser and the amount of wealth the purchaser owns.

Consider the value of an increase in enjoyment from the purchase of a good. To see why free markets do not necessarily maximize collective utility, notice that an exchange price p does not exceed the value of a transaction to a purchaser, as measured by the ratio of the utility gained by the purchase of a good and the utility lost for a decrement in income for its purchase. This implies:

$$p = \frac{\frac{\partial U}{\partial q}}{\frac{\partial U}{\partial I}} = \frac{MU_q}{MU_I},$$

where MU_q is the marginal utility derived from the purchase of a quantity q of the good, while MU_I is the loss of utility per dollar of income sacrificed.

We take for granted such calculations that drive our market decisions. The wealthy derive less enjoyment from purchases they make because they have relatively more income and can make more purchases. Their income and their greater consumption both experience diminishing marginal returns, and hence economic theory shows they derive less enjoyment from the last unit they purchase using their last increment of income earned.

Economic theory implies that market prices are not solely measures of our collective valuation. Instead, valuations are weighted toward those who can most easily afford the purchase price because of their greater income and lower sacrifices in utility for a relatively smaller loss in income through their purchase. Purchase of a good, service, or title by an individual in a free-market system is a combination of both an individual's intrinsic value MU_x and the conversion of income to utility MU_I. But while market transactions do not maximize collective utility, governments arguably seek the enhancement of the collective utility of the people they serve through alternative non-market mechanisms that are independent of the diversity of wealth within a people.

Businesses and corporations also perform a value calculation that maximize net surpluses generated in their purchases. These principles apply to market-oriented transactions within a nation and exchanges between nations. Voluntary exchanges of property rights are protected by well-established commercial and common law principles that are understood by the parties in each transaction. Participation in global commerce is thus an implicit acceptance of market principles. Violation of fair play in international markets can result in sanctions imposed by such groups as the *World Trade Organization*.

These principles of ownership and just compensation, and allocations of property rights based on the valuation and wealth of the transactors, are the underpinnings of the free-market system. They apply equally to exchanges among peoples in a nation and between peoples of different nations. In the absence of titles to property established by sovereignty, takings of common property, such as resources on the seabed in international waters, are governed by the UNCLOS, and require both prior approval and compensation at fair market value to all peoples of the world.

Peoples raised in well-ordered free-market nations implicitly understand these principles of consumer sovereignty and the voluntary exchange of property rights. The ability of governments to enforce protections of the right to transact gives rise to vibrant and creative economies in which participants have a free-market faith that the exchange of value will be fairly compensated. While, within a nation, criminal and civil sanctions enforce unlawful takings of property rights, in the international context any illegal takings can only be met with sanctions in the form of exclusions from future transactions and the potential expropriation of property of equal value owned by one sovereign nation but in the custody of another.

Within a free-market system, it has been understood since Pigou's work on externalities a century ago that those who take any resource, held either by private owners or in common, must compensate the harms they cause to avoid negative externalities. Should such takings be tolerated while compensation is ignored, the cost of production no longer reflects the value of the factors used to create it, nor the cost on others who are deprived of a resource without compensation. Compensation is necessary to reimburse those who sacrifice a resource, even if held in common. Hence, it is also necessary to ensure that the value of natural capital is properly reflected in production costs.

Such free-market principles differ from those that govern transactions between sovereign peoples. The principles enforced through public governance are designed to address the values within a people. Governments presumably transact in ways that balance the value to voters and distribute the benefits and costs imposed on present and future voters. Rather than transactions based on a combination of wealth and value that regulates private exchanges, governments transact to balance the interests of voters, ideally without regard to such factors as income and position. The various components of the Governance Pillar in the ESG paradigm are designed to protect these principles. ESG identifies a number of avenues to ensure the protection and expansion of human and economic rights. These include:

- International treaties and regulatory oversight
- Public and private sector practice
- Financial markets and corporate viability
- Public transfers and private equity
- Securities oversight and regulation
- Consumer rights
- Investment and risk management
- Public and corporate asset valuation and management
- Capital infrastructure and investment
- Corporate governance.

Some of these protections are statutory or regulatory, such as financial market regulations and oversight, securities and exchange laws, and banking and accounting principles. Others enshrine standards of practice that may govern commercial and market transactions, while still other protections are enshrined in treaties and constitutions.

Responsibilities and protections of rights at the sovereign level include adherence to regulations designed to protect economic and human rights, public and private sector practices, financial markets, public transfers, securities oversight and regulation, and consumer rights. The establishment and enforcement of these rights can profoundly affect the efficiency of market economies and the distribution of rewards and wealth within a nation.

GOVERNANCE OF DOMESTIC TRANSACTIONS

Economic justice and the distribution of wealth in democracies are determined at the ballot box. Ideally, they are made as choices among the peoples within sovereign nations according to Rawls' principle. While these practices ought to be designed to optimize economic efficiency and prevent conflicts that arise from misappropriated resources, they ultimately have profound economic and social justice implications as well.

Government has the responsibility to ensure that its free markets operate efficiently and fairly. The degree to which a well-ordered market economy can create and enhance value is proportional to its ability to promote efficiencies within the free market or alternative allocation systems. Principles related to the protection of property, the internalization of externalities, the preservation of human rights, and the encouragement of innovation and creativity are designed to create optimal balances between resource usage, rewards, and effort.

Oppressive governance measures that enhance the interests of one side of a transaction but discourage participation on the other side may ultimately deter otherwise worthwhile exchanges and hence hinder commerce. Likewise, insufficient regulation may create uncertainties that discourage transactions, while regulatory complexity may discourage transactions by increasing their costs. Each sovereign system establishes appropriate protections to optimize the balance between these competing effects so that markets can function sustainably and efficiently.

Each sovereign establishes these principles in ways that balance the rights and privileges of the various sides of transactions somewhat differently. Customs or societal ethics may afford greater protections beyond those that laws enshrine. At times these societal ethics may be so profound that sovereign laws may not even be necessary. In net, societal ethics and sovereign laws combine to create harmony and efficiencies that are unique to each well-ordered people. Each sovereign nation balances differently the rights and rewards to the various sides of a transaction.

Policymakers must acknowledge that any determination of an appropriate balance of rights within a people has distributional effects that may need to be addressed. Side payments may be offered to induce citizens to accept the outcome of market principles designed to enhance efficiency. For instance, a well-ordered society may choose to protect strongly the rights of an employer or a corporation, and hence limit employee protections. The greater efficiencies and profits that accrue to corporations if such a balance is established in their favor can presumably be redistributed, in the form of reductions in individual taxpayer obligations at the expense of higher corporate taxes. Because democratic governments establish principles based on the consent of the people, should the efficiencies gained be insufficiently distributed to address reduced rights of those harmed in the balance, these laws and principles may be unsustainable. Stable governance requires sustainable practices.

This principle of sovereign regulation then bifurcates sovereign administration and oversight of markets for the exchange of property rights into two categories. The first is the establishment of efficiency in transactions to spur economic efficiency and growth. The second is the establishment of income and property distribution and redistribution among citizens to promote economic justice and sovereign sustainability. Well-ordered societies are able to navigate these two functions to simultaneously enhance both efficiency and economic justice on behalf of their citizens. A sustainable people will also balance the needs of all generations over time to enhance intergenerational equity.

Governments also necessarily impose constraints and provisions on corporations so that corporations too can maintain the balance between efficiency and equity as they substitute their implicit contracts within their internal governance mechanisms for the more rigid costs of external arms-length transacting within markets.

Governance of International Transactions

These principles also apply to multilateral actions that affect sovereign nations. Without the establishment of principles of orderly exchange among sovereigns through treaties, uncertainty is created and otherwise viable exchanges may not be transacted. In the absence of effective global governance, nations may instead redress conflicts that arise from a lack of treaties to moderate competing interests through armed conflict and war, or conquest, invasion, and arbitrary resource expropriation.

But, while treaties, international laws, and international courts and tribunals may attempt to resolve potential conflicts between sovereign nations either before or after they occur, the tools of resolution available within a sovereign nation are much more substantial and extensive than between nations. Transactions between sovereign nations are restricted to voluntary agreements, and are limited to sanctions, confiscation of custodial property, and refusal to participate in future otherwise beneficial transactions.

A wide body of international law and conventions has been established through voluntary negotiation by signatories to agreements created through the United Nations. For instance, the Conference of Parties to the United Nations Climate Accord and the various rounds of negotiations, most recently the Paris Agreement, are attempting to assign responsibilities both to the cause of past depreciation of atmospheric resources and the prevention of future abuses.

We now regard the Montreal Protocol on ozone depletion as a great success in international collaboration. In that instance, global damage was not yet too extensive, and the duration of continuing damage spanned only a few generations. Remedies were of relatively modest cost, while the stakes of climate change and environmental sustainability were extremely high. In contrast, climate change negotiations have proceeded at a slow pace, even in the wake of rapidly escalating damage. In the absence of mechanisms to coerce parties to participate and to pay for damages they cause, such voluntary negotiations prove to be cumbersome, especially when there is both a political and a value of time incentive among some atmosphere-offending nations to delay the ultimate solutions necessary to resolve the issues. Ultimately, any global agreement is also difficult to enforce at best, and are far more challenging when they also invoke current and future generations.

Summary

The public sector has the responsibility of regulating free markets to create efficiencies and redistribute rights and wealth to ensure economic and social justice. Such regulation creates a level playing field for the private sector and maintains the public's trust in the integrity of both public and private institutions.

But, while tools available to the public sector within a sovereign nation can be profound, governments have little power to impose their will on other sovereign nations in any way but through voluntary exchanges.

For these reasons, while the private sector is typically not charged with social goals beyond their creation of profits, corporations have available to them avenues that may well exceed the power of nations to effect some types of social or economic change. The ESG movement attempts to empower corporations on the global stage. The next three chapters describe the abilities of corporations to contribute to convergence onto a more sustainable global path.

ESG Inquiries

To what degree does your institution engage in efforts to effect social change? Is such a goal consistent with your corporate strategy?

Does your organization partner with other similar organizations or associations to further the interests and efficiency of society beyond philanthropy?

Notes

1. Rawls, John. (1971). A Theory of Justice, Cambridge, MA: Harvard University Press.
2. Rawls, John. (1999). The Law of Peoples, Cambridge, MA: Harvard University Press.
3. https://www.un.org/depts/los/convention_agreements/texts/unclos/part11-2.htm, accessed November 30, 2022.

CHAPTER 23

Private Sector ESG Responsibilities and the Environment Pillar

A corporation is considered a legal person governed by a charter granted under the jurisdiction of a sovereign state. The corporation is afforded various rights, must operate in ways prescribed by law, and is responsible to its shareholders. Regulators have increasingly asserted that the corporation also has responsibilities to other stakeholders and to the people of a state that granted its charter. The basis for these responsibilities is analyzed next.

While ESG considerations in the private sector typically treat private profit-seeking corporations, many of these considerations also apply to public benefit corporations, not-for-profits, and non-governmental organizations. The various internal processes of institutions may not be devoted solely to the enhancement of profits. For instance, many argue that the corporation has a responsibility to operate in ways that are sustainable, represent strong and effective managerial principles, and practice effective governance processes that complement other corporate values.

THE FRIEDMAN DOCTRINE

In 1970, the eminent conservative economist Milton Friedman published in the New York Times what is commonly known as the *Friedman Doctrine*. Entitled "The Social Responsibility of Business is to Increase its Profits," Friedman offered a rebuttal to what is now known as *Corporate Social Responsibility (CSR)*.[1] His doctrine was a product of an era in which large American manufacturers were considered the economic engine that powered the world economy.

© The Author(s), under exclusive license to Springer Nature Switzerland AG 2023
C. Read, *Understanding Sustainability Principles and ESG Policies*,
https://doi.org/10.1007/978-3-031-34483-1_23

In the early days of the emergence of the post-World War II military industrial complex, Senator Robert Hendrickson in a 1950 senate hearing asked the General Motors Chief Executive Officer Charles Wilson whether holding GM stock would be a conflict of interest were he to be appointed the nation's Secretary of Defense under President Dwight D. Eisenhower. Senator Hendrickson asked,

> Well now, I am interested to know whether if a situation did arise where you had to make a decision which was extremely adverse to the interests of your stock and General Motors Corp. or any of these other companies, or extremely adverse to the company, in the interests of the United States Government, could you make that decision?

GM President Wilson responded,[2]

> I cannot conceive of one because for years I thought what was good for our country was good for General Motors, and vice versa. The difference did not exist. Our company is too big. It goes with the welfare of the country. Our contribution to the Nation is quite considerable.

This statement that the GM president believed the interests of the nation and his corporation were aligned is often now misquoted as "What's good for GM is good for the country."[3]

The ensuing discussion between Wilson and the senator made it clear that the country and Wilson's company were intertwined to such a degree that their interests were aligned. Over his extensive academic career, Milton Friedman often asserted that such alignment be taken as an article of faith. He asserted that the corporation has the sole obligation to generate profits by any lawful means. By doing so, the corporation offers its shareholders the greatest liberty and capacity to use their corporate dividends in any legal way, including promoting causes and positions consistent with their right to free speech and participation in a civil society, if they so wish.

Friedman advocated for a set of values for which he argued for decades. His writings, and those of the novelist Ayn Rand, were gospel for those who advocated that profits ought to be the sole shared objective of those who own public corporations. Friedman's strident tone in 1970 was a challenge to a book that appeared just before his *New York Times* opinion piece with a very similar title. Morrell Heald had just published a book entitled "The Social Responsibilities of Business: Company and Community, 1900–1960[4] in which he argued that the emerging principle of *Corporate Social Responsibility* arose from the excesses of monopolies during and following the **Gilded Age** of the late nineteenth century.

Corporate Social Responsibility did not argue that the modern corporation should not produce profits. Instead, it clarified that:

(1) Long-term profits should replace the immediate gratification of short-term profits.

(2) With greater profits comes increased monopolization and market share.

(3) The concentrated power of the modern corporation usurps the dispersed power of the businesses it obsoletes.

(4) While myriad obsolete small businesses must act within the social fabric of the corporation and strive to sustain themselves so they may realize their dual goals of sufficient profits and good citizenship, the modern corporation instead converges only on profit creation.

(5) Myriad businesses made obsolete are replaced by myriad shareholders, few of whom have any significant influence on the goals of the corporation.

(6) Directors and managers of the corporation, if compensated at least partly in stock, are motivated more narrowly than the motivations of the shareholders they represent.

(7) The charter offered by a state to a modern corporation is based on an implicit premise that a corporation creates value for society by paying the full cost of all the resources they consume and producing products society values on a sustainable basis commensurate with the indefinite life of the charter.

The CSR movement had its roots in the critique on the excesses of the *Gilded Age*. Even those who benefitted most from late nineteenth-century accumulation of wealth recognized the social dangers of wealth excessively concentrated in the hands of a few. Arthur Twining Hadley wrote in a 1907 paper entitled Ethics of Corporate Management[5] that Aristotle related a story of a man who, in a visit to Syracuse, was so clever in business that he earned twice the profit of other iron ore buyers. Dionysius advised that he ought to keep his money, but he'd best not stay in Syracuse too long. The Gilded Age industrialist Andrew Carnegie, in his *Gospel of Wealth*, drew the same conclusion.

Carnegie argued that a share of large profits earned in Gilded Age railroad, steel, and oil monopolies ought best to be returned to society in the form of libraries, museums, and colleges. Friedman would accept that philanthropy, if that is what the wealthy chose to do. But while Friedman argued that social responsibility was an option and not a requirement, Aristotle, Dionysius, and Hadley recommended our business leaders behave in a socially responsible manner as a matter of social sustainability within a State that issued corporate charters.

The CSR movement of the early 20th century argued that the twin goals of profitability and social sustainability are not merely a matter of pragmatism. Instead, they are an implied function of the very charter that society offers corporations. The modern corporation is designed to expedite raising of capital and stimulate corporate growth by providing a product the market demands. But corporations are also afforded the privilege to shed liability and discharge debt at society's expense, if the corporation cannot generate sufficient value, a right that few ordinary members of society can effectively exercise to the same degree. They are also afforded the opportunity to transact internally in a streamlined manner and much more informally and efficiently than through external markets. With extraordinary corporate rights, the CSR movement argues come social responsibilities.

In the absence of such a sense of responsibility, Hadley asserted that the excesses of the railroad monopolies in the Gilded Age created a blind sense of pursuit of monopoly profits for which these captains of industry felt no restraint but the letter of constitutional law. Their corporate charter did not specify that they had no right to harm the public. This arrogance that Carnegie hoped to ameliorate with philanthropy instead created a backlash of antitrust legislation.

Hadley argued that the larger the position and influence of a corporate agent in society, the greater the responsibility the corporation has to society, just as citizens with outsized influence have a proportional responsibility to their community. He observed,

> There is many a man who, in the conduct of his own life, and even of his own personal business, is scrupulously respectful of public opinion, but who, as president of a corporation, disregards that opinion rather ostentatiously.[6]

Hadley advised, "If a man intends to stand on his legal rights it is generally wise for him to keep as quiet as the circumstances admit." He added that blind devotion to short run profits should be tempered to prevent conflicts between the haves and the have nots. Hadley believed a corporation's lack of social responsibility and a failure to meet a modicum of public needs shakes the public's confidence in a corporation regardless of any market interest it may serve.

Hadley was writing in the aftermath of the banking panic of 1907 and the last example of corporate America's anemic attempt to correct a recession of their creation. The Great War and the excesses of the Roaring Twenties followed. In the wake of the Great Crash of 1929 and increasing frustration of the modern corporation in the onset of the Great Depression, Adolf Berle and Gardiner Means produced the first treatise in CSR. In their 1932 The Modern Corporation and Private Property, Berle and Means argued for a shift in corporate values from the creation of shareholder profits to stakeholder returns.

Berle and Means noted that the modern corporation transforms what had traditionally been private and independently owned rights to productive capacity in the hands of small entrepreneurs to a mere claim on the profits of corporate capitalism. Small business owners of such morsels of property, which they would otherwise have used to satisfy their individual and community interests, are transformed into a share of a corporation which only pursues profits. Gone are myriad opportunities to contribute to community, replaced instead with ownership of thousands or millions of shares, each with a right to a slice of corporate profits. According to this theory, the modern corporation generates wealth, but does little to distribute the opportunity for community development granted by the wealth they create.

In essence, a sense of private property that also embodied moral obligations to contribute to community was replaced by profits as mere property. No longer do newly created shareholders have the ability to lever their property in socially beneficial ways. This ability is surrendered when property is converted to shares and managed centrally to generate profits for one large corporation rather than across myriad small businesses and communities. Berle and Means observed,

> The economic power in the hands of the few persons who control a giant corporation is a tremendous force which can harm or benefit a multitude of individuals, affect whole districts, shift the currents of trade, bring ruin to one community and prosperity to another. The organizations which they control have passed far beyond the realm of private enterprise - they have become more nearly social institutions.[7]

Limited corporate liability exacerbates this corporate dilemma in ways far more profound than the privileges granted individual people. While individuals must ensure the value they create exceeds the debts they incur on an ongoing and sustainable basis, a shareholder is not liable for the debts of a failed corporation. A corporate charter granted by the people is considered a legal person with many of the rights of citizens, such as the right to contract, but without some of our human responsibilities.

For instance, while a person can be imprisoned, a corporation cannot. In 2020, Pacific Gas and Electric (PG&E), a California company, pled guilty of the manslaughter of 84 people in a 2018 forest fire for which it was criminally convicted, and was ultimately fined $4 million as its sanction.[8] Such a penalty was far less onerous than an individual may have faced for the same transgression.

Corporations are also granted other concessions. A corporation can be viewed as a self-contained institution that mimics free market principles internally without the need to resort constantly to formal exchanges of titles to property. For instance, one could imagine the various functions of a corporation as a set of transactions, each of which could alternatively be performed on a contractual basis within arms-length markets and under the auspices of commercial law. Assembly of an item on a manufacturing assembly line could be contracted out; The various auditing functions in an internal audit department could be performed by independent contractors; The process of preparing financial statements in anticipation of the need to raise funds through borrowing or the issuance of bonds could be performed by an investment bank paid by commission rather than in-house under the Chief Financial

Officer; A corporation is then an internalized collection of these various transactions.

Ronald Coase, who had won the Nobel Memorial Prize in Economic Sciences in part for his The Problem of Social Cost, was also awarded the prize for ideas described in his 1937 The Nature of the Firm.[9] Coase observed that the performance of myriad internal corporate functions by contracting rather than internal staffing would be cumbersome, more expensive, and less expeditious. Hence, the corporation acts as a locus of internal contracting to efficiently substitute for less efficient market transactions. The resulting efficiency gains by mimicking market functions internally through the hiring of agents for the corporation, called employees, results in synergies that create profits for the corporation.

These internal functions of the modern corporation are efficient because they rely on implicit internal contracts and policies rather than more explicit and repeated external contracting. The implicit contracts are in the form of assignments to internal tasks as necessary on an ongoing basis, all under the auspices of an explicit or implicit employment contract. The contract for the employment of such resources must follow general principles and laws set forth by the governments of well-ordered peoples. But, within the corporation, there is a great deal of latitude to expeditiously renegotiate these assignments, within the scope of employment contracts and the unalienable rights of employees granted by their sovereign governments.

In the 1970 book to which Friedman so stridently responds, The Social Responsibilities of Business, Company, and Community, 1900–1960, Morell Heald documented how the modern corporation was steadily transformed over the six decades following the Gilded Age, through the Panic of 1907, the Great War, the Roaring Twenties and the Great Crash, the Great Depression, World War II, and the rapid growth of American corporations following World War II. Heald concluded that the overarching power usurped by large American businesses began to rival the power of government, and indeed often acted much as private governments.

While not commenting on this emerging trend of increasing corporate power, Coase had argued that it is precisely the ability of a corporation to function as a substitute for the marketplace and deflect governmental oversight that is the source of corporate efficiencies. Commentators such as Heald, and Berle and Means and others before and after them, drew similar conclusions that such an ability to function beyond the direct checks and direct oversight of government brings with it responsibilities commensurate with this power, whether or not businesses acknowledge their increased social influence.

When this ability to operate opaquely and beyond the marketplace within which others compete, and with private channels of communications and interaction that substitute for public institutions, the ability to perform both for the good of and to the detriment of society are possible. Monopolies and their ability to fix prices, cause environmental damage from pollution, pesticides, or production of known defective products, and other externalities such as greenhouse gas emissions, demonstrate that the power of the corporation is sufficiently potent to affect peoples and their property for many generations to come. Corporations are hence able to enrich themselves by at times taking property that belongs to persons or the people. Heald argued that this corporate power justifies either constitutional checks or an assumption of increased corporate social responsibility commensurate with corporate power that at times rivals the power of government.

It is the concern by Friedman of a growing movement of Corporate Social Responsibility that motivated his strident reassertion that the sole responsibility of a corporation is to generate profits. However, his appeal to a simpler era was based on a rejection of the *stakeholder model* of corporate governance rather than a strong argument for a *shareholder model*. He did not address the growing power of corporations and their quasi-governmental capacities, nor the many ways that a corporation even with the best profit-generating intentions can profoundly influence society, the economy, and the environment. In other words, Friedman conflated the ideal of atomistic perfect competition with the aspirations of increasingly monopolistic and anticompetitive corporations.

Friedman's denial of the Corporate Social Responsibility movement and the stakeholder theory had proponents over the next half century following his assertion of the Friedman Doctrine. More recently, failures of corporate ethics that gave rise to the bankruptcy of Enron, the malfeasance of mortgage and finance companies that led to the *Global Financial Meltdown* of 2008, and the complicity of corporations in the opioid health crisis have eroded the people's trust in corporations and institutions.

These breakdowns of corporate ethics have led commentators and academics to increasingly embrace the theory of stakeholder value rather than mere advocacy for stockholder profits. The ESG movement that arose with society's growing concerns over the perils of global warming is an extension of the Corporate Social Responsibility movement, but with a greater immediacy framed based on economic justice rather than merely on a theory of corporate responsibility.

Meanwhile, proponents of the Friedman Doctrine, from principals in the cryptocurrency movement and Bitcoin miners to the corporate raiders who employed what some labeled as *vulture capitalism*, found the ensuing half century profitable, even if socially precarious. These two contrasting views, one of corporate responsibility to society and the other solely to profits, resulted in a retrospective among economists on the fiftieth anniversary of Friedman's provocative statement.

In 2020, the *New York Times* published the retrospective "A Free Market Manifesto That Changed the World, Reconsidered"[10] as a reflection of the successes and failures of the movement Friedman fomented. Commentators noted that Friedman's assertion of the superiority of corporations' sole focus on profits was an artifact of the assumption that markets are perfectly competitive and also perfect in the sense that all information was available and all externalities were internalized.

We have explored the inefficiencies and inequities that arise when the economic and societal consequences of corporate actions are not fully factored into their decisions. These intra- or intergenerational externalities can be corrected either through corporate actions or Pigouvian taxes. Markets also depend on perfect information to function in the Friedmann ideal. It is the internalization of externalities within the walls of a corporation the ESG movement hopes to enhance.

An essential element of the ESG paradigm is the enhancement of complete information as a prerequisite for well-functioning markets. As an example of the legal liability that can be created for failure to inform the public on matters of material interest, in 1972 a model of automobile called a Pinto, manufactured by the Ford Motor Company, stalled and was rear-ended by a vehicle traveling about thirty miles per hour. Its gas tank ruptured and ignited. The driver of the Ford Pinto died and her passenger, Richard Grimshaw, survived severe burns after dozens of operations.

Ford lost the ensuing lawsuit, *Grimshaw vs. Ford*, not because of Ford's attempt to produce a low-cost alternative to compete with the popular and inexpensive Volkswagen Beetle, but because the corporation did not inform its customers that the gas tank Ford chose was less safe than those in more expensive and accident-resilient models. Social welfare suffered because customers were not given sufficient information to balance the savings arising from a less expensive design with the welfare effect of product safety.

As another example, in the 1960s, scientists employed by the oil corporation Exxon discovered the acceleration of global warming because of the fossil fuel policies of the corporation. These internal studies were quashed by Exxon management for fear the research may spawn either greater regulatory or Pigouvian taxes on oil profits. In 2015, the publication Scientific American published an article entitled "Exxon Knew about Climate Change almost 40 years ago: A new investigation shows the oil company understood the science before it became a public issue and spent millions to promote misinformation."[11] These obfuscations and omissions of information obstruct the peoples' ability to make optimal decisions and hence violates a fundamental tenet of the perfectly competitive model.

Friedman's competitive ideal depends on access to full information, while corporations at times find it convenient to obscure information. The ideals of the perfectly competitive model also depend critically on the assumption that economic actors do not have sufficient power to control market prices. Public policy is designed to encourage such competition and discourage concentrated market power. But, a primary goal of large corporations is to expand market share to lessen competition and increase its pricing power.

While public policy attempts to reduce corporations' efforts to expand market power, the number of large corporations that dominate commerce has increased relatively unabated. As a corporation exercises its market power to increase prices to convert to producers' surplus for themselves the consumers' surplus their customers would otherwise enjoy, some marginal customers are priced out of the market. This rent-seeking behavior ultimately results in some consumer and producer surplus that is no longer capture by either side of the transaction. The resulting *deadweight loss* results in reduced combined social welfare that the competitive model attempts to enhance, as we demonstrated in Chapter 11 (Fig. 23.1).

While corporations have no comparative advantage in philanthropy from the wealth they create compared to shareholder philanthropy from the dividends they earn, corporations have significant comparative advantage in influencing other behavior for our collective good. Oliver Hart, the 2016 Nobel Prize-winning economist, noted in the New York Times retrospective that a company which legally manufactures a dangerous product can enhance societal welfare by making the product safer. In doing so, while present profits may fall, long-term social welfare may nonetheless rise.

By narrowing the purview solely to the generation of profits, the Friedman Doctrine ultimately abandons other values held by shareholders and stakeholders. Profits are certainly on one end of the spectrum of values, with pure altruism on the other end. In between are many actions a corporation can take that benefit their median shareholders or stakeholders in ways these stakeholders cannot replicate themselves as efficiently as can the corporation. To

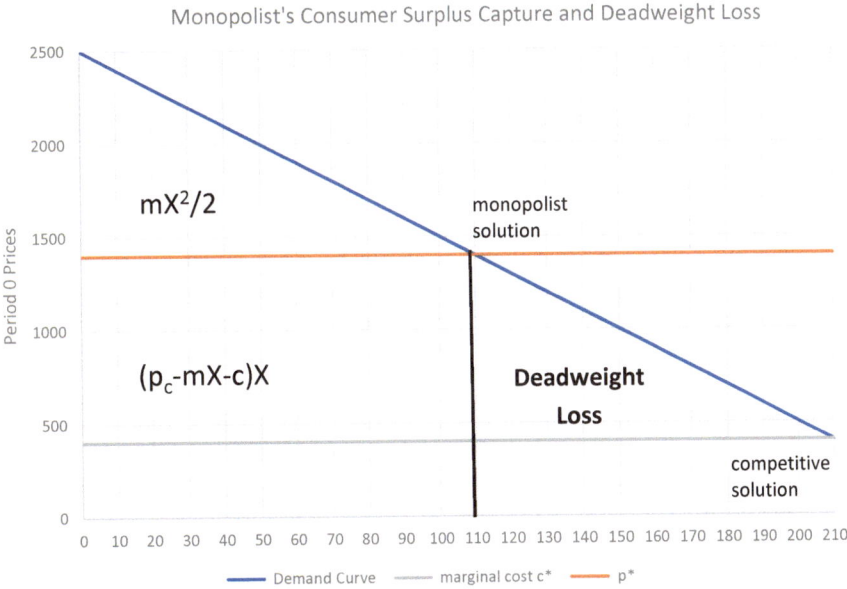

Fig. 23.1 Deadweight losses resulting from monopolistic strategies for various levels of period 0 output

then claim that only the profit extreme of this spectrum of corporate values be recognized as its sole motivation poorly represents the broad set of values that motivate shareholders and stakeholders.

The other glaring omission in the Friedman Doctrine is the lack of clarification of the significance of profits. Since his statement, corporations have been accused of focusing excessively on the quarterly profits that financial markets appear to value most highly, especially in light of the difficulty in communicating uncertain long-term profitability potential. Friedman skirts this short-termism by failing to differentiate between the type of profits he prizes. In a steady state economy, perhaps short-term profitability is the relevant measure. But in an economy that dynamically evolves, and in which many decisions have long-term and some have irreparable intergenerational effects on natural capital and sustainability, this distinction is important.

In summary, these artifacts of the Friedman Doctrine are what the ESG paradigm attempts to correct. They include:

- The need to ensure that market failures arising from externalities are corrected, either by market intervention or by the actions of corporations.
- The need to ensure markets remain competitive and that corporations do not engage in behavior to expand market share and violate the assumptions of the competitive model.
- The need for corporations to provide complete information regarding the actions they take that could affect their shareholders and stakeholders such as employees, suppliers, and those who purchase their products.
- The asymmetry between the ability of governmental institutions to deter antisocial outcomes by individuals and corporations through the courts. This asymmetry creates a social imperative for corporations to better align their actions in the public interest without using the corporate shield as protection from the consequences of antisocial behavior.
- The problems created when markets respond more profoundly to short-term profits verified in quarterly reports rather than long-term profitability that is more difficult to estimate. Such short-termism has the effect of increasing the corporation's discounting of future returns arising from present investments and hence worsens intergenerational efficiency.

Milton Friedman died in 2006, and could not defend his doctrine and the assumptions of perfect competition upon which it depends. Nonetheless, we can glean some insight from his original argument. He stated

... the doctrine of "social responsibility" involves the acceptance of the socialist view that political mechanisms, not market mechanisms, are the appropriate way to determine the allocation of scarce resources to alternative uses.

Friedman's concern is perhaps based on a fear that private markets would be replaced by central planning. He assumed that the problems ESG strives to reduce are failures of public policy rather than private corporations. The current ESG movement instead argues that some corporate policies may be more effective and better able to anticipate societal needs. Hence, corporate adopters of ESG may prevent the march toward greater governmental intervention and hence may even yield greater profits for corporations and society in the long run.

Certainly, corporations have an obligation to follow the practices established by their government of jurisdiction. The corporation is not an avenue to avoid these obligations, but is rather a collection of opportunities to follow these obligations in a more efficient manner. Many acknowledge that corporations have either a responsibility to society or to its bottom line to recognize values inherent in the ESG movement, but Friedman Doctrinists remain skeptical.

Beyond the requirement that corporations follow the letter of the law, the Friedman Doctrine presumes that corporations also have an obligation to provide to their shareholders with all information relevant to the pricing and profitability of their corporation, with the legitimate exception of corporate trade secrets. If the profitability of a firm can be adversely affected by such sustainability challenges as liabilities arising from climate change, efficient markets require that such information be revealed to stakeholders and the market.

In addition, corporations have a responsibility to reveal whether their implicit or explicit discounting of the future differs from the discount rate their shareholders or society maintain. These discount rates also certainly differ from the social discount rate appropriate for intergenerational efficiency. There may be avenues for public policy to narrow such gaps, but they would require either market intervention or corporate subsidization, both of which introduce other societal challenges.

The Friedman Doctrine also implies that corporations act in the long run best interest of its shareholders. For instance, Friedman may claim that a corporation has no obligation to pursue an ESG policy solely because it is a value shared by the corporation's shareholders. If the shareholders are representative of the customers of the corporation's product, sound strategic planning should anticipate whether customerage would suffer from reputational effects for a corporation that does not incorporate ESG values. The security price may also suffer if the growing share of investors who allocate funds based on ESG values reduce their demand for corporate equity that does not follow ESG principles. Such risks are described more fully in Chapter 26.

Social concerns have been heightened in the half century since Friedman provided his rebuttal to the Corporate Social Responsibility model that call for the substitution of stakeholders for shareholders in statements of corporate goals. In such a broader approach, the corporation's profound social role and its ability to act as a quasi-government within the bounds of the corporate veil produces a proportional responsibility to the peoples of a nation that affords corporations its charter and special privileges.

Corporations that accept either an obligation or recognize a competitive advantage in adhering to ESG principles respond in a number of ways in the Environmental, Social, and Governance paradigm. The incorporation of ESG values is not inconsistent with Friedman's faith in the ideal of the competitive economic model. They may actually enhance long run profits for the corporation, the industry within which it operates, nations, and the planet.

The Corporation and the Environment

Within the premises of the ESG Paradigm and Corporate Social Responsibility, corporations have an essential role in private sector aspects of environmental sustainability. The avenues for these roles include:

- The ecosystem and markets
- Lifecycle stewardship
- Green bonds and carbon markets
- Carbon Sequestration and Carbon Markets
- SEC overview
- Carbon footprint studies

Markets determine economic values based on the balance between supply and demand for products or funds at any given time. Since these determinants of market prices are derived based on the spectrum of values of current mortal market participants, they do not typically incorporate the values of generations not yet born. For instance, we have seen that a corporation which demands a cost of capital-based rate of return will consume fixed resources at a rate significantly higher than is socially perspective.

There are remedies to this discrepancy between the intragenerational and intergenerational discount rate. Government has the ability to create intergenerational fund transfers. Permanent financial capital funds arising from royalties set aside under the weak sustainability principle as natural capital is consumed can be used to either provide substitutes or repairs to fixed and natural resources such as the ecosystem and the atmosphere.

At the corporate level, tools can be imposed through resource royalties sufficient to create and enhance permanent funds. In addition, public policy can provide both Pigouvian taxes on environmental degradation or on consumption in excess of sustainable levels of natural resources in amounts sufficient to reestablish intergenerational efficiency. These interventions first demand adequate imposition of property rights to avoid any tragedy of the commons common property problems on a global basis. A mortal society is still left with a pattern of corporate decision-making that is responsive to the expected returns of markets rather than of generation-spanning peoples.

For the elements of corporate decision-making that do not suffer intergenerational implications, the risk adjusted rate of return determined by markets is correct. There remain many corporate decisions that have sustainability implications. If corporations do not recognize the intergenerational implications of these decisions, such as their use of fossil fuels, Pigouvian taxes can correct the market incentives.

The greater problem lies in inadequate corporate investment in activities that yield private returns which do not fully capitalize societal returns. For instance, an investment in alternative energy may not be feasible for a corporation given its cost of capital and any inherent market risk, even if the investment produces significant social dividends. In such instances, governments could subsidize their cost of capital, indemnify components of risk, or supply the necessary investment capital at a lower cost through public issuance of *green bonds*. Government recoups these subsidies over time through its superior ability to pool intergenerational risk or from the various social returns and multipliers and the taxes they generate arising from private investment in the long and very long run.

In addition, government can impose caps on corporate carbon emissions and allow corporations to trade their rights in a cap-and-trade system. Trading of carbon credits between firms needing to reduce their carbon footprint and offered by firms that do not use their entire carbon dioxide equivalent emission allocations, or for enterprises that create carbon sinks or sequestration.

To date, such carbon markets have proven relatively ineffective with regard to global warming, though. Governmental establishment of appropriate emissions limits in free market economies has also been insufficient, especially in cases when corporations play a disproportionate role in the influence of public policy. Even the monitoring of actual greenhouse gas emissions is challenging from a technical perspective, although recent innovations in orbiting satellite monitoring are beginning to yield vastly superior monitoring data. Finally, claims of carbon sequestration are fraught with a lack of standardization.

As an example, entrepreneurs have marketed tracts of forest land as natural sinks that absorb carbon dioxide. The appropriate economic measure is the effect of their claims at the margin. If that forest would not have been developed regardless of its placement in a carbon sequestration pool, then the inclusion provides no net contribution to sequestration. In the absence of effective monitoring of emissions and their net changes, marketers of carbon sinks also have incentives to exaggerate their claims. A recent study by the Guardian, Die Zeit, and SourceMaterial media investigative journalists determined that 90% of carbon offsets are ineffective.[12] The various activities to which a corporation commits are important to document, though, even if their current actions do not translate into enhancements of carbon dioxide equivalent sinks or natural capital preservation.

To improve the effectiveness of corporate climate stewardship, the U.S. *Securities and Exchange Commission* and the *Basel Accords* issued by the **Basel Committee on Banking Supervision** (BCBS) are increasingly requiring various types of corporations to report the extent of their carbon footprints and environmental liabilities in standardized ways. Communication of corporate patterns of activities provides stakeholders and markets with important information with regard to the inherent risk a corporation may face should public policy or sustainability markets evolve toward demands for greater sustainability in the future. In this regard, the carbon footprint of a corporation, or their exposure to activities that are harmed with further environmental degradation, creates an unpriced liability on a corporation which is a material matter that should be communicated to stakeholders and investors. Such risks are described in Chapter 26.

Corporations are also required in some jurisdictions to report their plans for recycling the resources they embody in the goods they produce. Such *lifecycle stewardship* places the responsibility for intergenerational resource optimization at the level which is least costly to administer. Consumer markets may also be levied with point-of-sale surcharges for the ultimate disposal of a purchased item at the end of its useful life, and disposal facilities can be maintained that refund this surcharge when the obsolete item is returned.

These various mechanisms are examples of ways in which intergenerational environmental externalities can be internalized at the corporate level through well-designed public policy. However, given the global and intergenerational aspects of such externalities, optimal public policy coordination at the global level is often necessary.

As we have seen with regard to the difficulties in negotiating standards to slow the rate of climate change, efficiency enhancement also frequently creates uneven distributional and equity burdens. There remain challenges of global coordination in mechanisms to induce optimal corporate decision-making, especially in the absence of necessary mechanism for transfers between national governments charged with putting such mechanisms in place.

These challenges do not reduce the need for greater environmental factors reporting by corporations, though. Without efficient and accurate reporting, neither governments nor markets can formulate optimal policies or prices.

Summary

With a growing inability to meet sustainability goals, there is an increasing assumption that corporations may play an essential role in natural capital repair and a movement toward a more sustainable future.

The sense that Corporate Social Responsibility, as extended by the ESG movement, is at odds with the insistence by some that the sole common denominator among corporate shareholders is the pursuit of profits. Instead, well-implemented ESG, through a partnership between the private and public sectors, can produce both better environmental results and enhanced corporate profitability over the long run. A number of avenues for active corporate participation in global sustainability are described in Chapter 25. In the next chapter, we describe a rationale for corporate concern of social and governance aspects of the ESG paradigm.

ESG Toolkit

A consensus is growing that corporate citizenship carries with it a corporate responsibility. This sense of corporate responsibility tends to increase with economic crises. Carbon footprints have accelerated with increased GDP, even as wealthier nations and corporations have attempted to reduce their footprint per unit of economic activity.

To what degree do corporations accept their degree of corporate responsibility?

Are the philanthropic gestures of corporations in proportion to the sustainability challenges they create?

How might corporate gestures be better aligned with their sustainability responsibilities?

How can corporations reduce the absolute level of emissions in carbon dioxide, methane, and nitrous oxide?

How does the level of an individual's corporate profits correlate with the carbon footprint?

Is this relationship tracked over time?

Notes

1. Friedman, Milton. (1970). "A Friedman Doctrine—The Social Responsibility of Business Is to Increase Its Profits", New York Times, September 13.
2. https://blogs.loc.gov/inside_adams/2016/04/when-a-quote-is-not-exactly-a-quote-general-motors/, accessed January 5, 2023.
3. https://www.hemmings.com/stories/2019/09/05/fact-check-did-a-gm-president-really-tell-congress-whats-good-for-gm-is-good-for-america, accessed December 2, 2022.
4. Heald, Morrell. (1970). The Social Responsibilities of Business, Company, and Community, 1900–1960, Routledge Publishers.
5. Hadley, Arthur Twining. (1907, Jan. 18). "Ethics of Corporate Management", The North American Review. **184** (607): 120–134.
6. Hadley, Arthur Twining. (1907, Jan. 18). "Ethics of Corporate Management", The North American Review. **184**(607): 120–134, at 128.
7. Berle, A. A. and G. C. Means. (1967). The Modern Corporation and Private Property, 2nd ed., Harcourt, Brace and World, New York, 46.
8. https://www.reuters.com/article/us-california-wildfires-pg-e/pge-pleads-guilty-to-84-counts-of-involuntary-manslaughter-in-california-wildfire-idUSKBN23N35T, accessed December 2, 2022.
9. Coase, R. H. (1937). The Nature of the Firm. Economica. **4**, 386–405.
10. A Free-Market Manifesto That Changed the World, Reconsidered. (2020). New York Times, September 11.
11. Hall, Shannon. (2015, October 26). "Exxon Knew About Climate Change Almost 40 Years Ago", Scientific American, October 26, 2015.
12. https://www.theguardian.com/environment/2023/jan/18/revealed-forest-carbon-offsets-biggest-provider-worthless-verra-aoe, accessed January 27, 2023.

CHAPTER 24

The Private Sector and ESG: Social and Governance Pillars

The ESG movement, and the Corporate Social Responsibility principle that preceded it, has long-standing roots dating back to the *Utopian Societies* initiated by the socially conscious nineteenth century industrialist Robert Owen in Scotland. Worker cooperatives that incorporated human values into corporate decisions spread to locations in England, Canada, the United States, Europe, and elsewhere in various forms of industries, from manufacturing to farming. These organizations were run on the principle that corporate success was designed to satisfy a variety of stakeholders, with social values beyond simply profits and dividends to shareholders.

Attitudes with regard to stakeholder values have come in waves, but the social responsibility tide has been rising with each successive and public failure in corporate responsibility. The excesses of the Gilded Age, the bank failures of the 1907 Great Panic, the excesses of the Roaring Twenties that culminated in the Great Crash of shareholder value in 1929, the Great Depression, environmental calamities of the 1950s and 1960s, the corporate excesses of the 1980s and 1990s, and the global financial meltdown of 2008 each created a greater clamor for Corporate Social Responsibility. Academic research and the Club of Rome's Limits to Growth in the 1970s and the development of the triple bottom line in the 1980s and 1990s all reassessed the degree of responsibility corporations must assume in proportion to their profitability.

By the 1990s, a number of companies began to espouse ESG values, even before ESG was fully articulated. Most influential may have been the formation by Royal Dutch Shell of its *Social Responsibility Committee* that issued a report a year titled *Profits and Principles—does there have to be a choice?*[1] This 1998 report perhaps inspired BP's 2001 slogan change from British

C. Read, *Understanding Sustainability Principles and ESG Policies*, https://doi.org/10.1007/978-3-031-34483-1_24

Petroleum, to Beyond Petroleum, with its introduction of a new logo that contained a stylized green and yellow sunflower to represent the Greek Sun god *Helios*. Despite BP's recognition by the American Marketing Association in 2008 for its green efforts, BP was criticized for relatively trivial alternative energy investment of $1.5 billion US relative to a $20 billion fossil fuel investment. Two years later, the Macondo Spill in the Gulf of Mexico further eroded its green corporate reputation, while, in 2022, BP admitted that it has renewed its focus on fossil fuel extraction in the wake of higher crude oil prices following Russia's invasion of Ukraine.[2]

Such examples of corporate greening and subsequent retrenchments have resulted in various industry organization responses. The **World Business Council for Sustainable Development** formulated a **Natural Capital Protocol** that helps guide corporate behavior and avoid the depreciation of trust in corporations arising from their failure to protect the Earth's resources. A number of private sector-led initiatives have also been created to coordinate corporate actions around ESG and better communicate collective efforts to the public in attempts to prevent eroding corporate goodwill arising from *greenwashing*.

GREENWASHING

Some accused BP of insincerity in its corporate responsibility to adhere to pledges made in its conversion to Beyond Petroleum. The term **greenwashing** was coined to describe efforts to project a sustainable corporate philosophy in an attempt to improve its reputation without a genuine commitment to substantial green policies. A corporation may provide the impression that it is performing based on stakeholder expectations by making pledges it fails to keep, or it may issue deceiving and disingenuous statements in attempts to deflect attention from its activities or instead direct attention to the environmental vulnerabilities of others.

Siano et al. (2016) label the first phenomenon, the failure to follow through with its environmental commitments as *decoupling*, while *symbolic actions* involve misleading statements designed to distort or deflect their record.[3] De Freitas (2020)[4] argued that such breakdowns and the emerging clarion call of greater corporate environmental sensitivity have created a greater impetus for Corporate Social Responsibility.

In addition, regulatory agencies such as the U.S. Security and Exchange Commission (SEC) and the Financial Conduct Authority in the United Kingdom have promulgated regulations for increased oversight designed not to regulate industry ESG activities by fiat but rather to ensure that financial statements and corporate communications to the public be accurate representations of material matters that impinge on the financial valuation of the firm.

Greenwashing, or its counterpart *greenwishing*, defined as either cynical or well-intentioned statements of unrealized climate change ambitions, is often designed to endear a corporation in public forums without any commensurate change in corporate priorities. Sincere corporate social responsibility can be profitable, especially as stakeholders increasingly embrace sustainability values, but unrealized climate change ambitions can damage entire industry sectors.

Markets depend on accurate information and representations, and the value across entire industrial sectors can be tarnished by a few bad actors. While some regulatory bodies now assert that they have a responsibility to ensure accurate corporate representation of matters that could affect their valuation in capital markets, the industry as a whole recognizes the negative externalities to related corporations if a company's actions cast aspersions on competitors as well. Reputational risk may cross corporate boundaries and hence institutions have been created to protect the integrity of the private sector that have increasingly extended beyond mere preservation of reputations.

To enhance both the corporate role and responsibilities in shared sustainability concerns, the *World Business Council for Sustainable Development* (WBCSB) was formed following the 1992 *United Nations Conference on the Environment and Development* (UNCED) in Rio De Janeiro to incorporate emerging directions emanating from the UN that may affect the private sector. It developed a set of best practices through its *Principles for Responsible Investment* working group in 2006 to guide corporate social responsibility. Corporate participants are expected to commit to the incorporation of ESG values into their corporate actions and communicate their ESG policies and actions to stakeholders, and collaborate in the development, implementation, and reporting of activities related to the PRI principles.

The *Global Reporting Initiative (GRI)* had been formed a few years later in 1997, in the aftermath of the 1989 *Exxon Valdez* oil spill. Following the spill, stakeholders and the public clamored for greater corporate transparency, which resulted in a voluntary environmental code of conduct called the *Valdez Principles*. GRI was formed to ensure at first that corporations could adhere to a voluntary set of environmental accountability principles but increasingly engaged in developing corporate principles assembled as the *GRI Sustainability Reporting Standards* across the spectrum of ESG values. GRI stated:

> The tide is turning, and these stakeholders are understanding that the risks they face are long-term and many are related to ESG concerns. Nowadays investors are changing the way they view sustainable investment, to include environmental, social and governance factors.[5]

In doing so, the GRI saw its scope as expanding beyond CSR and into creation of well-functioning investment and finance markets that correctly price ESG values.

In 2011, the *Sustainability Accounting Standards Board (SASB)* was formed to create ESG metrics by which its corporate participants can gauge their performance. They also develop standards for material financial disclosures related to corporate environmental sustainability. SASB defined standards across 77 industries in the areas of sustainability related to human and social capital, governance and corporate business models, and the environment.

The GRI and related groups such as the SASB improve market and investment efficiency by reducing negative externalities that damage commerce when some firms misrepresent their corporate actions, and by correcting the free-rider problem by encouraging broader corporate participation in such voluntary corporate associations.

Sustainability Initiatives and Coalitions

Private or Non-Governmental Organizations have instituted a number of programs to address sustainability. They include:

- World Business Council for Sustainable Development (WBCSD) formed following the Rio De Janeiro summit of nations and corporations and has now grown to more than two hundred major corporations with substantial international operations. They include BP, Dupont, and Nestle, and span the spectrum of consumer production to the energy sector.
- Business for Social Responsibility is an organization that networks and shares best practices among corporations interested in sustainability and resilience when faced with climate change.
- The UN Global Compact now includes 80% of nations and more than 13,000 participants to promote principles regarding the universality of human and labor rights, anti-corruption, and environmental protection.
- The Business Roundtable is a business advocacy group that has increasingly directed its attention toward the attainment of sustainability and economic resilience for economies that serve all Americans and stakeholders.
- Within the Financial Services Sector, Asset managers and owners have formulated Principles for Responsible Investment (PRI). These six principles for Responsible Investment are aspirational and voluntary investment principles that provide avenues to incorporate into investment practices viable ESG concepts.
- Insurers have assembled since 2012 to create Principles for Sustainable Insurance (PSI). Their Four Principles for Sustainable Insurance contribute to improved management of entities challenged with ESG issues. The coalition seeks to foster a resilient, inclusive, and sustainable society.
- International bankers have produced the Principles for Responsible Banking (PRB). Their Six Principles for Responsible Banking offer a platform for sustainable banking by embedding sustainability into strategy and bank portfolios across the various banking markets

These initiatives hope to remedy the damage that can occur to broad corporate sectors if individual corporations improperly assess reputational risk. Private sector associations, government, and global bodies such as the United Nations increasingly recognized the need for proactive action that aligns corporate and societal interests. The private sector recognizes the various risks of failure to respond to the growing clamor for corporate sustainability responsibility.

While corporations play an essential role in stewardship of the environment, the stakeholder model and a risk management approach also have implications on a number of other aspects internal to the corporation. These include:

- Public, corporate, and employee policies
- Employee and executive compensation and incentives
- Privacy and information governance
- Diversity, Equity, and Inclusion (DEI)

The ESG paradigm identifies these various avenues as critical in the enhancement of both equity and efficiency within the stakeholder approach to corporate value creation and social responsibility. These ESG policies are designed with two objectives. The first is to ensure that the policies within the veil of the private sector corporation are communicated to stakeholders, including boards of directors, shareholders, regulators, and the public. This element of transparency is consistent with leading theories of corporate governance.

Such articulation of policies also reduces operational and reputational corporate risks. Communication of policies ensures consistent coordination across institutions and ensures that institutions speak with a clear and concise voice. In doing so, not only are internal processes aligned along a consistent framework but stakeholders understand these policies and respond accordingly.

The second advantage of an insistence on consistent and clearly communicated corporate policies is that legal risks are reduced. Such risks are described more fully in Chapter 26. At the state level, governing policies and principles take the form of laws and can be enforced criminally or civilly. Civil rights and responsibilities often explicitly define the relationship between the state and the people. However, actions within corporations are more opaque and less well-defined. If applied inconsistently, or if internal policies violate laws, corporations may create avoidable legal risks. By articulating and communicating policies, codifying employee rights and responsibilities, and ensuring that entities within the corporation must reflect shared values within civil society, the risks of civil suits against the employer by employee or union members or groups are reduced. This reduction in legal liability creates more harmonious internal processes, minimizes operational risk, and enhances profits, but it also limits potential legal risks and reputational damage.

Two social criteria for the corporation in the ESG framework are of particular relevance. Well-defined Diversity, Equity, and Inclusion (DEI) criteria both reduce corporate liability and contribute to a reduction in people risk and an improvement in institutional harmony. DEI also reduces operational risk by ensuring that diverse perspectives are included in corporate deliberations and that the outcomes of such deliberations are fair from the perspective of various stakeholders.

The Corporation and Governance:

Corporations are increasingly recognizing the need to communicate their sustainability values to their shareholders and stakeholders. Internally, the corporation must also establish governance principles that advance corporate sustainability. Factors include:

- Corporate asset valuation and management
- Capital infrastructure and investment
- Investment and risk management
- Role of Corporate Governance in ESG Policies
- The Public Process Necessary to Formulate Intergenerational Equity

The modern corporation in many respects mimics social conventions and public policy but within the veil of its corporate structure. Modern stakeholder theory then implies that the corporation might be best served by the adoption of certain acceptable values that are also prized by the people of the society that grant corporate charters.

It is conceivable that a group of managers or directors who are homogeneous in education, wealth, age, experience, and race could make well-informed decisions that equitably and efficiently represent the values of stakeholders who are diverse in these various factors. However, such decisions are less likely to be equitable not because of any assumption of ill will on the part of directors and managers but rather because the perspectives brought to a boardroom are narrow. Groups that are diverse are more likely to interject a broader set of factors and perspectives into the decision-making process in ways John Rawls constructed with his *veil of ignorance* approach to social policy-making.

The other criterion is the alignment of incentives for stakeholders. Under the Friedman Doctrine, the sole stakeholders are a corporation's shareholders. Any broader stakeholder perspective is incorporated into corporate policies and strategies only if it also yields higher profits. Opportunities for profits arising by addressing stakeholder preferences are more likely if management and boards are representative of the diversity of stakeholders.

Under the ESG paradigm, the reward structure of senior and critical managers and directors should be aligned with the criteria of both shareholder profit maximization and stakeholder's value optimization. Under the Friedman Doctrine of shareholder profit maximization, executive and board compensation is typically only tied to stock performance. Incentive compatibility is enhanced when senior managers' and board members' compensation are at least partly in the form of stock. If a corporate goal is to manage and enhance long-term profits, these rewards should be in the form of a right to purchase stock at some time in the future based on the prevailing price when issued. Under such an incentive scheme, directors and managers have a greater incentive to ensure that stock prices rise in the medium and long-term as corporate strategies are realized.

Corporations that adopt a broader stakeholder Community Social Responsibility approach in which diversity, equity, and inclusion are valued are articulating a broader set of economic justice goals beyond mere maximization of short term profits. The establishment of corporate goals can combine to create a corporate culture that is more sustainable over time. These can also include sustainability and climate change mitigation measures, in diversity and inclusion, in a lifecycle approach to the products the corporation produces, in a corporate carbon footprint, and in managerial and human resource processes that optimize expertise and minimize undesired turnover.

The optimization of corporate resources, especially in enhancing sustainable corporate and human resources, provides for a healthy and potentially more productive corporate environment. In turn, this productivity translates into greater corporate value and stock price. The formulation and communication of a broader set of sustainability values also engender the corporation to external stakeholders such as consumers, suppliers, regulators, and analysts, and especially ESG investors who believe there is a positive correlation between strong corporate ESG values and the market valuation of the corporation.

Corporate Asset Valuation and Management

There is a saying within corporations. *If you can't measure it, you can't manage it*. Corporations are ultimately held accountable to their stakeholders and to markets based on their performance. Publicly traded corporations are required to regularly report their enterprise value, in terms of balance sheets that compare corporate assets and liabilities, and income statements that compare annual or quarterly revenues and costs. Public policy has imposed standards of accounting reporting, called *Generally Accepted Accounting Principles* (*GAAP*) to ensure these measures are reported consistently and in ways that lend themselves to intercorporate comparisons.

These measures attempt to account for the value of investments made in physical capital, human resources, research and development, and other items for which the path of productivity may be difficult to estimate. For instance, standard depreciation principles assume a physical investment's productive capacity declines over time. However, such accounting principles may not always be truly representative of corporate values from an economic perspective. For instance, while the liabilities of a firm are often contractual and prone to measurement with greater accuracy, asset valuation is often uncertain. This is especially true when risks impinge upon a firm, most often in their valuation of assets.

Goodwill is one such asset. It anticipates that the value of a firm in financial markets is an unbiased and accurate measure of the firm's true value. Since market exchanges are the mechanisms we use to value assets, if a market determines a stock is of high value, its price adjusts to meet these expectations.

If such a price is in excess of the value of a corporation's assets net of its liabilities, then *GAAP* principles state this excess as the value of an *intangible asset*. This value would be impossible to extract were corporate assets sold at their GAAP value and liabilities paid off with proceeds. The difference in value between tangible assets and liabilities represents the tangible value of the enterprise but does not include the goodwill associated with its intangible assets.

If market valuation depends on the effective communication of these financial assessments, then accurate and standardized corporate valuations are critical. If corporate managers and directors have the sole short-term goal of corporate value maximization, then they tend to manage to tangible values rather than true values of assets and their ability to generate future revenue flows, or to alternative values that their stakeholders may also hold. In doing so, they may sacrifice potential goodwill in the long-term to enjoy short-term expediency.

Corporate Reporting and Regulation

Market valuation of corporations then critically depends on accurate reporting of corporate finances and *forward guidance* by senior corporate managers regarding the prospects for future net revenue generation and the possibility of current asset degradation. Access to capital then depends critically on the accurate and unbiased reporting of financial and other material data. If this process is short-circuited, corrupted, or too narrowly or cynically defined, markets lose confidence in reporting, and market reliability and resiliency are diminished.

The integrity of financial markets and the accuracy of market information is critical for the ability of markets to raise capital in support of corporate investment. Reliable financial and prospective information reduces uncertainty and risk and hence increases corporate value and returns. This reduction in risk and enhancement in value then affords access to a superior opportunity to raise capital for investment.

Financial regulatory authorities then play a significant role in enhancing investment value and optimizing capital availability to stimulate economic growth. For instance, the U.S. *Securities and Exchange Commission* (*SEC*), among others, has adopted the critical role of maintaining financial market integrity by ensuring accurate corporate reporting so markets have complete information. Such entities standardize the reporting and communication of financial and other information that may affect the value of securities. They also ensure that information which can affect market valuations are made available simultaneously to all those who may trade on new data to prevent some traders from having an informational advantage over others. In doing so, they attempt to prevent *insider trading* and ensure all stakeholders have equal and accurate access to relevant corporate information.

Corporations wish to maintain the competitive advantage they create through their strategic investments. Some of the information within the corporation constitutes legitimate **trade secrets** that must be protected if society hopes to benefit from the uncertain corporate innovation investments. There is no expectation these trade secrets that constitute the success of such investments should be revealed to competitors. However, beyond corporate trade secrets and their proprietary strategies to expand their profitability and market share, corporations are increasingly expected to report material items that will affect market valuation, either positively or adversely. ESG values and opportunities are now broadly considered items of market concern.

ESG and Corporate Governance

Corporate boards are the conduit for both oversight and direction of internal corporate strategies to generate profits on behalf of shareholders and to provide assurance that the corporation conveys critical information so that its stakeholders can best determine the prospects of the corporation on an ongoing basis. For instance, in the United States, publicly traded corporations must issue to the SEC 10K forms annually and 10Q forms quarterly that report on corporate finances and describe factors that may enhance or impair these results. Governing bodies and senior management must approve these statements.

The SEC and similar regulatory bodies must then ensure accuracy and completeness in the information contained within such statements. Regulators do not specify the actions of corporations but instead strive to provide the public with the information it needs to properly assess corporate value. The type and extent of information that is necessary to convey to financial markets is thus proportional to the dynamic nature within which corporations operate. An environment that is unchanging may necessitate less extensive reporting compared to dynamic environments in which factors that could impinge on corporate valuation are evolving rapidly.

For these reasons, regulatory authorities, from the U.S. SEC to its *Federal Reserve Board* and the *Financial Stability Board*, and the equivalent *Basel Committee on Banking Supervision (BCBS)* in Europe gauge evolving stakeholder expectations and foment reporting regulations to ensure the public has access to relevant information that is standardized and hence easily benchmarked and compared. Recent changes in actual or proposed reporting items by these agencies related to ESG are described in greater detail in Chapter 26 on corporate risk management. Such risk management and reporting requirements must necessarily adapt to evolving industry norms and stakeholder expectations and have increasingly directed greater reporting on corporate sustainability policies.

One can argue that such reporting is activist in that it is responding to the agendas of those who may wish to amplify corporate responsibility, especially in terms of sustainability and climate change. However, such reporting of ESG matters can materially affect market valuation and hence the efficiency of markets to raise capital. As such, regulators are increasingly broadening requirements for standardized reporting on ESG matters in the belief that these material factors influence stock prices. Such reporting is consistent with both the shareholder and the stakeholder model of corporate responsibility. Statements confined to documentation that a corporation acts in ways consistent with ESG values are merely accurate communications of ESG-related actions consistent with the principle of good corporate reporting.

INVESTMENT AND RISK MANAGEMENT

Economists claim there is **no free lunch**. In the context of investment, it implies that one cannot have both a high rate of return and low risk. One can certainly invest in securities such as a U.S. government bond and be assured that, since the U.S. government has never defaulted on their promised return of periodic payments and of the face value of the bond upon maturity, their risk is low, and confined primarily to an uncertainty about future interest and inflation rates. There is even a type of U.S. bond called a *treasury inflation protected security (TIPS)* that is indexed to inflation and hence removes the inflation risk. While some downside risk is never entirely unavoidable, it is limited to the regret of purchasing a bond if the return on other investments subsequently rises, government bonds are deemed relatively *risk-free*.

Financial securities instead attempt to garner a higher than risk-free rate of return by effectively managing greater risk. Financial markets are an excellent mechanism to disseminate and price such risk and adjust the offer price for securities to compensate for the degree of risk an investor absorbs in the purchase of a security. The primary measure of appropriate returns in financial securities is called the **Capital Asset Pricing Model** (*CAPM*). It states that the rate of return r a security offers over and above the risk-free rate of return r_f and the overall market rate of return r_m is proportional to the degree of risk the security imposes on the investor compared to the relative risk of the overall market. This comparative measure of risk is called a beta coefficient, and uses the Greek symbol β:

$$\beta = \frac{r - r_f}{r_m - r_f}.$$

For instance, if $\beta = 1$, the return r on a given investment offers the same premium over the risk-free rate of return as does the return of the market as a whole. The beta coefficient is thus a measure of relative risk or volatility. The greater the volatility of returns, the greater the beta and the greater the market expectation of a higher return to compensate for a security's greater risk. If all securities are priced according to such an analytic approach, the market will deem which securities are of higher value and hence permits well-performing corporations the opportunity to raise more capital for new investment.

To enhance corporate value, governance boards, and management then have an important role in the reduction of uncertainty and risk as reported to regulatory agencies and disseminated to the public. By managing risk, the rate at which the market discounts firm returns falls, and hence valuations and returns to shareholders improve. The corporation has the incentive to manage effectively and optimally the risk inherent within the firm. In turn, regulatory oversight bodies must then ensure that this risk is properly conveyed to both maintain investor confidence in financial markets and ensure that each security is priced accordingly.

Governance boards then have the dual responsibility of ensuring optimal risk management and the creation of shareholder value within the corporation through the communication of any relevant factors to stakeholders and shareholders. Regulatory agencies such as the SEC then provide the infrastructure and produce the reporting mechanisms to ensure maximal accuracy and hence minimal unnecessary uncertainty. In doing so, regulatory oversight of ESG reporting enhances financial market efficiency and the ability of corporations to raise investment capital.

For these reasons, while reporting on such matters relevant to ESG and sustainability may seem counterproductive to some, and may demand resources internally by the firm, they are nonetheless an essential element of a firm's internal auditing function. Accurate ESG reporting plays a crucial role in the long-term ability of a corporation to maintain their capital value and raise new capital as market conditions and expectations evolve.

Summary

Modern corporate strategy must evolve to both play a role in and profit from emerging societal trends and greater urgency in the need to attain sustainability. Regulators are also increasingly expecting corporations to articulate their ESG strategies to stakeholders so that markets are well-informed about evolving corporate strategies in a corporate world threatened by the evolving needs of sustainability. Adoption of ESG reporting standards also creates some opportunities for leading-edge companies to effectively raise capital and invest. The next chapter describes some of these opportunities, while Chapter 26 describes how corporations are incorporating ESG, climate change, and sustainability into their risk management models.

ESG Toolkit

Leading corporations and non-governmental organizations are increasingly forming associations to share best practices and develop sustainability standards.

Has your corporation joined other similar organizations to share best practices?

Are private entities able to succeed in establishing sustainability when national leaders could not?

How can it be profitable for a corporation to act more sustainably in a fiercely competitive corporate environment?

NOTES

1. https://www.climatefiles.com/shell/1998-shell-report-profits-principles-cho ice/, accessed February 18, 2023.
2. Worland, Justin (Feb. 10, 2023). "BP's Green U-Turn Shows Exactly Why the Energy Transition Is So Hard," *Time* Magazine. https://time.com/6254378/ bp-oil-profits-climate-energy-transition/, retrieved February 13, 2023.
3. Siano A, Vollero A, Conte F, Amabile S. (2017). "More than words": Expanding the taxonomy of greenwashing after the Volkswagen scandal. Journal of Business Research. 71: 27–37. https://doi.org/10.1016/j.jbusr es.2016.11.002.
4. de Freitas Netto, S. V., Sobral, M. F. F., Ribeiro, A. R. B. et al. (2020). Concepts and forms of greenwashing: A systematic review. Environmental Sciences Europe. 32: 19. https://doi.org/10.1186/s12302-020-0300-3.
5. https://globalreportinginitiative.medium.com/from-disclose-to-disclose-what- matters-the-growing-role-of-investors-in-sustainability-566cbf85ab32, accessed December 10, 2022.

CHAPTER 25

Sustainability Scenarios and Private Sector ESG Opportunities

As accurate ESG reporting is increasingly considered an essential responsibility for publicly traded companies by corporate associations, regulatory agencies, and a diverse set of stakeholders, major corporations are realizing the opportunities presented by adherence to ESG principles. This chapter describes the various opportunities corporations have in enhancing ESG values and long-term corporate profitability.

OPPORTUNITIES BASED ON RESILIENCE AND ADAPTATION

Dynamic societal change forces reactions and accommodation by some and presents opportunities for others. In an effort to prolong a status quo, corporations may choose to invest in the preservation of past practices and slow adaptation to change rather than the potentially costlier in the short-term and more uncertain embracement of the future. The *Taskforce For Climate-Related Financial Disclosure (TCFD)* is an international assembly of corporate leaders and regulators that has embraced a sustainable future and is increasingly enlisting more major corporations into their vision by providing pathways for success and for collaboration across like-visioned institutions.

The TCFD is responding to a societal and investor trend. Given the public interest in a successful transition to sustainability against the natural tendency to preserve the status quo, incentives for change and disincentives for preservation of past practices may be necessary. When these benefits to society exceed the benefits to an entity within society, the positive externalities that arise in the transition may warrant assistance and perhaps even corporate subsidization just as government routinely discourages negative externalities.

Such encouragement of the public good aspect of sustainability creates an essential role for public policy to ease the burden of transition. Economic encouragement can come in the form of transitional subsidies, product purchase guarantees, subsidized or guaranteed lending, or indemnification of insurers willing to encourage mitigation over adaptation.

SUSTAINABILITY IN THE PRIVATE SECTOR—CLIMATE POLICY AND GOVERNANCE

The private sector has a unique role in climate change and its mitigation. Almost all anthropogenic emissions of greenhouse gases arise from actions taken within the private sector. In addition, actions necessary to adapt to or mitigate climate change also work primarily through the private sector. Finally, private sector institutions all strive to optimize their profits and returns while they manage the risks and liabilities they create, including the climate risks described in Chapter 26.

Corporations are designed to manage and indemnify risk through diversification and through risk management techniques for which they have an advantage both from their *economies of scale* and *economies of scope*. At the same time, sustainability in general and climate change in particular is an international challenge which goes well beyond nations and private sector entities. Sustainability is riddled with intergenerational externality challenges, while climate change also invokes intragenerational externalities arising from unpriced emissions and actions. Corporate acts with significant externalities and with global consequences are difficult to integrate into public policy at the best of times, and typically evolve very slowly, especially in comparison to the rapid acceleration of consequences.

The 1987 Montreal Protocol demonstrated that nations could move relatively quickly to develop risk mitigation- and evidence-based policies when the stakes are less dramatic and costs to prevent damage are relatively modest. Scientific conclusions quickly resulted in policy to mitigate ozone depletion and forged a set of principles for consensus on the need to mitigate, the development of alternative technologies, the development of a global plan, and the recognition of and ability to move funds from the MDCs to the LDCs so they too may develop their economies.

To illustrate the dilemma even with the less intractable issue of ozone depletion, Diane Dumanoski (1992) wrote in an article entitled Global Solutions Sought on Pollution[1] that:

> Environmental problems are spilling across international borders much faster than the world is developing ways to deal with them, according to a new study by the Worldwatch Institute.... The study by the Washington-based group urges world leaders to undertake a major reform of longstanding international institutions such as the World Bank and the United Nations and to adopt treaties to address specific problems. (But government negotiators) ... seem more inclined

to tinker with existing laws and institutions than to undertake a needed over-haul (while …) developing countries are disinclined to strengthen international institutions.…

The recognition as a matter of international law of responsibility for acts by private entities began in 1941 with the *Trail Smelter Arbitration*. In that case, Canada was deemed responsible to U.S. citizens for actions by the *Consolidated Mining and Smelting Company of Canada* for their emissions of sulfur dioxides arising from smelting operations in Trail, British Columbia, Canada. This case established as a matter of international law the premise to *use your own property so as not to injure that of another*, translated from the Latin *sic utere tuo ut alienum non laedas*.

This principle was subsequently applied not only to attach responsibility to nations for acts of their corporations but also to acts of a nation itself, for instance, in the decision of the International Court of Justice to hold France responsible for damage to New Zealand and Australia arising from nuclear detonation tests held by France in the South Pacific. Causation is sufficient to establish national responsibility. Such a Polluter Pays Principle is a consequence of the legal principle that claims of ignorance or assertion of sovereignty do not absolve a corporation or a nation of its legal responsibility to remedy the consequences of its actions.

This premise of responsibility, under the legal pragmatism that only a nation can properly govern the corporations it charters, was enshrined in the United Nations *Stockholm Declaration on the Human Environment* in 1972. It affirmed the right of a nation to afford its corporations access its resources but also stated in its Principle 21[2]:

> States have in accordance with the Charter of the United Nations and the prin-ciples of international law … the responsibility to ensure that activities within their jurisdiction and control do not cause damage to the environment of other States or of areas beyond the limits of national jurisdiction.

These principles led to the 1985 United Nations Framework called the *Vienna Convention* that led to success by the Montreal Protocol to provide a pathway for remedy to ozone depletion:

> To this end the Parties shall, in accordance with the means at their disposal and their capabilities: … Adopt appropriate legislative or administrative measures and cooperate in harmonizing appropriate policies to control, limit, reduce or prevent human activities under their jurisdiction or control should it be found that these activities have or are likely to have adverse effects resulting from modifications or likely modifications of the ozone layer.

The original Montreal Protocol created a joint responsibility but without the financial transfers necessary to permit a path of development for LDCs that the MDCs had enjoyed. A series of meetings followed. When 90 nations met in London to enforce the Montreal Protocol, an observer, Professor Joel A. Mintz noted,

> Faced with strong evidence that increasing stratospheric ozone depletion poses a serious and growing threat to human health and the world environment, the parties to the Montreal Protocol instituted important modifications in several of the Protocol's central provisions. These modifications include: 1) adjustments strengthening existing measures for the control of substances covered by the original Protocol; 2) control measures for ozone-depleting substances not originally regulated; 3) establishment of a multilateral fund to assist developing countries in meeting Protocol commitments; 4) provisions for further investigation of specific scientific, technical, and legal matters.[3]

Pope John Paul II also weighed in through his 1990 New Year's Day speech:

> ...World peace is threatened not only by the arms race, regional conflicts and continued injustices among peoples and nations but also by a lack of due respect for nature .... The fruits of the Earth are divinely intended as a common heritage for all mankind.... Responsibility for protecting the environment and for limiting damage already inflicted, such as (among others) depletion of the ozone layer. . . lies with the entire human community - individuals, states and international bodies.... The right to a safe environment is ever more insistently presented today as a right that must be included in an updated Charter of Human Rights.[4]

While commentators on the morality of international treaties, and even some declarations of the United Nations itself demonstrates that nations face a responsibility for the current and past transgressions of its corporations and governments, treaties that articulate this responsibility clearly and fund obligations accordingly are difficult to negotiate, especially in the present instance of the mammoth costs of global warming remediation.

Such acceptance of responsibility may remain fleeting, but attribution science and economics have proceeded regardless. The economist William Nordhaus, who won the 2018 Nobel Memorial Prize in Economic Sciences for his work in the economics of climate change, first began to publish in the field of sustainability with his Yale colleague (and fellow Nobel prize winner) James Tobin in 1972. Their paper Is Growth Obsolete?, published in the same year as Limits to Growth, questioned the premise of maintenance of growth in Gross Domestic Product as the basis for sound economic policy. They observed that GDP is an imperfect measure of the welfare of society and that it implicitly prescribes population growth as a means to realize economic growth. They note[5]:

> Possible abuse of public natural resources is a much more serious problem. It is useful to distinguish between local and global ecological disturbances. The former include transient air pollution, water pollution, noise pollution, visual

disamenities. It is certainly true that we have not charged automobile users and electricity consumers for their pollution of the skies, or farmers and housewives for the pollution of lakes by the runoff of fertilizers and detergents. In that degree our national product series have overestimated the advance of welfare. Our urban disamenity estimates given above indicate a current overestimate of about 5 per cent of total consumption... There are other serious consequences of treating as free things which are not really free.

Nordhaus has been criticized for more recent statements in which he concluded, from an anthropocentric benefit/cost perspective, that a global temperature increase limited to 2.0 degrees Celsius is an acceptable threshold. His conclusion is partly based on his focus on efficiency and aggregate effects rather than on economic justice and the burden some LDCs may face. It also reflects an anthropocentric approach that does not fully measure damage to the entire ecosystem.

Despite concerns that his analyses tend to employ market-based parameters, such as a market-determined discount rate, Nordhaus has nonetheless interjected into the discussion the use of economics to augment good science. The Nobel Memorial Prize committee noted that the award was made for Nordhaus' work that "significantly broadened the scope of economic analysis by constructing models that explain how the market economy interacts with nature."[6]

Modern attribution theory has arisen from an acceptance among legal precedents, academics, and some government leaders for the need to apportion responsibility for sustainability. With the growing collaboration between scientists, economists, and other social scientists, attribution theory measures the size of past greenhouse gas emissions by nations, in proportion to the wealth of these nations. It acts as the basis for discussions regarding both the attribution of causes of global warming and the financial and technical capacity to remedy past acts by corporations and countries. Such an analysis must begin with a study of cumulative emissions. The next two figures show both the annual and total accumulated carbon emissions by countries since the onset of the Industrial Revolution (Fig. 25.1).

The largest national emitters of carbon dioxide to date are the United States, the European Union, and China, each with between 249 and 422 Teratons of carbon dioxide equivalent emissions since the beginning of the Industrial Revolution. The annual increase in these emissions represents the rate of change of aggregate emissions. In this measure, China and the United States remain the largest current contributors to global warming, with annual emissions of 11.5 and 5 Teratons of carbon dioxide respectively in 2021 (Fig. 25.2).

India has recently overtaken the European Union in annual emissions and, at current trends, is on a path to exceed the United States as the second largest contributor to carbon dioxide emissions annually by the end of the decade. However, because carbon dioxide persists in the atmosphere for centuries,

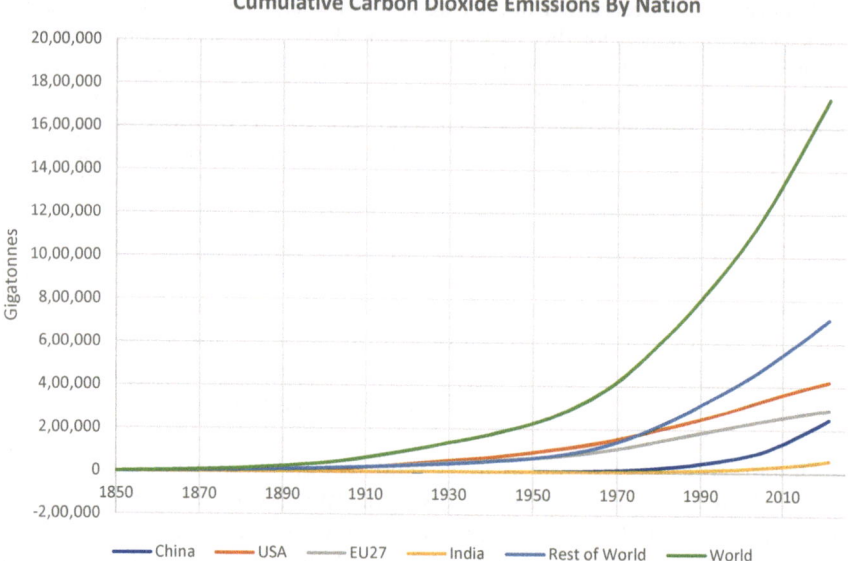

Fig. 25.1   Cumulative Carbon Dioxide emissions by nation since 1850

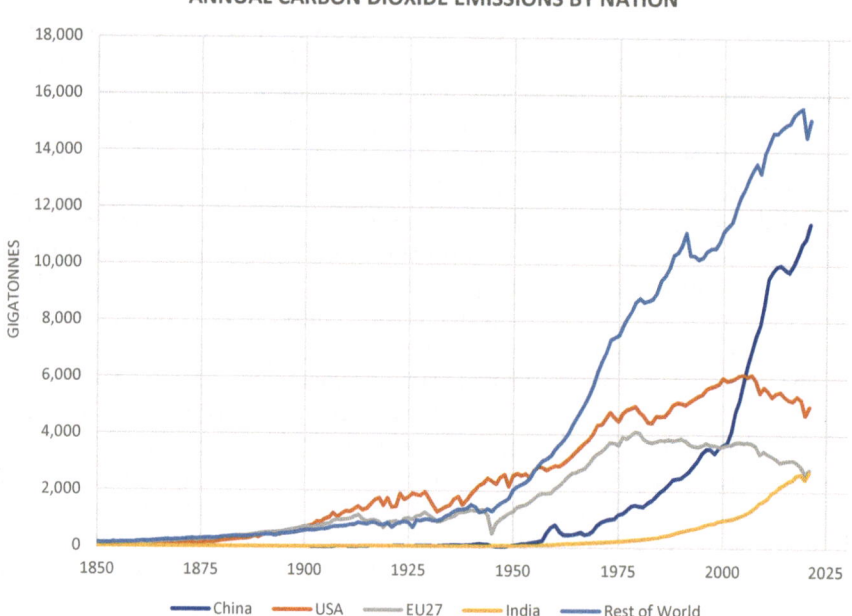

Fig. 25.2   Carbon emissions by nation since 1850

India's aggregate effect on global warming remains much smaller than the accumulated and persistent emissions of the United States, China, and the European Union.

The challenge in corporate and national responsibility is that an entity which decides to reduce its emissions captures for itself only a miniscule benefit while it absorbs the entire cost of its mitigation. While emissions represent a profound negative externality, their mitigation creates a public good which the market does not reward. Such an economic phenomenon inevitably results in attempts to free-ride instead of the actions of others. Within a people, government legislation can cure free-rider problems. Prevention of global free-riding requires treaties.

A second confounding aspect of corporate or national responsibility is the accelerating nature of the carbon emissions externality. Since the onset of the Industrial Revolution, economic growth in MDCS was furthered by fossil fuel production and consumption for decades without penalty for the externalities it caused. The aggregate effects of global warming have only recently accelerated to the degree we see today.

Tipping points and positive feedback loops that increase the marginal costs of emissions are only now being triggered. Hence, early adopters of fossil fuel-based economies may argue that their environmental costs were small when they acted over past decades. LDCs then wishing to further their economic development unimpeded by the current pricing of externalities must face higher costs because of emitters that came before them. Various attempts to resolve these arguments of convenience have created an international stalemate, despite some promising signs at the recent Conference of Parties 27.

Economic theory breaks the aspects leading to this impasse into two parts. First, any externality must be internalized at its current cost, regardless of history, to ensure economic efficiency. The carbon tax, the Polluter Pays Principle, and past failures to pay for externalities must then, from an equity viewpoint, be internalized to prevent the artificial enrichment of past recipients of climate abuse at the expense of those wishing to develop in their wake. Permanent funds should then be used to distribute income to nations that are precluded from developing at the artificially high level enjoyed by MDCs.

Second, the current emission cost must be sufficient to ameliorate costs imposed on those who now and in the future are harmed by global warming, including the degree to which humanity collectively values our shared ecosystem. While each type of emission requires a different calculation, the **carbon tax** is the most relevant internalization of the carbon dioxide emission externality. David Gordon Wilson, an engineering professor at the Massachusetts Institute of Technology, proposed such a Pigouvian tax in 1973 to ensure that the price of carbon dioxide emissions reflects the true social cost of carbon (SCC). It matters little from an efficiency basis whether consumers or producers bear the tax, only that the market price reflects that tax.

These principles are widely accepted by economists. Even the conservative economist Milton Friedman wrote in his memoir, *Free to Choose*, that:

> The preservation of the environment and the avoidance of undue pollution are real problems and they are problems concerning which the government has an important role to play. ... Most economists agree that a far better way to control pollution than the present method of specific regulation and supervision is to introduce market discipline by imposing effluent charges.[7]

In 2021, over 3600 economists, including 28 Nobel Prize winners, signed a letter of support for a carbon tax that Friedman implicitly endorsed through the market efficiency principles he espoused.[8]

About a third of the world's nations and more than a hundred cities have committed to such taxes by 2050, although few countries have put carbon taxes in place as of 2022. In 2015, a group of Stanford scientists estimated its cost at $220 per US ton, or $243 US per metric ton of carbon dioxide emissions.[9] Few governments have accepted such a high price at the political level, although Canada has articulated a goal of a carbon tax at $170 Canadian per ton by 2030.[10] Sweden's carbon tax is currently 1200 Swedish Krona, equivalent to about $118 US per metric ton in 2022. However, some jurisdictions have determined that such a tax is actually a tool for economic growth. The province of British Columbia in Canada imposed a significant carbon tax and subsequently determined it had negligible adverse economic impact and may even have a small positive effect on employment.

From a global perspective, differential private sector carbon taxes are problematic in that they may confer a greater level of profitability to those nations that make the least effort to incorporate carbon externalities into market prices. Even within the European Union, various member countries differ in their belief or capacity to address global warming.

In addition, while various standards for measurement of product carbon content have been proposed, one single set of standards accepted by all nations has yet to emerge. In the absence of a holistic approach that incorporates carbon content at each level of production and distribution, entities wishing to price carbon have developed their own approaches.

The European Union has initiated a process that taxes imports of goods based on their carbon content.[11] This approach, while laudable, will create challenges in measurement and application, and will likely be challenged at the level of the *World Trade Organization* by nations that believe this provision places EU countries at a competitive disadvantage in their markets.

Some companies also incorporate an internal carbon price into their financial analyses of various projects. In doing so, for the purposes of calculating their projects' *internal rates of return*, they increase the potential costs of projects that include the additional risks and liabilities carbon-intensive activities may incur in the future. Examples of this *risk premium* range from $60 per ton of carbon dioxide emitted in proposed projects of ExxonMobil, while Microsoft imputes a risk premium of $6 per ton.

Some nations have also introduced carbon trading. This system establishes a property right up to a certain threshold of emissions. It also allows those that can emit below their allotment to sell excess rights to corporations that are challenged to meet their permit allocation with the technologies available to them.

This cap-and-trade technique potentially provides capital for those who can then invest to enhance their ability to abate emissions. The market-determined carbon price also provides an incentive for others to reduce emissions to avoid the price. Such markets have been prone to significant volatility though, especially in times of macroeconomic crisis or energy sector volatility, and hence are not an ideal mechanism to impose the costs of emissions in the long term.

Summary of Various Emission Reduction Public Policies

Carbon taxes—a common ad-valorem or mass-proportional tax levied on the level of emissions of carbon dioxide equivalent gases from the combustion of a hydrocarbon purchased at the retail level, or as a value-added tax at each level of production. The size of the tax is calculated in relation to the externalities created by the combustion of fossil fuels.

Emissions trading regimes—a method to impose a collective quota on emitters across industries that produce pollutants. The advantage of the methodology is that it penalizes polluters but also incentivizes those polluters that can invest in technologies that result in discharge reductions. Innovative companies can then market discharge levels no longer met by selling discharge rights and hence incentivizing innovation.

Power Generation Portfolio Standards—the imposition of utilities to provide a specified share of its power through renewable sources.

Corporate Average Fuel Economy Standards—require manufacturers to attain an average fuel efficiency across their portfolio of vehicles manufactured.

Fuel tax—a tax designed to increase the supply cost of fossil fuels for which their combustion created disproportionately large emissions. The effect is to shift the use of fossil fuels that produce a reduced level of emissions per quantity of heat produced.

Bans on land use with a deep carbon footprint—burning of forests and peat bogs can be discouraged by regulation or statute to discourage the release of greenhouse gases they sequester.

Green Procurement Standards—statutes and codes that may require those who do business with local, state, and federal governments and agencies to meet certain standards for the use of green technologies or the substitution toward sustainable energy sources as a way to enhance green practices and develop the economies of scale necessary for their profitability.

VOLUNTARY CLIMATE CHANGE REPORTING

The *World Resources Institute (WRI)* and the *World Business Council for Sustainable Development (WBCSD)* have included in their corporate social responsibility measures guidance on how to account for and report corporate greenhouse gas emissions. The resulting *GHG Protocol* identifies a number of additional dimensions for corporate social accounting and reporting. These include a *Policy and Action Standard* that develops a standardized protocol for measuring GHG policies.

The WRI also specifies ESG opportunities, through their *Project Protocol* to quantify the benefits of reductions in emissions, a *Corporate Value Chain Standard* to guide how corporations can best identify ways to improve their sustainability footprint, a *Product Standard* to educate corporations on emissions over a product life cycle, and a *Mitigation Goal Standard* to provide a standardized method to develop, measure, and communicate standards and performance under their standards.

More than 90% of all U.S. Fortune 500 companies and a growing number of national regulatory agencies have since adopted these standards and embraced some of the opportunities WRI describes. The emissions reporting standards are particularly effective since they offer one concise set of standards for corporations to adopt and because they measure both the direct and indirect emissions that arise from corporate decisions.

Under the protocol, Scope 1 emissions arise from the activities directly under a corporation's control, Scope 2 emissions arise from the production of energy purchased by the corporation, and Scope 3 emissions include any additional emissions that occur upstream or downstream from suppliers, distributors, and product users. These Scope 3 aspects of emissions are the most inclusive, but only about half of the participants have created net-zero targets for all or part of their Scope 3 emissions.

A number of national governments and regulatory agencies are also promulgating their own standards. The European Union has been most active in the development of such taxonomies as part of their 2030 Green Deal. Their taxonomy establishes six objectives:

The Taxonomy Regulation establishes six environmental objectives.[12]

1. Climate change mitigation
2. Climate change adaptation
3. The sustainable use and protection of water and marine resources
4. The transition to a circular economy
5. Pollution prevention and control
6. The protection and restoration of biodiversity and ecosystems

Such efforts are laudable. However, at national level, they are often the result of a great deal of political compromise. They must also endure significant lobbying by corporations, interest groups, and, in the context of the

European Union, disagreements between nations, typically drawn along differences in national income and wealth. Nonetheless, the resulting standards have the ability to affect financial markets and introduce the same sort of volatility observed in carbon markets.

For these reasons, corporations often have significant vested interests, typically in proportion to the degree to which their balance sheets contain fossil fuel or depletable assets or, alternately, sustainable assets. Nonetheless, by imposing such standards across broad regions such as the European Union, these values are most effectively communicated to consumers who can then respond on a consistent basis. This is particularly helpful when countries within regions trade significantly with each other.

There is obvious interest among corporate associations concerned about the ramifications of sustainability standards imposed by financial regulators who oversee the accurate reporting to stakeholders of corporate liabilities. *Central banks* charged with the maintenance of stable prices, consistent employment, and economic growth have also become increasingly interested in the mitigation of climate change. Mark Carney, the former governor of the Bank of Canada and of the Bank of England, two influential central banks globally, has asserted that bank oversight in the interest of financial stability necessitates a concern about sustainability as an essential element of monetary policy over time.

A significant challenge among regulators is credibility. The *Financial Conduct Authority* (FCA), the UK financial regulator of corporate conduct and financial consumer protection, found in a 2019 review that the sustainable label was applied to a "very wide range" of financial products, some of which did not appear to have "materially different exposures" when compared to products not marketed as sustainable. This raised the specter of greenwashing, that is, marketing that portrays products or activities as producing positive environmental outcomes, when this is not actually the case.

Another key contributor is the *Institutional Investors Group on Climate Change* (IIGCC), which was launched in 2001 and now encompasses over 300 members, mostly pension funds and asset managers, in 22 countries and represents over US $40 trillion in assets under management. Sustainability policies are also of increasing importance in discussions across the group of nations, with the *IIGCC*, global economic leaders as part of the World Economic Forum, and various non-governmental monitoring groups increasingly articulating standards.

The uniting principle for such engagement is the large amount of international activity that depends crucially on sustainability. In a recent World Economic Forum report, they noted that in the analysis of 163 industry sectors representing half of the value generated in the world's supply chains depended either moderately or highly on nature's resources, from water quality to pollination and disease control, healthy soils, and a stable climate. These supply chains represent $44 trillion US annually in economic value, with

China's exposure at \$2.7 trillion, European Union exposure at \$2.4 trillion and \$2.1 trillion in exposure annually in the United States.[13]

SUSTAINABLE FINANCES

Major corporations and investors increasingly demonstrate growing awareness in the ways in which sustainability issues, especially climate change, affect the value of stocks, bonds, loans, derivatives, and insurance premiums. *Green finance* is the funding of investments that can enhance sustainability, including the costs of conservation, alternative energy, waste, and lifecycle management, conservation, and ecosystem loss mitigation. The most developed financial markets are seeing unprecedented growth of activity in investments in these activities.

Some of these investments are for identifiable projects. For instance, *green bonds* provide investment capital for specific infrastructure projects to build out sustainable energy. *Social bonds* may be used for innovations aligned not on specific physical projects but rather to address other ESG issues such as economic justice for the economically disenfranchised or for underrepresented entities such as minority- or female-owned businesses or for low-income entrepreneurs. *Sustainability bonds* are designed to simultaneously advance both social and environmental goals.

To assist in the development of viable markets for these new financial instruments, the *United Nations Environment Programme's Finance Initiative (UNEP FI)* and its governmental and corporate partners have collaborated in a series of workshops for potential investors, analysts, and others. Their objective is to assist in sharing best practices and creating opportunities for green, social, and sustainability investments and bonds so public and private sector issuers can be matched with entities that can benefit from these funds. The UNEP FI expects that such bonds will be issued on national stock exchanges.[14]

Such ESG-related bond funds experienced substantial growth in demand over the COVID-19 era and are expected to continue to grow. At the same time, regulators are developing and imposing reporting standards to ensure funds are used to promote their specified purposes, with the requisite interest payments falling to the degree to which societal goals are met. Those funded by such bonds must regularly report *Key Performance Indicators* to ensure this alignment between fund objectives and project performance.

The European Union and their Basel Accord specify criteria for mandatory ESG-related items on quarterly and annual corporate reports for large corporations that employ more than 500 people. For instance, the United Kingdom mandated greenhouse gas emission reporting and diversity, equity, and inclusion (DEI) success since 2006. Climate risk as defined by the TCFD mandated as of 2022.

ESG reporting practices are spreading. China has also introduced mandatory ESG reporting, while Canada also requires diversity reporting in addition to mandated reporting of TCFD standards performance for those who received COVID-19 governmental support. In the U.S., major banks and bank holding companies have been notified to expect new regulations from the U.S. Federal Reserve, and the Securities and Exchange Commission is also promulgating new regulations on ESG matters.

SCENARIO ANALYSIS

Traditional financial analysis has generally relied on past financial success as a predictor of future financial returns. For instance, *Modern Portfolio Theory*, the *Capital Asset Pricing Model*, and *Options Pricing Theory*, the primary tools of financial pricing analyses, all include the variance of past financial returns to predict the correct asset price. The principles of such financial analysis extend to much of traditional finance theory.

The defining premise of sustainability, and especially climate change, is that society is evolving and is typically not in a steady state sustainable equilibrium as financial analyses typically assume. Hence, from one period to the next, prices change, resources are consumed, and externalities are suffered. When, as in climate change, uninternalized externalities compound and future generations are increasingly challenged, the economy is not in any sort of a steady state that allows analysts to rely on the past to predict the future.

With such evolving volatility over time, one approach to determine the state of future resources and the atmosphere is to rely on good scientific data. Analysts can then construct scenarios for various future assumptions based on possible paths of public policy. We can also perturb these assumptions in one direction or another to test how changes in these scenarios will affect our conclusions.

Researchers, government planners, and leading edge corporations have been developing scenarios to determine the range of possible inputs so they may generate outputs for their planning models. These plans may specify budgets, simulate the value of financial securities, or, in the case of the sciences, help define possible outcomes when systems are so complex that formal modeling cannot always the full range of possible outcomes.

Financial experts are becoming increasingly sophisticated in such scenario analyses in ESG-related areas over the past few decades. The new component with regard to private and public sector risk management is the integration of increasingly sophisticated scientific analyses into the financial and public policy process.

The Taskforce on Climate-Related Financial Disclosures provides guidance for using scenario analysis as a way to ensure that inaccurate and backward-looking assumptions do not force models to draw false conclusions. The process of scenario development affords policymakers an opportunity to challenge their assumptions and determine how the resulting strategies may depend critically on small changes in assumptions.

For instance, scientists and social scientists associated with the Intergovernmental Panel on Climate Change have produced five operational scientific scenarios regarding the global average temperature over the balance of the century, depending on various assumptions of public and private sector efforts to mitigate forces that drive greenhouse gas emissions and climate change. Many of the assumptions are based on potential changes in human and corporate behavior that gave rise to global warming, and still, many other drivers trigger tipping points. Not only are our human actions that drive greenhouse gas emissions relevant, but collective efforts to mitigate emissions or even reverse past climate damage through carbon sequestration depend critically on public policy and corporate initiatives.

Given the vast number of possible physical, economic, political, and social factors that contribute to climate change, it is not surprising that scenario analysis is complex. This complexity is magnified because we must attach to each moment in the future for some uncertainty. Such uncertainty then compounds itself because uncertainty in one period contributes to additional uncertainty in the next. Another way to imagine such compounding is that it is straightforward to judge the most likely weather this afternoon, more difficult to model the weather in a few days or a week, and wrought with uncertainty beyond that. Models become increasingly unpredictable as uncertainty compounds with each additional period. Yet, climate change scenarios span not days, weeks, or quarters, but instead span decades and generations, far in excess of traditional public policy time horizons.

Scientists have been developing increasingly sophisticated climate change scenarios for about three decades. Important variables such as atmospheric physics and meteorology are combined with the technologies humans employ to meet our wants and needs, our expected economic growth, and population growth. These latter three variables are interrelated in complex ways that also depend on cultural norms and hence may be regionally and culturally specific.

While model developers have become adept in their understanding of the complexities of the physical world, human economic development and innovation remain far more difficult to predict. Hence, while a nation may be able to outline policies for a generation or two, climate change models become increasingly uncertain even a decade into the future.

Typically, in climate change modeling, the physical variables of global temperatures, climate change, sea level rise, and storm strength can be described with some confidence if atmospheric scientists know the pattern of anthropogenic emissions. Hence, scenario analyses often then begin with the assumptions of human behavior and then demonstrate the resulting implications. Human variables are typically ranked from best to worst case, with worst case reserved for *business as usual* including an acceleration of emissions from

LDCs as their wealth and economic activity increases and they rely on fossil fuels to the same extent as MDCs had in the past.

Various scenarios are defined by evolving **Representative Concentration Pathways** (RPCs) that are issued by the Intergovernmental Panel on Climate Change. These RPCs are differentiated by the increase in net insolation, in watts per square meter, under various emissions and tipping point trajectories (Figs. 25.3, 25.4).

The International Energy Agency (IEA) also issues a streamlined set of scenarios that include a **Stated Policies Scenario** (SPS) and a **Sustainable Development Scenario (SDS)**. The SPS considers only existing and announced policies, while the SDS includes actions necessary to remain within the 2.0° Celsius threshold agreed upon in the Paris Agreement.

The IEA approach is less sophisticated and not as nuanced as the IPCC scenarios. The IPCC includes both physical and social scientists who perform what are called **integrated assessments**. Such **integrated assessment models (IAMs)** thus permit feedback between physical variables, such as global temperature increases and accelerated societal reactions.

The World Health Organization also models the physical implications of various emissions trajectories in an effort to understand the atmospheric science and its implications on the planet. Human costs associated with

Fig. 25.3 Assumptions in global warming scenarios (Courtesy of the International Panel on Climate Change)

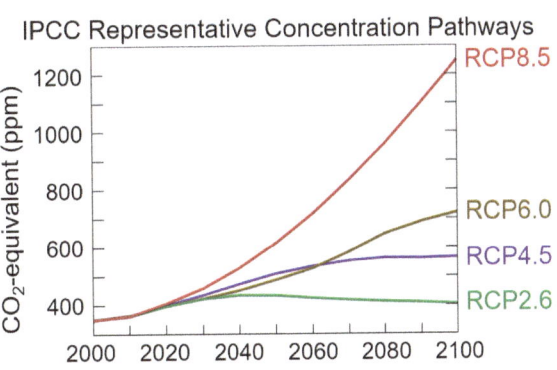

Pathway Name	Rise in Temperature by 2100	Trend
RCP 1.9	~1.5°C	very strongly declining
RCP 2.6	~2.0°C	strongly declining
RCP 4.5	~2.4°C	slowly declining
RCP 6.0	~2.8°C	stabilizing
RCP 8.5	~4.3°C	rising

Fig. 25.4 Pathway scenarios based on various assumptions of responses to global warming

higher sea levels, increased storm intensity and frequency, reduction in coastal barriers, and shifts in weather patterns require an additional level of analysis by experts. The analysis draws upon the expertise of property and casualty insurers, agricultural economists and engineers, policymakers, and even legal and political markets who can best gauge scenarios for governmental response to such physical calamities.

A number of entities rely on these scenarios generated from integrated assessment models to determine the effects on particular nations, sectors of the economy, corporations, or groups of people. The results can be sobering once total climate change damages and liabilities are summed across peoples, sectors, financial markets, and nations. Concerned activists from across the political and development spectrum are increasingly recognizing the extreme costs of *business as usual* if humans are unable to mitigate global warming.

As a result of these analyses, there is increased awareness that the goal of net carbon-zero by 2050 must be attained, but scant political consensus also exists on how and when to make the biggest changes necessary to ensure we remain on the net-zero by 2050 path. With troubling implications arising from even the most optimistic scenarios, the 2021 Glasgow COP26 Conference garnered net-zero commitments by 135 countries that constitute 90% of global GDP and 88% of current emissions. Yet, few nations are on track to fulfill their carbon commitments agreed upon in the Paris Climate Accord in 2015 and reiterated in 2021.

THE RACE TO ZERO

These scenarios and pledges have inspired private and public sector corporations, agencies, and entities to create their own *United Nations Race to Zero* pledges. The pledge has subscribed businesses, governments, financial institutions, cities, and states to each develop strategies so they too may be net-zero by 2050. Corporate participants were asked to develop plans and formulate targets that specified actions by dates to meet the 2050 target. The *Net-Zero Tracker*[15] reports that 133 countries representing 80% of the world's population and 83% of its emissions have pledged participation. These include more than 40% of 2001 of the world's largest corporations.

The world's leading corporations, cities, and countries each specify their own target year and trajectory, but their plans differ substantially. For instance, India and China do not specify net-zero carbon emissions until after 2050, while other countries pledge net-zero well before 2050. In addition, a number of countries specify that they shall be net-zero as early as 2028.

There has also been some backtracking by nations since the Paris Agreement and the Glasgow Conference. The policies of a nation in global commitments are only valid so long as committed political leaders remain in office. Regime changes are much more common among governments than corporations closely wed to preferences for corporate stability on behalf of their stakeholders. In addition, there may be differences in the capacity to limit emissions by corporations compared to the capacity of nations to limit emissions through our consumption decisions. Hence, the challenges among nations are more varied and prone to definitional differences and politics.

Corporations too are at the whim of financial markets upon which they depend to raise the capital and make their necessary investments to transition to a cleaner economy. They function at the juxtaposition of evolving climate goals and new technologies. Ultimately, corporations must adopt new technologies but with each technological change comes uncertainty and potential delay until new technology costs decline.

Governments and global bodies can play a critical role in the indemnification of the risk of new technology embracement and subsidize the inevitably higher costs to first adopters of new and sustainable technologies. Financial markets are also an essential component in these necessary transitions. However, financial markets abhor uncertainty and risk. The importance of financial markets in the mobilization of investment capital to meet climate change challenges necessitates a partnership between the private sector, public sector organizations, and governments.

In addition, some large financial units such as pension funds are increasingly investing for the long-term physical and financial health and happiness of their clients. Such pension funds, in combination with insurance funds may, with the creation of proper incentives offered by government, be able to mobilize a good share of the $61.5 Trillion US the Jacobson study described as the necessary investment in our sustainable energy future. These institutions would then also be able to reap the rewards of revenue that increasingly flow to sustainable rather than fossil fuel energy producers.

Organizations such as the TCFD, the UN Race to Zero, and the Net Zero Tracker area offer transition assistance to accompany its development of standards and goals. Three quarters of companies and many nations specify a strategic transition plan that is developed by these agencies. The agencies help formulate corporate, agency, and national plans for:

1. Realistic trajectories for emissions reductions so that aggregate emissions can be estimated and deviations from trajectories can be tracked.
2. Complete strategies for environmental pledges that include realistic assessments of challenges and a thorough vetting of engagement potential for all the various stakeholders, including individual targets for these various subgroups.
3. Realistic rather than aspirational plans for the adoption of new technologies. For instance, it is enticing to assume one can attain net-zero through the purchase of carbon credits through carbon markets or the hope that the cost of carbon sequestration comes down. Both these assumptions are speculative rather than practical.
4. Realistic consideration of equity aspects. Especially for nations, but also for corporations with significant liability for past practices, a feasible

plan may need to anticipate the degree to which distributional effects come to bear. If MDCs must provide financial assistance for transitions in LDCs, wealthy countries will be asked to bear the brunt of both the financial costs and the technology transfer necessary for LDCs to attain energy sustainability. It is natural to presume that corporations who amassed great wealth partly through unpriced emissions may likewise bear a proportionate liability toward a net-zero world.

5. Realistic assessment of broader ESG factors beyond simply climate change.

Any transition in society invokes significant renegotiation of relationships between members of society. Sustainability requires consideration of significant economic justice issues. The role of low-income members of society or indigenous peoples and the predicament of LDCs must be considered. In addition, climate change mitigation has implications for the entire ecosystem and those who depend on it. The context of change is then much broader than the accounting of costs to businesses, corporations, and individuals.

Our market institutions measure actions and consequences in terms of monetary gains and losses, even if losses to marginalized populations and to the ecosystem as a whole may be far more significant, and less evident, in terms of financial measures. As an example, if it is determined that alternative use of forest land or cropland is necessary from a carbon sequestration perspective, net-carbon-zero policy may nonetheless have profound implications on people dependent on the land. This difference in private and public returns and costs necessarily requires the imposition of public policy to create or incentivize socially optimal responses to ESG challenges.

Increasingly, regulatory agencies are playing a role in the definition of standards to communicate to the public the actions and transition strategies of corporations. To avoid greenwashing, and ensure accurate financial reporting, agencies that offer oversight of publicly traded securities are increasingly attempting to ensure that corporations treat reporting of sustainability matters in the same way they report their financial results and forward guidance. Groups such as the TCFD have developed appropriate reporting standards.

Regulatory agencies ensure that corporate communications to the public exhibit ***materiality***, which is a measure by which financial disclosures properly depict both the potential returns and the risks of corporate actions and their effects on financial variables such as profitability. Much like the triple bottom line, more accurate reporting of the social and economic damages arising from climate change enshrines the concept of ***double materiality*** in which corporate actions and strategies can align to improve both corporate profits as well as benefits to the environment and society.

Summary

Institutions and corporations have become increasingly aware of the need to address climate change and sustainability risks and the evolving regulatory and public policy landscape that intersects with corporate goals. On the horizon are such regulatory impositions as carbon taxes and the application of the Polluter Pays Principle that will create risks for some corporations but opportunities for those most adept at redefining corporate strategies.

There are also emerging number of private sector boards and associations that provide significant guidance and support on ways to transition to the new sustainability reality. Increasingly, major corporations are making pledges to attain carbon neutrality, for instance, and are embracing ESG and Corporate Social Responsibility to meet the emerging reality.

In addition, new financial instruments are helping to pave the way. Guaranteed by large global funds to absorb some of the uncertainty in emerging expectations for sustainability, such green, social, and sustainability bonds provide access to low-interest capital for long-term projects in the best interest of our intergenerational society. These new capital instruments provide opportunities for those corporations that recognize the inevitability of the need to attain the path of carbon neutrality upon which nations agreed at the Paris Conference on Climate Change.

ESG Toolkit

Associations of private sector institutions are increasingly formulating an agenda and a set of standards that are expected to enhance long-term profitability and resiliency.

Are differences in corporate strategy explained primarily by differing regards for the short and long-term and by differences in the weighted average cost of capital?

Can corporations find it profitable to follow sustainability practices under the premise that their stakeholders are willing to provide a premium in support of such policies?

Notes

1. Dumanoski, Dianne. (Mar.15, 1992). "Global Solutions Sought on Pollution," *Boston Globe* p. 8.
2. Declaration of the United Nations Conference on the Human Environment (Stockholm Declaration), U.N. Dor. A/Conf. 48/14IRev.I (1973), U.N. Publ. No. E.73.11a14 (1974).
3. Joel A. Mintz, Joel A. (1991). "Progress Toward a Healthy Sky: An Assessment of the London Amendments to the Montreal Protocol on Substances That Deplete the Ozone Layer," 16 YALE J. INr'L L. 571, at 578.
4. William D. Montalbano, William D. (1989). Pope Warns of Global Ecological Crisis; Environment: Pontiff s Peace Message Attacks Consumer Greed and

Pillaging of Resources, L.A. TIMES, Dec. 6, page A6, accessed December 13, 2022.

5. Nordhaus, William and James Tobin. (1972). "Is Growth Obsolete?" Yale University. https://www.nber.org/system/files/chapters/c7620/c7620.pdf, accessed December 13, 2022.

6. https://www.nobelprize.org/prizes/economic-sciences/2018/nordhaus/facts/, accessed December 13, 2022.

7. Friedman, Milton; Friedman, Rose (1990). *Free to Choose: A Personal Statement*, New York: Harcourt. pp. 213–218. ISBN 978–0-156–33,460-0.

8. https://www.econstatement.org/all-signatories, accessed December 13, 2022.

9. https://news.stanford.edu/2015/01/12/emissions-social-costs-011215/, accessed December 13, 2022.

10. https://www.canada.ca/content/dam/eccc/documents/pdf/climate-change/climate-plan/healthy_environment_healthy_economy_plan.pdf, accessed December 13, 2022.

11. https://www.nytimes.com/2022/12/13/world/europe/eu-carbon-tax-law-imports.html, accessed December 14, 2022.

12. https://finance.ec.europa.eu/sustainable-finance/tools-and-standards/eu-taxonomy-sustainable-activities_en, accessed December 13, 2022.

13. https://economictimes.indiatimes.com/news/international/business/half-of-world-economy-dependent-on-nature-wef-study/articleshow/73371574.cms, accessed December 13, 2022.

14. https://www.unepfi.org/training/training/green-bonds-training/#:~:text=Local%20green%2C%20social%20and%20sustainable,on%20their%20national%20stock%20exchanges, accessed December 13, 2022.

15. https://zerotracker.net/, accessed December 13, 2022.

CHAPTER 26

Private Sector Sustainability and Climate Change Risk Management

All enterprises manage risk. The ability to effectively balance risk with returns allows firms to generate profits and organizations to produce value. The challenge is to effectively model and measure risk so that returns adequately compensate managers for the risks that occur. While risk managers often have the luxury of past history to guide their prescriptive management policies, risks related to sustainability and climate change create new challenges to traditional risk management because the future shows little resemblance to the past. This chapter reviews the various types of risk and provides insights into their management, with a special focus on the particular forces that must also be considered when addressing sustainability and climate change.

A better understanding of the risks allows managers to ensure that the benefits expected are worth the costs and risks accepted. If these risks are well-understood and all costs have been addressed, benefits should then be sufficient to ensure that the profits of corporations and the welfare of society are enhanced by the decisions we make. This approach is particularly relevant as we better understand challenges to sustainability and risks arising from climate change.

All organizations assess and manage risk to some degree. While they may differ in their capacities to identify, classify, and manage risks, corporations stand to benefit from an enterprise-wide awareness of sustainability and climate change risks. In doing so, organizations are offered the opportunity to employ scientific and systematic principles that afford them a pathway in our planet's inevitable transitions as it experiences climate change.

C. Read, *Understanding Sustainability Principles and ESG Policies*, https://doi.org/10.1007/978-3-031-34483-1_26

All else equal, humans and our institutions prefer something of value with certainty over that same opportunity even with an equal and symmetric chance of loss or gain. To see why risk matters, even when it could be equally upside or downside, consider the St. Petersburg Paradox that puzzled mathematicians in the eighteenth century. A member of the famous extended and intergenerational family of mathematicians, Nicolas Bernoulli, asked why people would not be willing to pay an almost infinite sum for the right to play a game with an infinite expected return.

This casino game asks what one would pay for the right to an uncertain payoff. The game works as follows: If I flip a coin and it comes up tails, the game ends and I pay you $2. If it comes up heads, I flip again. If it then comes up tails, the game ends and you receive $4. If instead, it comes up heads, I flip again and pay you $8 if the coin comes up tails and the game ends. Otherwise, I continue flipping until tails occurs.

In this game that ends when the first tail is flipped, there is a 50% chance it ends after one flip, a 25% chance it goes to two rounds, etc. The expected value of this game when these rounds are added up is then:

$$\text{Expected Value} = \frac{1}{2} * \$2 + \frac{1}{4} * \$4 + \frac{1}{8} * \$8 + \frac{1}{16} * \$16 + \cdots$$
$$= \$1 + \$1 + \$1 + \$1 + \cdots = \infty$$

Nicolas' cousin, Daniel Bernoulli solved the paradox and showed why human nature abhors risk. In doing so, he demonstrated mathematically what another contemporary, the mathematician Gabriel Cramer, had stated intuitively:

> The mathematicians estimate money in proportion to its quantity, and men of good sense in proportion to the usage that they may make of it.

Since the return to any human activity is not boundless, our willingness to invest in any opportunity is likewise bounded. People quizzed seem willing to offer about $25 for the opportunity to engage in the coin-flipping game, despite its infinite expected payoff. In other words, risk matters.

The economist John Maynard Keynes described the cost of risk in another way. Consider a number of green investment opportunities a corporation may have, ranked from best to worst. If it knows with certainty the amount of its available financial capital, the corporation can gauge its expected investment reward. However, if capital fluctuates in value, the corporation suffers from losses to a greater extent than it may enjoy equal gains. Daniel Bernoulli's quick demonstration of the effects of diminishing returns, and subsequent work by the economist Irving Fisher in the early twentieth century demonstrates why risk matters, and what we must do to manage risk and optimize returns.

Keynes demonstrated the inherent cost of risk through a simple example. Imagine a firm with $3 million of investment capital. It has four potential projects under consideration, each of which cost $1 million. The first investment generates net revenue of $1.4 million, the second $1.3 million, the third $1.2 million, and the fourth $1.1 million. In the absence of risk, with $3 million to invest, the firm stands to net a profit of $0.4 million + $0.3 million + $0.2 million by allocating its capital to the three most profitable opportunities. These returns sum to $0.9 million for the $3 million investment, which translates into a 30% return.

Let us next calculate the returns should available financial capital instead rise or fall by $1 million, with equal probability. Then, the return will be either ($0.4 + $0.3 + $0.2 + $0.1) million if available capital rises by a million, or ($0.4 + $0.3) million if capital falls by the same amount. The average return then falls from $0.9 million when there was no uncertainty to an average of $0.85 million when there is equal upside and downside risk. In a world of risk and diminishing returns, expected returns fall when risk exists, even if the risk is symmetric and equally likely to be upside as downside.

Insurance companies understand such costs of risk and indemnify companies from these risks for a premium that is proportional to the cost risk imposes on an organization. In fact, there are a number of techniques a firm can employ to insure itself against fluctuations in the value of their investment capital or assets values, or uncertainties in their liabilities. Mitigation of risk affords a corporation an opportunity to avoid the risk premia required for market protection.

The *World Resources Institute* manages *Corporate Ecosystem Services Review* that identifies both corporate vulnerabilities and opportunities. It lists various attributes of traditional risk categories. These sustainability and climate change risks include[1]:

Operational	• Risks occur in the use of increasingly scarce or vulnerable water and coastlines that may arise as freshwater becomes more scarce, hydroelectric reservoirs are drained after a prolonged drought, or businesses in coastal regions suffer from rising sea levels or increased storm severity
	• These threats increase opportunities to use water more efficiently or preserve wetlands
Regulatory and legal	• Corporations face regulatory risks in user fees, fines, new regulations, or government and citizen lawsuits that arise from insufficient corporate environmental sensitivity
	• Corporations can reduce these threats by participating in the development of public policy and by encouraging incentives to reduce their environmental footprints and restore ecosystems
Reputational	• Retail-oriented corporations are increasingly subject to organized campaigns on behalf of concerned consumers, for instance, for corporate practices that degrade forest, land, or ocean ecosystems
	• Corporations can respond to such threats through better communication of their ESG values as a way to demonstrate their corporate commitments and improve their brands

(continued)

(continued)

Market and product	• Corporations face market risks as stakeholders instead patronize sources that ensure their products have attained various eco-certifications or follow new sustainable procurement policies
	• Corporations can demonstrate that their products have reduced ecosystem footprints, that they participate in new technologies such as carbon sequestration, or that they manage their natural assets to maintain sustainability
Financing	• Corporations face risks as banks are increasingly required to document vulnerabilities of their commercial loan portfolios to ESG-related risk such as climate change
	• Such scrutiny creates opportunities for improved loan terms if borrowers can demonstrate how they reduce their environmental footprints or better manage the ecosystem

Challenges to sustainability and from climate change have increased the inherent risk facing all organizations. Corporate and institutional success depends on the abilities of risk managers to balance rewards with their intended risks. These risks come not only from within an organization but from the markets within which they operate and their industry. In the case of such mammoth global environmental phenomena as climate change and challenges to sustainability, risks arise as the planet transitions toward a more uncertain future.

Regulators and stakeholders alike are increasingly charging boards of directors, senior leadership, and risk managers with the identification of economic forces often beyond easy corporate control. These managers protect their organizations and stakeholders to the extent possible from various traditional and climate change-inherent risks. Effective climate risk management then requires a corporate culture that is forward-looking to a greater extent and is cognizant of market forces that can profoundly challenge corporate profitability.

Certainly, the corporation's investors are sensitive to the level of corporate returns. They are also aware of the risk of these returns, not only as measured by the market based on past and recent performance but also based more prescriptively on how corporate returns may be affected by sustainability challenges and climate change.

The risks associated with large societal and economic movements may even challenge the liquidity of a corporation and hence induce liquidity risks. Those who lend financial capital to the corporation or borrow from the corporation are also concerned about how risks arising from climate change and other challenges may impinge upon the value of their loans.

It is essential for the corporation to acknowledge and mitigate such climate risks. These risks are most significant for a corporation that is exposed to properties vulnerable to the effects of climate change. While many of these risks affect financial returns, there is cause for concern for many non-financial firms as well. Sustainability and climate change risks are becoming increasingly significant to the operation of institutions that must raise funds, expand in markets, rely on a secure supply chain, and address the ethics and interests of their various stakeholders.

Finally, institutions and corporations may have a strong environmental ethic but may lack a consensus among their directors in the ways in which a corporation should manage sustainability and climate change risk, among other risks. Knowing that institutions are not united in their regard for sustainability and climate change, governments are increasingly mandating requirements on firms, and especially publicly traded companies, to incorporate and communicate values contained within the Environmental, Social, and Governance (ESG) paradigm.

As a consequence, leading-edge corporations must not only master the traditional tools of risk management but must also extend the risk management paradigm into degrees of climate and sustainability risk rarely seen in the absence of global depressions and meltdowns.

Those well-versed in the tools of effective risk management will find their risk management toolbox can be extended to incorporate sustainability and climate change. While some aspects of these risks may be more pronounced under global climate change, the overall risk management paradigm remains the foundation.

The *United Nations Framework Convention on Climate Change* facilitates a coordinated set of standards and actions among all nations of peoples and provides guidance on corporate strategies that span borders. Such broader actions engage the corporate sector as a critical stakeholder. *Michael Bloomberg,* the former New York City mayor and principal of Bloomberg Global News, an international subscription service that provides data and analytics for corporations and governance alike, chairs an assemblage of private sector participants from across corporate sectors under the auspices of the Taskforce on Climate-related Financial Disclosures (TCFD). This entity, formed by the *Group of 20 Financial Stability Board (G20 FSB)*, was charged with the description of various climate risks that must be addressed by publicly traded and major companies in the private sector.

The TCFD provides guidance on ESG-related matters and their reporting. The TCFD produced a set of recommendations regarding climate-related financial reporting and has updated the status of progress toward their recommendations in 2022.[2] In his 2022 update to the Bank of International Settlement's Financial Stability Board chair Klaas Knot, TCFD chair Bloomberg acknowledged ever-increasing public interest in sustainability matters over the five years of annual updates since their first 2017 report.

Bloomberg reaffirmed the 2017 goal to ensure that financial asset pricing risk is reduced through the application of clear recommendations for climate-related financial disclosure. The task force noted:

The Task Force' recommendations provide a common set of principles that should help existing disclosure regimes come into closer alignment over time.

Preparers, users, and other stakeholders share a common interest in encouraging such alignment as it relieves a burden for reporting entities, reduces fragmented disclosure, and provides greater comparability for users. The Task Force also encourages standard setting bodies to support adoption of the recommendations and alignment with the recommended disclosures.

More than 3800 companies have pledged to align their reporting to TCFD standards, which have most notably been implemented in part by proposed rulemaking by the U.S. Securities and Exchange Commission, the *International Sustainability Standards Board*, and the *European Financial Reporting Advisory Group*.

The risks that corporate entities face are not unrelated to those faced by other institutions and by humanity and our ecosystem. Elements of adaptation, mitigation, and climate change resilience combine to affect the vulnerability of eco-, human, and organizational systems to the various climate change risks. There are a number of components to these risks. These include:

- the stranding of assets such as fossil fuels that remain in the ground once backstop technologies are broadly adopted,
- the physical risks related to property and casualties arising from climate change, and
- indirect risks to supply chains and operations.

In addition, corporations must:

- incorporate and manage legal risks,
- avoid damage to their reputation, product, and other market risks, the cost of obsolescence or depreciated technological assets as economies experience structural changes,
- protect against impairment as human capital depreciates when new combinations of skills are required in emerging economies.
- anticipate the shift from an unsustainable economic equilibrium to a sustainable one that creates risks during the transition but also creates opportunities proportional to inherent corporate and public sector resilience and a capacity to adapt.

OPERATIONAL RISK

The predominant sustainability risk is the threat to human systems and the ecosystem arising from climate change. There are various avenues such as climate risk imposed on corporations and other private entities. Some are inherent risks arising directly from the actions they take. Others result from the risks imposed should corporations violate the expectations of stakeholders.

Still, others are imposed through the broader social and regulatory contract they implicitly accept as members of society and the broader economy.

Some operational risks are related to transition costs as a corporation adjusted to the new reality of climate change. These include the stranding of assets and human capital.

Stranding of Assets

The economist Joseph Schumpeter originated the notion of *creative destruction*. Transitions from one technology to a newer, more efficient, and potentially more sustainable capacity inevitably requires a change in the way a society values assets. An earlier technological era that prized pools of coal, crude oil, or natural gas, may not be valuable when a new and sustainable technology prevails.

These assets are not stranded in the physical sense but are stranded economically once they have little value in the new economy. Corporations that own physical assets which become obsolete once a new technology is adopted are disadvantaged by progress, even if society inevitably benefits from the transition to more sustainable natural capital.

While creative destruction recognizes that economic resources flow to their best use, inevitably this incessant quest toward greater efficiency and evolution toward a more sustainable planet creates winners in new technologies but also losers who own obsolete technologies or resources and refuse to develop strategies to transition to a more sustainable technology.

Stranded Human Capital

Perhaps the greatest challenge is the need to overcome both a corporate culture and a workforce immersed in and comfortable with the current corporate status quo. The corporation recognizes the costs of stranded assets and obsolescence, but must also navigate the human costs of repositioning itself for increased sustainability.

In the long run, such a reorientation of a workforce can be accomplished. However, while economic efficiency typically focuses on long-run values, in the short run, significant internal resource redistribution costs are incurred as managers must shift away from increasingly obsolete technologies and write off stranded assets. It may also be costly for the corporation to retrain existing employees, and some employees may be sufficiently adverse to retraining that their career prospects are hampered. The fear of adopting an uncertain and unsustainable future at the expense of sacrificing a potentially less ideal but more familiar present often inures resistance.

Politics too tends to align itself to the preservation of the known present rather than the adoption of a less certain future and may thus protect employees who may be displaced by change. Internal pockets of resistance may frustrate transitions and the raising of financial capital necessary for investment in a more sustainable and potentially productive future. Corporations and government may need to make sizeable investments to overcome such frictions while, at the same time, they sacrifice known and historically reliable resource rents and profits.

Physical Risks

Scientists have unambiguously concluded that global warming will have a displacing effect that imposes costs on some, regardless of global ambitions to limit warming to 1.5° or 2.0° Celsius. Polar regions shall warm considerably and the meteorological patterns shall shift, with storms becoming more severe. Some of these exposures, for instance, the duration of warm arctic summers, will lengthen summer seasons and accelerate permafrost melting and the loss of entire arctic villages as a consequence. Tropical cyclone seasons will extend and storms may be of greater magnitude, which will accelerate property and habitat loss with coastal erosion.

Meanwhile, droughts extend for more years, forest fire seasons may continue year-round in some regions, and the most acute changes with high regional specificity may have a chronic influence on human and economic decision-making. The scale of these intense or persistent effects depends on which of the various Stated Policy Scenarios (SPSs) are realized. While scientific models can gauge with increasing confidence the extent of damage that may occur, humankind's success is dependent on the sociological and political responses that act as the primary drivers of climate change, the decisions, and the policies of humans.

Attribution science has grown out of necessity over the past three decades as stakeholders increasingly recognize that extent of physical damage and the inherent risks to our economies arising from climate change and lack of sustainability imposes real and substantial costs to the human and ecosystem. From a risk management perspective, it is increasingly relevant to know to what degree the severe flooding in Pakistan, for instance, in the spring and summer of 2022, and the increased frequency of droughts and forest fires in the western United States, are due to a regular but low probability event or an increased frequency of climate calamities because of global warming.

Without the attribution of increased probabilities arising from contributing factors, society will be unable to effectively insure and mitigate against such events in the future. These disruptions could be substantial, not only in real human and ecosystem costs but also in the insurance industry's capacity to indemnify against future calamities.

Indirect Risks to Supply Chains, of Legal Liability, and to Systems and Operations

The nature of risks also depends on the degree to which corporations are exposed to markets or are insulated from them. If a vertically integrated corporation controls the flow of resources such as fossil fuels, from exploration to well development, the wellhead and distribution to refining, marketing, distribution, and retailing, it is better able to manage the risks to its entire supply chain than the various risks a less integrated and more supply chain dependent firm must face. For a vertically integrated corporation, much of the supply chain is internal, but for other corporations, each link in the chain may be exposed to sustainability shocks and market regulations and oversight. While their ability to manage cascading risks to their supply chain differs and can be mitigated to varying degrees, both types of corporations face costs associated with climate change.

While these costs may be isolated to each link in the supply chain for independent contractors, the vertically integrated firm also faces a greater legal liability given their almost complete control of all aspects of their industry. Under the theory of proportional responsibility in civil torts, this vertically integrated firm can find no refuge in deflecting greater responsibility to other parties.

The degree to which vertical integration prevails in an industry also has ramifications on the complexity of its internal systems, and may indeed create risks in themselves as organizational complexity can grow to an extent that risk is magnified. A more complex system has a correspondingly greater number of failure points. Communication of corporate informational becomes less effective with increased corporate complexity as well. While the public and the legal system consider a corporation a single monolithic legal entity, it is in fact increasingly problematic for one department of a large organization to fully understand the actions and consequences of another department. For such a complex organization, legal liability increases rather than dilutes with increased size, despite the actual likelihood that such organizations have failures to communicate within.

Policy (Regulatory) Risks

A fast-evolving issue such as sustainability typically results in equally accelerating regulatory and oversight regimes, even if regulations are sometimes slow to evolve. Cancelation of obsolete public projects, evolving regulations that may or may not grandfather existing operators and agreements, and legislated transitions from one technological regime to another all impose significant risk. *Clean(er) Coal* investments considered innovative a decade ago may be obsolete as coal-fired electrical plants are under climate change pressure to shut down. Entire industries may opt to physically move to more lightly regulated jurisdictions as emissions requirements tighten. Some industries may find reporting requirements costly from an operational perspective or prohibitive from a legal perspective. Organizations that are optimized to compete within one competitive landscape may be placed at a competitive disadvantage when forced to compete in a new and sustainable competitive landscape.

Legal and Reputational Risks

Legislation only slowly responds to rapidly evolving social concerns such as global warming, but legal theories and actions can evolve quickly and may be able to punish past behavior retroactively. Legal sanctions that are imposed when a corporation violates the expectations of a people are often severe.

The most severe violations that assault the sensibilities of a society result in criminal proceedings. The conviction of manslaughter for the death of a human arising from the negligence or intentional acts of a corporation can result in significant fines and damages. Rarely would convictions result in imprisonment because only in exceptional circumstances will a chief executive be also indicted individually and beyond a capacity as an agent of the corporation. Certainly, shareholders have a limited liability that does not go beyond risk to their investment.

Legal risks more typically flow from violation of civil law. These transgressions impose sanctions in proportion to the damage caused, with punitive damages sometimes imposed if the action is considered particularly egregious and insufficiently deterred by mere damages. On occasion, the people may be a party to a corporate lawsuit for violation of either a statute that specifies civil damages or for torts that are simply considered legal wrongs that must be corrected because they imposed harm on individuals or a class of individuals.

An individual or a group of individuals assembled as a class may pursue a civil claim, with the assistance of attorneys who are often compensated with a commission as a share of the damage award. At other times when a government agency determines that a transgression damages a sufficiently broad swath of society, the government itself may litigate a civil suit. However, such civil suits are often insufficient to deter egregious behavior because the award is typically limited to the damages incurred.

To see this, consider a decision a corporation makes to enhance efficiency because its benefit B exceeds the damages D induced. It may recognize that the cost C of a civil action with a potential legal sanction that imposes damages D but with only a probability p of detection or civil liability is then:

$$C = p * D$$

Hence, with less than a certain probability of detection of a civil transgression, a corporation may impute a legal risk C that is less than the damages D the corporation may cause if its acts may go undetected with a probability p. This calculation implies that a corporation may be induced to take actions with insufficient social concern. While such calculations appear unethical, they are nonetheless the basis for occasional corporate legal strategy. Reputational effects aside, legal liability often extends not from transgressions that shock the sensibilities of civil society, but rather from business decisions that benefit the corporation by imposing an externality on another entity.

Obviously, takings of property or a right from another within the market-place of well-defined and regularly traded property rights is theft. But, beyond the types of exchanges that are transactional, many other corporate actions occur that are legal even if they cost other parties. Our resource decisions today that cost future generations are one example. The failure of a corporation to reveal to the public the long-term costs of global warming for their fossil fuel extraction today is another example.

Such legal risks may or may not be imposed ultimately on the violating firm. Liability depends both on the successful determination of damages and on proof based on the preponderance of the evidence that a corporation was responsible for the damages it causes. Such a preponderance is not easy to establish under any circumstances and is incredibly difficult to establish in matters such as a failure to act in a sustainable manner. However, while civil suits are difficult to prove, they may nonetheless impose costly reputational costs to a corporation. In addition, standards can evolve quickly at times, such as the presumption of *strict liability*, that can immediately and retroactively shift responsibilities to corporations with climate change exposure.

Reputation Risk in the Environment of Climate Change and Challenges to Sustainability

One type of business or strategic risk that deserves special discussion, especially for organizations that have a high public profile, is risk to corporate reputations. While the failure to enhance or protect its reputation can be considered a part of its corporate strategy, some modern organizations are particularly vulnerable to reputational risks arising from challenges because of a lack of Environment, Society, and Governance (ESG) policies.

Equity values of some major corporations have suffered because of the exposure of their balance sheet to fossil fuel resources in the wake of concern over global warming. Some firms associated with fossil fuels, such as natural gas and oil companies or internal combustion engine automobile manufacturers, have reduced their reputational risk by diversifying their corporate strategies into alternative energy or electric automobile production. The recognition and management of reputational risk is particularly important for organizations associated with greenhouse gas emissions or assisting in the mitigation of climate change.

Inherent in the stakeholder approach to corporate social responsibility is the notion that firms are offered the privilege of a corporate charter because they can efficiently fulfill the needs of the people they serve. Violation of the sensibilities of the people in the exercising of privileges of the corporation imposes a significant risk. The issue of consideration and communication of corporate values to conform to societal expectations is at the heart of the ESG paradigm. Consumer and investor expectations from the bottom up, and guidance from the top down by boards of directors are designed to maintain and enhance corporate reputations.

A significant harm arising from the mismanagement of sustainability risk is from the depreciation of the esteem held by the corporation among its stakeholders. The supply chain, staff, stakeholders, and shoppers of a corporation's product have the discretion to contract with a range of corporations. The product one buys is increasingly an amalgam of a physical good or service and the sense of values shared between stakeholders and the corporations they patronize.

Corporations increasingly realize that they can enhance patronage and increase brand loyalty if they more closely align with the sustainability values of their stakeholders. Such an alignment allows corporations to differentiate their product from others and enhance product loyalty. Conversely, trust and loyalty can suffer if a corporation devotes insufficient effort to enhance its reputation or damages its reputation by engaging in short-term expediency at the expense of long-term sustainability. Reputation risks arising from gaps between the values of a corporation and its stakeholders, or from cynical misrepresentation of corporate values are most problematic for products in competitive markets and for which there is a diversity of retail customers and choices.

Reputational risks may also translate into risks of reduced demand for corporate production in factor and especially in final product markets. Consumers concerned about sustainability increasingly demand products that have low carbon footprints and are willing to pay a premium price to align their values with corporate product climate pledges. Alternately, any savings arising from corporate opacity can be small compared to the risk of sacrificed reputation.

Maintenance of corporate reputations in an uncertain and evolving environment is one of the most challenging aspects of risk management when challenged by sustainability and climate change. Successful risk management requires both the mitigation of existing operations and the creation of new policies and strategies. Since the maintenance and enhancement of reputations are more ephemeral than the creation of new technologies, firms face a large degree of uncertainty in the adaptations and effective communication necessary to preserve reputations.

On the surface, the least costly reputation risk management technique is to attempt to preserve reputation in ways that require only minor adjustments to corporate strategies. Such a tactic runs the risk of *greenwashing* in which the perception of mitigation is attempted rather than the more costly sincere mitigation. For instance, natural gas companies may argue that their product is green without explaining that their statement only refers to the fact that their product is damaging to the climate to a lesser extent than coal extraction and combustion. In realms such as climate change in which much of the issues are intertwined with complicated science, there is a great deal of latitude to enhance reputation through misinformation.

A thoughtful and strategically sophisticated corporation has both a comparative advantage and a greater need to understand the implications and costs of mitigating the damage that their actions may impose on society. The corporation is also in the best position to estimate their costs should they proactively adopt new and necessary technologies to mitigate climate change. Institutions may then have an incentive to quash their analyses to delay inevitable but costly transitions or to avoid the wrath of public opinion. Oil companies that have prevented the dissemination of their research which acknowledges the effect of fossil fuel combustion on global warming may provide short- or medium-term respite from the obsolescence of their industry, but at a great reputational cost once their strategies are exposed.

It may then make sense for large corporations that have benefited substantially from an increasingly obsolete industry to use their immense resources to proactively adopt sustainable new technologies. However, to do so not only makes obsolete much of their capacity and past investment, but it also requires a major shift in its corporate reputational culture.

Obsolescence Risks and Backstop Technologies

Transitions are costly in that the corporation's people and processes are optimized over decades or generations to perform a particular function. These people and processes cannot quickly or inexpensively garner the human and institutional capital to shift entirely toward another product or technology that may employ vastly different processes. While the corporation may have significant financial resources at their disposal, they may in fact have inferior internal resources compared to another corporation that is fully immersed in a new and sustainable technologies on an ongoing basis. These various risks combine to induce many corporations in obsolete industries to manage their risks by delaying rather than adopting new technologies and product lines that mitigate climate change or enhance sustainability, and hence risk their reputations.

Corporations tend to move through a lifecycle. New firms embrace emerging technologies and can make investment bets on the potential for technological adoption by the economy. Their industrial portfolio may be heavily concentrated in such a new technology, but they have garnered little income to be threatened by their failure. These are the speculative firms that are primarily funded by venture capital seeking high returns but understanding the high risks involved. Since they have little history or status quo to preserve, these firms may tolerate a high level of risk and be much more willing to adopt new and emerging technologies.

More sophisticated firms must balance the risk of threats to existing cash flows with the risk of failure to adopt emerging technologies. This balance biases such entities to the maintenance of their status quo. For instance, we have seen that new sustainable energy sources exhibit the lowest cost energy technologies in terms of their levelized cost of energy production, including amortized investment costs. Solar, wind, and geothermal energy are half the cost of competing fossil fuel-based energy technologies. Yet, we do not see the major energy-related corporations, such as oil companies and utilities, transitioning rapidly to alternative energy.

The price of solar, wind, and geothermal energy, when one includes the new investments necessary to bring them to market, represents a transition cost. Since the investment costs of legacy fossil fuel-based technologies have already been made and these investments have little alternative value in a world that has transitioned to sustainable energy, the market will not fully transition to these backstop technologies until their levelized costs, including investment returns, are comparable to marginal costs of legacy technologies, excluding sunk investment costs.

In other words, *obsolescence* is measured not solely based on the superior technology but also on the corporation's capacity to adopt new technologies and strategies and in a corporation's attachment to and investment in *legacy technologies*. Sometimes, only when these past investments are depreciated in effectiveness to require new investment are decisions made to fully transition toward state-of-the-art technologies and practices.

The implication of this economic reality is that legacy corporations may innovate only slowly and will attempt to delay their transitions to entire industrial sectors or more innovative technologies until a point is reached that they must make new investments to replace old technologies. This confounding of necessary transitions is exacerbated when there exists significant regulatory uncertainty and when large legacy corporations with significant sophistication and resources are able to influence public policy and exercise their advantage to delay transitions.

These *obsolescence risks* that delay the stranding of capital and discourage transitions to otherwise superior technologies must be carefully managed. Corporations unable or unwilling to adjust the technologies they employ to enhance sustainability run a higher risk of product and technology obsolescence. Some corporations are unwilling to transition to new technologies and leave existing technologies stranded, while others are prepared to retrain employees, shift suppliers, realign with product markets that maintain sustainability values, and adopt new and sustainable technologies to produce more sustainable products.

SUSTAINABILITY AND CLIMATE CHANGE TRANSITION RISK

Sustainability and climate change are not without cost. Those who invest in the preservation of the status quo avoid the costs of transition to an uncertain future and manage to garner the short to medium-term benefits of preserving their status quo. Meanwhile, successful technology transitions are often well-received by the marketplace, but they are not without some risk and displacement as well. Corporations unwilling to accept the risks of change must instead accept the risks of obsolescence and potential irrelevance.

Groups such as the TCFD have accepted the challenges and inevitability of sustainability and construct pathways to facilitate corporate transitions to a more sustainable future. The TCFD and enlightened corporations recognize that to manage transition risk, corporations must be presented with and embrace well-articulated transitions to sustainability. Public sector agencies, regulators, and government must also articulate a sustainability vision and navigate necessary market and corporate interventions toward sustainability. TCFD outlines various measures that corporations can take to reduce policy and legal risk, manage transition risks, adopt to new technologies, and enhance their reputation among stakeholders who value sustainability.

Governance Risk and the Corporate Risk Culture

As improved tools for risk management are developed, enterprises are increasingly viewing effective sustainability and climate change risk management as an element that contributes to both their corporate culture and ability to meet their mission. Risk management is rarely viewed by astute organizations as a mere set of accounting constraints imposed by regulators. Instead, risk managers are essential for organizations to implement their mission, maintain discipline in their strategies, and, increasingly, provide a conduit for individual departments within an organization to better understand their interdependences. The risk management backbone ultimately contributes to the organization's success and its reputation in the marketplace.

The appropriate culture for effective enterprise risk management begins at the organization's highest levels. The various major corporate financial breakdowns that have caused reevaluations of risk in publicly traded companies either began or ended in the boardroom. While in rare cases executive fraud was to blame, most often the lack of maintenance of sound risk management standards was either implicitly encouraged or condoned by members of the board of directors or chief corporate officers.

In response to these failures, organizations and regulators reassessed the ability of boards and executives to claim that they knew nothing of the transgressions, lack of oversight, or ineffective risk management that gave rise to major corporate failures. In the wake of the collapse of the energy trading firm ENRON in 2001, the U.S. Congress imposed a new set of board and chief executive officer accountability standards that are collectively known *as Sarbanes–Oxley provisions* (collectively called SOX) after the Congressional sponsors of the reform. Since the imposition of these laws, chief executives and members of the board of directors must periodically attest to the accuracy of the financial results a listed corporation must publish.

In addition, the Securities and Exchange Commission imposes on publicly traded corporations certain expectations of board responsibilities. The result is a reevaluation of board responsibilities and an increased separation of the traditional relationship between boards and chief executives.

Increasingly spelled out in policy are various new provisions for boards. When public stakeholder expectations evolve rapidly in response to ESG, sustainability, and climate change issues, governance should respond equally rapidly. These responses include:

- Greater specificity of board responsibilities
- An emphasis on board member qualifications, terms of office, and compensation
- Strong conflict of interest avoidance, transparency, and disclosure requirements
- Improved board structure and practices
- A reorganization of board and committee structures
- Increased board responsibility in regulatory compliance
- Incorporation of the internal audit function and risk management at the board and audit committee level
- Requirements for the independence of audit committee members

Astute boards of directors also develop an explicit *Risk Appetite Statement* that articulates the types and levels of risk that will confine corporate actions to achieve its strategic goals.

Many of these more explicit risk functions are overseen by either a separate risk committee or an audit committee charged with risk oversight. Publicly listed corporations typically have an internal *Chief Risk Officer* (CRO) who reports to the board's audit or risk committees and also serves a dual role of reporting directly to the corporate chief executive officer.

RISK OFFICER, RISK COMMITTEES, REPORTING STRUCTURES

Sustainability and climate change knowledge and challenges are evolving rapidly and are often beyond the scope of traditional corporate departments to track and respond. Risk managers then find themselves at the forefront of corporate sustainability, climate change, and ESG incorporation into their corporate culture and risk appetite.

The increasingly well-defined relationship between the Risk or Audit Committees and the Chief Risk Officer is critical for effective state-of-the-art enterprise risk management. The vast majority of major publicly traded corporations and the most effective of our various other organizations now employ chief risk officers. They realize that the CRO is the most effective element in both the development of best practices and the ongoing adherence to these practices to:

- Enhance market value
- Offer an early risk warning mechanism
- Keep abreast of evolving sustainability and climate change standards and expectations.
- Mitigate potential losses and risk exposures
- Ensure that a corporation can properly set safe capital buffers
- Propose mechanisms to transfer risk, for instance, to external insurance companies
- Reduce the cost of insurance by demonstrating to indemnifiers effective internal processes.

The CRO works closely with senior management and the board-appointed oversight committee to develop and implement an enterprise-wide risk management framework. The CRO and the risk committee also develop and implement policies consistent with the organization's risk appetite and culture, and builds into the framework a series of measures that indicate organizational risk exposure and potential mitigations. These frameworks and systems put in place act as the backbone of the organization's risk management program.

The CRO and the board also play an essential communications role. They act in concert to communicate to stakeholders the corporate risk policy and profile and respond to evolving concerns of shareholders in areas such as ESG, sustainability, and climate change. These stakeholders are not limited to the shareholders but also include the board of directors, regulators, analysts and rating agencies, partners and contractors, and the organization's employees. As such, the CRO and the Risk Committee act as leaders to provide a vision and mission for the organization's enterprise risk management culture and process.

Sensitivity Analysis and Stress Testing Frameworks

The factors described above to devise strategies to mitigate sustainability and climate change risk are typically adaptations of traditional analytic models of risks facing an enterprise. Another increasingly common and standardized way to measure enterprise risk is to perform analyses on various facets of the enterprise and perturbate certain important parameter assumptions to determine how sensitive these changes are to corporate performance and risk measures.

Stress Testing Definitions and Principles

Effective stress testing is a form of scenario analysis that explores the resilience of the balance sheet and capital sufficiency based on various conceivable challenges. It has a few different dimensions:

- *Losses*—First, effective stress testing must begin with appropriate modeling of the potential of losses from climate change. Risk managers must gauge the sensitivity of balance sheet asset value as various risks are modeled. For instance, a significant drop in the asset value of homes in a flood-prone region may generate significant capital losses for insurers or owners of mortgage-backed securities. However, such a profound sustainability driver will certainly affect some assets significantly but may not affect other unrelated securities. Some regions may also be affected while other regions go unscathed. Finally, even among corporations in the same industry, some may have more exposure and sensitivity than others. As a consequence, the sensitivity of a model for a given stressor is highly asset mix- and situation-dependent. Risk managers can perform regressions that establish a correlation between the stressor and its effect on the various instruments and their mixes within a given enterprise.

- *Revenues*—Second, enterprise revenue is equally important in the profit loss analysis as are asset values. However, while such losses are more standardized because enterprises within a given industry sector or region often share similar forms of assets, the pattern of revenue generation is typically enterprise idiosyncratic. Within an insurance company, for instance, much of their income may come from premiums and investments, but these earnings vary based on their relative exposure to climate change. Profitability of such corporations may differ substantially even under common stressors such as reduced economic growth, threats of climate change on property, or an increase in business risk over the economic cycle. As with asset loss analysis, the effects of changes in climate change scenarios will have differential effects on each enterprise, even within the same sector. These effects are also more difficult to identify, measure, and track, given the highly idiosyncratic and varied nature of the effects and on how each enterprise has traditionally navigated each risk. Historical and regression-based approaches can be employed to assist the risk manager in incorporating these risks, but the level of predictability and reliability tends to be lower for such stressors that are emerging rather than historical.

- *Balance Sheet*—Successful sensitivity analysis of asset losses and revenue shocks combine to affect the balance sheet and capital adequacy. Stressors may influence enterprise financial statements both from a stock balance sheet today but also flows of net income on the cash flow statement, and hence also future balance sheets. To measure these multi-year effects, stress analyses must necessarily span a long sequence of quarterly and annual balance sheets. For sustainability and climate change challenges, evolving needs and expectations are profound and can span decades rather than corporate quarters.

Stress analyses aside, the composition of balance sheets needs to be reoptimized over time, especially when exposed to the most rapidly evolving elements of climate change. A thorough stress test must then anticipate such rebalancing as an enterprise is expected to react to changes in the stressor variable. This necessary realism then requires stress testing to be dynamic and evolve in subsequent years of the simulation to incorporate wise business decisions.

This dynamic nature requires a great deal of sophistication of the risk manager charged with meaningful stress analyses. Fortunately, organizations such as the Intergovernmental Panel on Climate Change trace and communicate the best available science on sustainability. However, while the science is increasingly clear, there remains significant uncertainty with regard to global governmental policy responses to greenhouse gas emissions and the capacity of corporations to incorporate good science and translate it into sustainability strategies.

THE STRESS TESTING TAXONOMY

Effective stress testing must follow a definitional taxonomy so that the results can be meaningfully described and assessed across potentially diverse audiences. While stress testing unavoidably requires a number of factors idiosyncratic to each enterprise, such formalized scenario analyses also share a number of dimensions. These include:

- The inclusion of a baseline scenario that is the best-guess estimation of financial results based on the current understanding of market conditions and operations.
- The adverse or stressed scenario that imposes on the model conditions that can plausibly be expected. Such adverse conditions arising from sustainability and climate change risk can vary, depending on the goal of the stress test.
- A dynamic balance sheet that models the size, term structure, and composition of the balance sheet as it evolves and is modified in response to the effects of climate and sustainability stressors over the relevant time horizon.
- A time horizon that is sufficiently long to span the reasonably expected duration of the stressor and the period necessary to fully accommodate residual effects once the stressor reverts to an equilibrium level.
- The inclusion of feedback effects. Shocks and responses can arise from and be induced by the initial stressor. A first-round effect is typically a significant aspect of climate change, while the second-round or subsequent effects explore enterprise liquidity, implications of climate or sustainability responses arising from the first-round stressor, and management actions as a response to the evolving financial conditions.

- A stress testing framework that includes the design of the relevant scenario to be tested, and mechanisms to explore the effects of a stressor on the balance sheet, business operations, portfolio, department, or product.
- A historical scenario that replicates past shocks to allow the risk manager to determine whether the stress test model responds in a manner consistent with the actual enterprise realization of the historical shock. This technique is a method to "back test" the model to determine model validity.
- A multifactor stress test that models the implications of two or more economic shocks.
- A reverse stress test that defines a set of outcomes, such as illiquidity or insolvency, and explores the range or magnitude of a shock or shocks that could induce such an outcome.
- A scenario analysis that imposes hypothetical conditions on a financial system or business line to determine the effects of a future event.
- Sensitivity analyses that gauges the correlation with predicted outcomes as the scale of the stressor is changed.
- A solvency stress test that assesses the effect of an economic or sustainability shock or scenario on enterprise capital.
- A worst-case scenario that imposes the most negative plausible shocks over a defined time horizon on a financial model to explore the degree of serious financial vulnerabilities.

QUALITATIVE AND QUANTITATIVE APPROACHES

Ideally, a stress test employs a complete and well-specified financial risk model to produce results that can be easily interpreted and generalized. However, every enterprise has unique characteristics that may be substantial and material in such a test. In addition, models that fully represent the range of possible operational or financial outcomes arising from many shocks may be so complex that model elements may interact in unpredictable or unintended ways.

Perhaps the greatest challenge in sustainability is that there are few precedents for climate change risks given the future will necessarily be quite different than the present and recent past. While a very well-specified and quantitatively sophisticated and accurate model is the goal of any risk manager, even the most well-designed model remains an abstraction. The simplification of complex financial or operational details allows for meaningful and intuitive interpretation, but may inadvertently miss subtle yet important scientific and economic consequences.

For these reasons, while the goal may be to maximize the degree to which quantitative relationships are captured in the model, risk managers and decision-makers must recognize that the results still require human judgment for meaningful interpretation. In addition, some scenario results may seem implausible. Such could be the case because the model failed to capture details at intermediate stages in the time horizon when human intervention should have occurred.

One goal of a stress test is to attain some standardization so that the results of the test can be compared across similar enterprises. In such cases, certain elements of the operations may not be fully included. For these reasons, enterprise risk managers and decision-makers may need to impose upon the model various qualitative adjustments to ensure the model properly captures the nuances of the enterprise and its operations or to ensure the model produces results that are more consistent with past history.

A stress test that is too complex and quantitative, or is insufficiently detailed or relies excessively on qualitative adjustments, can provide erroneous or meaningless results. The goal from a model design perspective is to strike a balance between simplicity and complexity, and in quantitative results and qualitative adjustments so that risk managers and decision-makers can understand the model and its implications and can use the results to enhance their intuition and credibly influence their responses.

Measurable and Immeasurable Dimensions

Related to the goal to ensure the appropriate level of quantitative rigor is the need to rely on measurable variables. However, wholesale reliance only on the measurable while omitting less quantitative but still relevant information can also create a bias.

There is first the danger of attempting to quantify variables that defy easy quantification. This can lead to erroneous measures. The second is to omit important but intangible aspects. This failure to properly consider immeasurable factors, especially if the measurable factors point the model in one direction, while the immeasurable factors induce the opposite result, may lead to outcomes that are highly biased and hence dangerous from a risk management perspective.

Parametric and Macroeconomic Stress Testing and Scientific Extrapolations

A stress test explores the effects of various types of stressors based on the intended goal of the test. The simplest stress test is the parametric test that explores the effects of an isolated market parameter on financial results.

Such modeling serves three purposes. First, corporate governance bodies and risk managers are made more aware of sustainability or climate change trajectories that may affect their enterprises. Second, such partnerships allow enterprises to better prepare for and mitigate resulting risks. Third, regulatory agencies are better able to manage systemic risks and avoid cascading failures by fostering increased sophistication and preparation for emerging climate change and sustainability risks.

Scientific extrapolations of broad environmental or technological phenomena may result in changes to patterns of investment, relative growth or decline of markets, property valuations, obsolescence risk, and even macroeconomic variables. For instance, massive investments in alternative energy to combat scientific extrapolations of global warming may affect the availability of loanable funds, interest rates, inflation, and emerging markets.

In such cases, mitigation of systemic risk may require three partners. They include:

- the scientists and technologists who forecast broader environmental phenomena and predict the pattern of future infrastructure investment,
- the regulatory agencies, central banks, investment banks, and treasuries that must manage systemic economic and global risk, and
- the various enterprise risk managers that must respond to mitigate these risks or respond to these opportunities.

CREDIT RISK

Few organizations have the luxury of unconstrained funding sources that are forever sustainable and abundant. Most organizations borrow from time to time to invest in new activities or ensure smooth cash flow or may instead lend to provide operating and investment funds for other entities. The level of risk associated with an organization may affect the terms of its borrowing, and hence affect its profitability and viability.

In addition, an institution that invests in corporations that face excessive climate risks that threaten their profitability or viability may itself become financially vulnerable. Such related parties may not be fully repaid for the funds they may have lent. These external credit risks affect organizations' balance sheets. Risk management tools can be used to manage these sustainability and climate change-related credit risks.

CREDIT PRODUCTS AND CLIMATE CHANGE

The types of credit products are as varied as the entities that participate in credit markets. Corporations contract in commercial paper for short-term lending and borrowing, project financing to the point that a project begins to generate revenue, corporate bonds of a duration that may extend to ten or twenty years, long-term mortgages on real estate, and asset-backed credit which depends on the value of an underlying and perhaps climate change sensitive assets such as stock in a company or a portfolio of mortgages.

The duration of these instruments is typically related to the pattern of cash flows the lending can support. For instance, an insurance company collects premiums on an ongoing basis and invests the premiums in assets that can provide a return commensurate with its expectation of property and casualty claims. It is prudent for such a corporation to also ensure that it includes in its credit portfolio some assets that can be liquidated more quickly should unusual losses arise.

Such exercises in the matching of income and obligations typically rest with assessments of the likelihood of occurrence of future events. An insurance company that collects premiums on policies in flood-prone coastal areas may regularly study the occurrence of past property and casualty losses in such regions. This extrapolation of past risks into the future can be fraught with error should past losses underestimate the potential for future losses, especially as climate change accelerates.

Such mismatches between past modeling and future potential damages pose not only a risk for the insurance company itself but also a risk for those who have paid premiums for property and casualty indemnification and for entities that may have invested in or extended credit to the corporation.

Many corporations *self-insure* against risks for assets they hold on their books. Hence, while these risks and not strictly credit risks, they are associated with additional corporate risks that may not be adequately quantified when sustainability and climate change risk increase over time.

CREDIT RISK MEASUREMENT

A corporation traditionally employs various models to determine the degree of risk a counterparty or a liability may impose on its balance sheet. These models take into account the following factors:

- Probability of Default (PD). The PD is an estimate typically based on past observations of the probability that a credit will default on its ability to pay its debt. A variety of techniques may provide such an estimate. They may be based on past observations of the frequency of payment defaults, scientific models of event probabilities, or credit agency or scoring ratings from external or internal analysts.
- Loss Given Default (LGD). The LGD measures the share of the loss that must be absorbed by the corporation as a proportion of the credit exposure should a default occur. This exposure takes into account the type of debt and the degree to which other creditors may compete for a share of collateral.
- Exposure At Default (EAD). The EAD is the expected value of the exposure at the time the default occurs. For instance, this amount may be the remaining balance owed on a property subject to climate change losses. This amount may be adjusted downward if there is some residual value of the collateral at the time of the loss, or upward if the assumption of a share of the property incurs additional costs, such as recovery or liquidation expenses, maintenance or repair costs, or holding costs.

Given these parameters, an Expected Loss (EL) can be calculated as the product of the probability of default, the loss given default, and the exposure at default:

$$EL = PD * LGD * EAD$$

While these expected losses are often calculated based on past observations and history and augmented based on evolving losses and probabilities, for instance, as climate change accelerates, a corporation must also be cognizant of risks in excess of these expected losses. Such *Unexpected Losses* measure the difference between expected losses and potential loss should a rare event occur. Since *black swan* events may occur rarely, such as a hundred-year flood, they may not be factored into the probability of default given their rarity. Rare events may nonetheless be so catastrophic, despite their low probability, that the exposure to such unexpected events should nonetheless be quantified.

Unusual black swan events are sometimes quantified based on a Value-at-Risk (VAR) calculation as described below. Such estimates for instance, of the expected loss of an event that has only a very small probability of occurring may nonetheless attribute a significant liability should a black swan event occur. While such VAR calculations are common, they may also provide deceptive results. For instance, it is difficult to assess the worsening probability of future events as conditions change.

As a hypothetical example, a hundred-year flood may occur every decade if climate change accelerates. In addition, if one sets a 1% threshold over a given duration and determines that a $1 million loss will occur 1% of the time, it is possible that an even more catastrophic loss could occur perhaps 0.5% of the time. The VAR describes a potential loss at some threshold but gives no indication of potentially far greater losses that may occur with only a slightly lower probability.

A more useful measure would assess the probabilities of all catastrophic events that have a likelihood up to the VAR threshold to calculate an equivalent probability-weighted loss for rare events that occur no more frequently than the threshold. Such a measure would provide more information than the VAR and would be more relevant as the likelihood of extreme events increases with increasing sustainability and climate change risk.

Credit Risk Management and Stress Testing

Such credit risks evolve over time. Obviously, risk declines as a credit reaches maturity for a couple of reasons. First, balances are typically paid down and the total outstanding amount owed typically declines as the maturity date is approached. Second, the probability of a default or catastrophic loss is lessened as the remaining duration to maturity decreases. For instance, consider the probability of a hundred-year flood. Such a flood may not occur at all in a century, or it may occur once, twice, or more times. By definition, a 100-year flood occurs with a probability of 1% in a given year.

This process is defined based on the statistical model presented by Siméon Denis Poisson in 1837.[3] The probability of zero such occurrences, labeled p_0 over a given time period t, given an arrival rate λ, in this case, 1% is then:

$$p_0 = e^{-\lambda t} = e^{-.01t},$$

and the probability of at least one such event is $1 - p_0$. As an example, the probability of at least one one-hundred-year flood in the next fifty years is 39% while the probability of at least one-hundred-year flood in a hundred years is 63%. An increase in the duration of concern increases the probability that an extreme event will occur.

Given their exponential nature, these probabilities are very sensitive to even small changes in risk. For instance, climatologists have increasingly cited instances of 100-year flood that are becoming much more common. Reliance on past data may not well reflect future climate risk factors.

MARKET RISK

Every organization has elements it manages that are within its control and are also subject to external forces for which it must respond. No organization in a free-market economy is insulated from the marketplace. Publicly traded corporations are valued based on the market valuation of the excess of their assets over liabilities. Their balance sheets, and often even their cash flows, depend on how the market values these assets and liabilities. Institutions that do not engage in frequent market exchanges may still be vulnerable to changes in the value of securities, resources, donations, or obligations such as loans or covenants that they may retain on their books, even if these balance sheet items do not represent an important element of their strategic plans.

Market forces that affect these assets and liabilities in turn affect the institution as a going concern. Such risk can arise when securities market discount an asset in an organization's portfolio. When the organization's comptroller must mark the value of a security held in the company's portfolio to its market valuation, called *mark-to-market*, the corporation may experience a balance sheet (paper) loss. If assets lose sufficient value, or liabilities likewise increase due to changes in market valuations, corporate equity or working capital may be eroded to dangerously low levels. Risk management tools can be used to better understand and control such market risks.

For instance, a nonprofit organization may hold on its balance sheet securities that decline in value as carbon taxes are imposed on end users. Global warming can also increase the liabilities of the publicly traded corporations these securities represent. Management of the nonprofit organization can use risk management tools to anticipate and hence better manage such risks.

Market risk is traditionally backward-looking. The patterns of volatility of past returns indicate the expected distribution of returns and their probabilities. If returns are distributed normally, then one can calculate the return at a given threshold. As an example, one may want to know the expected loss at the 5% probability threshold on a given day or week. The threshold return, called the Value-at-Risk is then the loss at the specified probability threshold within a specified duration (Fig. 26.1).

However, the vexing aspect of risks arising from sustainability and climate change is not that these risks are the typical random shocks that have historically and predictably affected corporate valuations. Rather, they may arise from discrete new events that tend to be concentrated on the downside of risk rather than symmetrically around a median point. In addition, the risks not only defy history bur are also evolving rapidly and systematically in the case of climate change. Hence, reliance on traditional tools such as Value-at-Risk that rely on past risk history may lead to inappropriate risk management complacency.

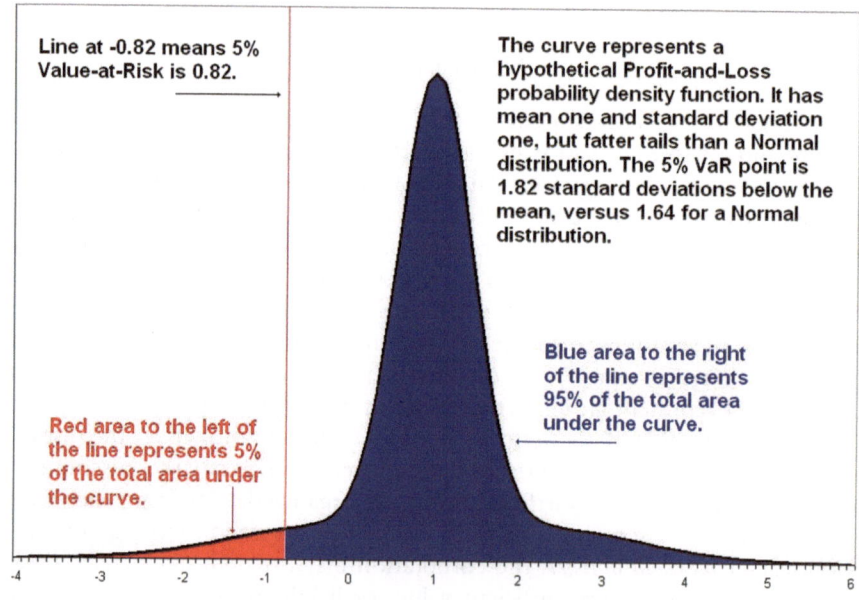

Fig. 26.1 The value-at-risk methodology to assess the expected costs of a calamity (By Original uploaded by AaCBrown [Transferred by Matanya]—Public Domain, https://commons.wikimedia.org/w/index.php?curid=14530760)

MARKET RISK PORTFOLIO
MANAGEMENT AND STRESS TESTING

A corporation has an advantage in managing these risks by aggregating such risks in a broader portfolio. If the various credit elements of the portfolio have probabilities and magnitudes of potential losses that are uncorrelated with each other, the overall risk of the portfolio falls just as a stock portfolio can reduce risk with diversification.

Effective sustainability and climate risk management can then decrease if managers ensure that the risks which they accept in their portfolio are uncorrelated. However, credits can share some component of risk because they are all subject to the same large macroeconomic forces that affect all instruments. For instance, corporate credit of all forms may be adversely affected if a recession threatens the market valuation of these various credits. Climate change may also increase the broad level of risk among credits. These systemic risks may affect the entire system of credits and hence may be unavoidable. They may also worsen as climate change accelerates.

However, climate change does not affect all entities equally. In fact, climate change may even benefit some nations and corporations and hence reduce their risk of default or loss of resource value. These differential risks across credits are called idiosyncratic because they take into account specific aspects of each credit on an individual rather than a systemic basis.

Such *idiosyncratic risks* may require more elaborate models that separate credits into various categories and then exposes stresses on the portfolio in a way that permits assessment of the ways such stresses affect each element in the portfolio.

These idiosyncratic risks are challenging to categorize and estimate at the best of times. The accelerating rate of climate change challenges even the most elaborate risk models that routinely impose known stresses on a portfolio and estimate their implications. Risk managers cannot be complacent. While stress testing is a valuable tool, even this most sophisticated technique fails to capture risks as conditions evolve in ways that are unexpected and may have not been observed historically.

STRATEGIC RISKS UNDER SIGNIFICANT
CHANGES IN SUSTAINABILITY REGIMES

The changes necessary to accommodate profound shifts in sustainability and climate change policies are of such a scale that the vast majority of private sector firms will find some accommodation necessary. These accommodations may be small, such as a shift in the mode and cost of the energy they consume. Other industries may suffer existential threats. Meanwhile, corporations communicate their priorities and strategies to financial markets as stakeholder expectations evolve.

Inherent in the strategic planning process is an analysis of a firm's current regulatory, market, and technological environment and anticipation of its operating environment from a few years to a few decades in the future. Strategic plans rarely anticipate rapid change and are often revised only periodically, usually on a five-year timeline. A corporation that traditionally revised its strategy but once every half decade may need to reevaluate strategies more frequently in the future. Rapid technological or sustainability regime changes necessitates much more rapid revisions of strategic plans. However, the analysis necessary for successful strategic planning is extensive at the best of times. When there exists a significant regime change with a great deal of uncertainty with regard to the new environmental and economic landscape, strategic planning also becomes highly speculative. These uncertainties create a significant amount of *strategic risk*.

External Risks

Private sector entities actively manage risk as they innovate, design new processes and products, construct novel marketing strategies, and often invest heavily in market and strategic research. Corporations function within an environment that increasingly challenges the control of forces that must be anticipated so their attendant risks are managed effectively. Changes in sustainability regimes introduce significant external risks of a scale that exceed what risk managers typically finesse. Beyond the traditional market, credit, regulatory, legal, and liquidity risks, firms experiencing the forces of major environmental or sustainability regime changes must also accommodate changes to their physical environment.

Environmental and physical risks arising from climate change often span sectors and regions and may impinge the greatest risks on entirely unrelated sectors. Climate change is a consequence of combustion of fossil fuels from which economic progress in MDCs was derived. The greatest benefactors have been the United States, Europe, Canada, and China, among others. These nations have been in temperate or subarctic regions that have escaped the worst consequences of global warming so far.

MDCs also have the greatest capacity to mitigate the costs of climate change. Sectors and nations most prone to physical risks to their environment have significant exposure to the effects of rising sea levels, changing weather patterns, and rising temperatures. Nations facing greatest sensitivity to sea levels or already face extreme temperatures, drought, or flooding tend to be most prone to the physical effects of global warming. Patterns of economic development in the most prosperous nations have often established significant resiliency to such physical shocks, or have the capacity to enhance their resiliency.

This leaves LDCs, their economies, and sectors most dependent on and sensitive to environmental and climate change risks and are most prone to the costs and displacements of major environmental shifts arising from global warming. Not only are they most profoundly affected but they also have the least capacity to manage and mitigate the costs.

Physical risks in such sectors and regions are profound. The 2022 Conference of Parties 27 has for the first time acknowledged that these extreme physical risks are disproportionate in cause and effect. A discussion has begun that recognizes the inevitable need for income redistribution to increase the capacity of the most vulnerable and least able to manage these risks.

Summary

Modern corporations in the most normal of times face risks. Indeed, these risks allow firms to command returns in excess of the modest steady-state risk-free rate of return that prevails in the economy. But while corporate risk managers have used models of past behavior to manage current and future risk, climate change and sustainability challenges necessitate new approaches.

The operational risk of our internal operations, the regulatory and legal risk as regulators are quickly modifying expected ESG standards, and the physical, obsolescence, and stranded asset risk as the patterns of our operations evolve under climate change must all be acknowledged and managed. If corporations do nothing to respond to ESG challenges, and if some resist the inevitable evolution toward ESG, corporations suffer reputational risk. Risk management is especially challenging if it is unclear how or how fast policymakers will adjust to mitigate our rate of decay of natural capital.

ESG Inquiries

Has your institution's risk management protocol adjusted to accommodate the fast-evolving climate change environment?

Does your institution face any reputational risk should it fail to respond to evolving stakeholder expectations?

Has your organization categorized its balance sheet based on its vulnerability to evolving climate change or expectations?

Appendix—Regulatory Trends Arising from Sustainability and Climate Change Risk

The Federal Reserve Bank Recommendations to Institutions on Climate Risk Reporting

This chapter closes with developments on the regulatory front. Challenges related to sustainability and climate change are not always resolved efficiently within traditional free markets. Scientists have described the human forces that give rise to climate change as arising from non-market implications of corporate actions. Unpriced pollution and insufficiently priced emissions arise because of such non-market externalities.

Likewise, sustainability is challenged when one generation does not fully integrate the costs it may impose on resources consumed that are lost to future generations. These non-market and intergenerational externalities are increasingly attracting the attention of governments and regulators charged with internalizing the externalities to create greater market efficiency and intergenerational equity.

Such regulatory risk can be assessed and mitigated through an organization's business and strategic risk management processes. This chapter closes with a brief description of significant regulatory avenues that the U.S. *Federal Reserve Bank (FRB)* and the *Basel Committee on Banking Supervision (BCBS)* have imposed on banks that often find themselves on the leading edge of climate risk.

In December 2022, the U.S. Federal Reserve Bank published for comment its intentions to address climate-related risk management. Their concerns rest primarily with the transitional risks associated with a shift in economic priorities from a fossil fuelled economy to one that relies on sustainable energy and practices. The U.S. *Office of the Comptroller of the Currency* (OCC), the *Federal Depository Insurance Commission* (FDIC), and the Taskforce on Climate-Related Financial Disclosures (TCFD) as a committee of the Financial Stability Board have also been proposing potential ESG-related financial disclosures to protect consumers, investors, and the economy.

While the Federal Reserve Bank oversees commercial banking, it has also assumed the responsibility to address systemic financial failures such as occurred following misguided investments in sub-prime mortgages in the first decade of the 2000s. Because the Federal Reserve has both the ability to affect bank reserves and hence liquidity and purchase assets that can become toxic due to significant economic shocks, it recognizes its important role in stimulating and coordinating economy-wide policy in anticipation of climate change transition and risks.

The Federal Reserve has proposed regulations that encourage improved procedures in corporate governance, policies and procedures, strategic planning, risk management, data and risk measurement and reporting, and scenario analyses around issues related to ESG, sustainability, and climate change. They propose a framework for *safe and sound* management of the financial risk that results when firms are faced with climate change risk exposure, either directly on their balance sheet or indirectly in their ability to adapt to climate change and sustainability challenges.

Their general principles recommend improvements in the frameworks for:

FRB Guidance on Governance

The Federal Reserve recommends boards create and maintain sufficient sophistication to assess potential climate risks within the corporate risk appetite and are able to anticipate the evolution of these risks over time. Corporations are expected to manage these risks by assigning roles and responsibilities that may evolve as necessary and providing sufficient resources. They also note that management must regularly communicate to internal stakeholders how climate change may affect the corporation's risk profile and report their analyses regularly to their boards of directors.

FRB Guidance on Policies, Procedures, and Limits

Management is required to incorporate into their policies and procedure detailed descriptions of how the corporations under their purview consider climate change risk and how these considerations shall affect operations.

FRB Guidance on Strategic Planning

The Federal Reserve considers climate change risk exposures as an essential element in strategic planning. These considerations should be incorporated into its business strategy, risk appetite, financial and capital planning, and operational plans. These corporations should address the effect of climate change risk on regions, stakeholder expectations, reputational risk, and the potential physical and financial harm on low- and middle-income and vulnerable communities. Corporations under their purview must clearly communicate to the public their climate-related strategies and commitments, and be consistent with risk appetite statements and internal strategies.

FRB Guidance on Risk Management

Corporate managers should ensure the development and oversight of processes to identify, measure and monitor, and control within their risk management processes their exposure to climate risk. They shall do so by:

- Developing processes to identify merging and material climate risks under various time horizons and scenarios.
- Considering stakeholder insights from the entire corporation.
- Developing processes that permit the measurement and monitoring of climate risks and communicating their results to internal stakeholders in the areas of physical and transitional risks, tools to effectively communicate risks and vulnerabilities, aligning the corporate risk appetite, promulgating appropriate metrics, and incorporating climate risk into internal audit and risk management frameworks.

FRB Guidance on Data, Risk Measurement, and Reporting

Corporations should make available timely, relevant, and accurate data to permit across the institution sound and rapid decision-making by:

- Incorporating climate risk data into institutional dashboards and risk monitoring processes.
- Ensuring the institution maintains effective risk data aggregation and reporting capabilities
- Monitoring the evolving state-of-the-art in risk and data measurement, the methodology of modeling and reporting, and incorporating their conclusions into their climate risk management processes.

FRB Guidance on Scenario Analysis

The Federal Reserve observes that "climate scenario analysis is emerging as an important approach to identifying, measuring, and managing climate risks… [and] an effective climate scenario analysis framework should provide a comprehensive and forward-looking perspective to apply alongside existing risk management practices when evaluating the resiliency of strategies and risk management to the structural changes arising from climate risks." They instruct institutions under their purview to:

- Create processes for climate scenario analysis processes that are tailored to the appropriate risk profile, size, type of business activity, and complexity.
- Develop standards to oversee, validate, and ensure quality control for the analyses in proportion to the risk they impose.
- Define analysis framework objectives commensurate with climate risk management strategies

The Federal Reserve defines the management of risk areas according to the following categories:

- Credit Risk. Climate-related transitional and physical credit risks that are managed across market sectors, geographies, and concentrations and include how correlations across exposure across classes may evolve.
- Liquidity Risk. Liquidity risk management and buffers should incorporate climate risks.
- Other Financial Risk. Institutions should monitor interest rate risk and other model assumptions to contrast how volatility increases or predictability declines because of climate risk.
- Operational Risk. The institution should conduct climate risk assessments across all business operations and consider counterparty, business continuity, and evolving legal and regulatory risks.

- Legal/Compliance Risk. Institutions should determine how climate risks and mitigation measures may affect legal and regulatory factors within which the institution operates.
- Other Nonfinancial Risk. The corporation should oversee the implementation of strategic plans and determine how financial condition and operational resilience are affected by an evolving operational environment. Institutions should consider:

  - The extent that the corporation's operations and finances may be damaged by reputational, liability, and litigation risks.
  - The degree to which measures to implement to mitigate these material risks is adequate in the face of evolving climate risk.
  - Methods to monitor on a regular basis service providers' performance and reassess their selection based on various risk elements to which the institution is exposed.

Internationally, the Basel Committee on Banking Supervision has also produced a set of principles. These Basel principles, entitled *Principles for the effective management and supervision of climate-related financial risks*, adopted similar guidance for entities under their purview and the purview of national regulators. They partitioned their guidance into the following categories:

a. The governance process (BCBS Principles 1, 2, 3, 6)
b. The internal control framework (BCBS Principle 4)
c. Management monitoring and reporting (BCBS Principle 7)
d. Liquidity and capital adequacy considerations (BCBS Principle 5)
e. Credit risk management (BCBS Principle 8)
f. Market risk management (BCBS Principle 9)
g. Liquidity risk management (BCBS Principle 10)
h. Operational risk management (BCBS Principle 11)
i. Scenario analysis (BCBS Principle 12)
j. Supervisory Responsibilities (BCBS Principles 13–15)
k. Responsibilities, powers, and functions of supervisors (BCBS Principles 16–18)

Basel Principles for Effective Management and Supervision of Climate-Related Financial Risk

Principle 1: Banks should develop and implement a sound process for understanding and assessing the potential impacts of climate-related risk drivers on their businesses and on the environments in which they operate. Banks should consider material climate-related financial risks that could materialize over various time horizons and incorporate these risks into their overall business strategies and risk management frameworks.

Principle 2: The board and senior management should clearly assign climate-related responsibilities to members and/or committees and exercise effective oversight of climate-related financial risks. Further, the board and senior management should identify responsibilities for climate-related risk management throughout the organizational structure.

Principle 3: Banks should adopt appropriate policies, procedures, and controls that are implemented across the entire organization to ensure effective management of climate-related financial risks.

Basel Principles on Internal Controls

Principle 4: Banks should incorporate climate-related financial risks into their internal control frameworks across the three lines of defense to ensure sound, comprehensive, and effective identification, measurement, and mitigation of material climate-related financial risks.

Capital and Liquidity Adequacy

Principle 5: Banks should identify and quantify climate-related financial risks and incorporate those assessed as material over relevant time horizons into their internal capital and liquidity adequacy assessment processes, including their stress testing programs where appropriate.

Risk Management Process

Principle 6: Banks should identify, monitor, and manage all climate-related financial risks that could materially impair their financial condition, including their capital resources and liquidity positions. Banks should ensure that their risk appetite and risk management frameworks consider all material climate-related financial risks to which they are exposed and establish a reliable approach to identifying, measuring, monitoring, and managing those risks.

Management Monitoring and Reporting

Principle 7: Risk data aggregation capabilities and internal risk reporting practices should account for climate-related financial risks. Banks should seek to ensure that their internal reporting systems are capable of monitoring material climate-related financial risks and producing timely information to ensure effective board and senior management decision-making.

Comprehensive Management of Credit Risk

Principle 8: Banks should understand the impact of climate-related risk drivers on their credit risk profiles and ensure that credit risk management systems and processes consider material climate-related financial risks.

Basel Principles on Comprehensive Management of Market, Liquidity, Operational, and Other Risks

Principle 9: Banks should understand the impact of climate-related risk drivers on their market risk positions and ensure that market risk management systems and processes consider material climate-related financial risks.

Principle 10: Banks should understand the impact of climate-related risk drivers on their liquidity risk profiles and ensure that liquidity risk management systems and processes consider material climate-related financial risks.

Principle 11: Banks should understand the impact of climate-related risk drivers on their operational risk and ensure that risk management systems and processes consider material climate-related risks. Banks should also understand the impact of climate-related risk drivers on other risks and put in place adequate measures to account for these risks where material. This includes climate-related risk drivers that might lead to increasing strategic, reputational, and regulatory compliance risk, as well as liability costs associated with climate-sensitive investments and businesses.

Basel Principles on Scenario Analysis

Principle 12: Where appropriate, banks should make use of scenario analysis to assess the resilience of their business models and strategies to a range of plausible climate-related pathways and determine the impact of climate-related risk drivers on their overall risk profile. These analyses should consider physical and transition risks as drivers of credit, market, operational, and liquidity risks over a range of relevant time horizons.

Basel Principles on Prudential Regulatory and Supervisory Requirements for Banks

Principle 13: Supervisors should determine that banks' incorporation of material climate-related financial risks into their business strategies, corporate governance, and internal control frameworks is sound and comprehensive.

Principle 14: Supervisors should determine that banks can adequately identify, monitor, and manage all material climate-related financial risks as part of their assessments of banks' risk appetite and risk management frameworks.

Principle 15: Supervisors should determine the extent to which banks regularly identify and assess the impact of climate-related risk drivers on their risk profile and ensure that material climate-related financial risks are adequately considered in their management of credit, market, liquidity, operational, and other types of risk. Supervisors should determine that, where appropriate, banks apply climate scenario analysis.

Basel Principles on Responsibilities, Powers, and Functions of Supervisors

Principle 16: In conducting supervisory assessments of banks' management of climate-related financial risks, supervisors should utilize an appropriate range of techniques and tools and adopt adequate follow-up measures in case of material misalignment with supervisory expectations.

Principle 17: Supervisors should ensure that they have adequate resources and capacity to effectively assess banks' management of climate-related financial risks.

Principle 18: Supervisors should consider using climate-related risk scenario analysis to identify relevant risk factors, size portfolio exposures, identify data gaps, and inform the adequacy of risk management approaches. Supervisors may also consider the use of climate-related stress testing to evaluate a firm's financial position under severe but plausible scenarios. Where appropriate, supervisors should consider disclosing the findings of these exercises.

Regulators worldwide are promulgating their own financial oversight principles along lines similar to those published by the U.S. Federal Reserve Board and the European Basel Committee on Banking Supervision. Following the Global Financial Meltdown of 2008, regulators understand the interrelatedness of financial markets that can suffer a contagion that is initiated by only a small fraction of financial institutions. In turn, such contagions can wreak havoc even on the commercial banking sector and stock markets and push national economies into recession.

The regulations of the FRB, the BCBS, and others are designed to allow institutions to anticipate climate risks and communicate their vulnerabilities so that the market can properly price these risks and protect itself from sustainability and climate change risks.

NOTES

1. Hanson, Craig, Janet Ranganathan, Charles Iceland, and John Finisdore (2012). *Guidelines for Identifying Business Risks and Opportunities Arising from Ecosystem Change, Version 2.0*, World Resources Institute, https://files.wri.org/d8/s3fs-public/corporate_ecosystem_services_review_1.pdf, accessed December 10, 2022.
2. https://assets.bbhub.io/company/sites/60/2022/10/2022-TCFD-Status-Report.pdf, accessed December 10, 2022.
3. Poisson, Siméon D. (1837). Probabilité des jugements en matière criminelle et en matière civile, précédées des règles générales du calcul des probabilités [Research on the Probability of Judgments in Criminal and Civil Matters] (in French). Bachelier, Paris, France.

CHAPTER 27

An Extension of Classical Economic Analysis to the Normative Realm

We have explored to now the problems created by people and corporations insufficiently aware or affected by the true costs of decisions with intragenerational and intergenerational implications. We have also described various ways to institutionalize and internalize such external effects of our decisions. However, our ambition to modify the conventional neoclassical model of economics to incorporate values that extend beyond the individual or corporation does not solve all such externalities. To see why, let us return to the earliest political economists who increasingly expressed hope and despair over the newfound ability of humankind to dominate our environment.

The earliest references from the age of Aristotle to the study of our oikos, our *house*, or our environment acknowledged that such economics refers to management within the greater world around humankind. Aristotle's many Greek deities were gods of the environment that dominated humans. Little had changed in the quality of human life and our interactions with the broader ecosystem in the 2100 years that intervened between Aristotle's authoring of the Nicomachaen Ethics and Adam Smith's *Theory of Moral Sentiment*. By 1757, the Industrial Revolution had not yet dawned to produce its fantastic growth in wealth, population, and greenhouse emissions. Economics was still the realm of mortals who functioned within an ecosystem rather than dominated it.

Adam Smith is regarded as the first economist, but Smith and those who soon followed him are more typically regarded as ***political economists*** who understood that optimization of the human condition could not be constructed solely within markets divorced from the normative world of governance and democracies. Smith, Malthus, Ricardo, Mill, and other contemporaries recognized that economies represent an aggregation of people rather

© The Author(s), under exclusive license to Springer Nature Switzerland AG 2023
C. Read, *Understanding Sustainability Principles and ESG Policies*,
https://doi.org/10.1007/978-3-031-34483-1_27

than of mere markets. In a political economy, corporations and individuals, and the markets that offer a forum for these entities to interact, had not emerged as the central elements we consider today.

Instead, in political economy, there is a greater emphasis on our collective good, on the effect of power on the nature of transactions, and on the importance of equity. Granted, Smith in his subsequent 1776 *Inquiry into the Wealth of Nations* touted the efficiencies that could arise from mass production and enhancements in markets, but even in this lesser of Smith's two great works, by his own estimation, the prevailing theme remained the common good. Ricardo further introduced the importance of natural capital in a way that is unique among productive capacities, while Malthus lamented the dismal prophecy of exponential population growth in the face of arithmetic resource availability. Mill ended this era of political economy by advocating for mechanisms for greater economic justice in the face of growing economic inequality.

Mill's *Principles of Political Economy*[1] was one of the last major works toward the end of an era that 183,088 placed humankind within the greater context of our environment. He lamented:

> If the earth must lose that great portion of its pleasantness which it owes to things that the unlimited increase of wealth and population would extirpate from it, for the mere purpose of enabling it to support a larger, but not a better or a happier population, I sincerely hope, for the sake of posterity, that they will be content to be stationary, long before necessity compel them to it.

Mill was concerned about the lack of a steady state economy and environment that could serve the ecosystem indefinitely. In retrospect, he was founding a new approach to optimization of the human condition that is called ecological economics today.

Soon after Mill's lament, economics took two forks. One emphasized the political in political economy by noting that power often dominates markets while concentration of power within capitalism ultimately depreciates humankind. Karl Marx founded a movement based on these concerns. Despite some strong analysis and insightful conclusions, his theories failed to prevail among most economists. Instead, a classical economics emerged that incorporated a value statement no less profound that Marx had proposed. In classical economics and the neoclassical version that constitutes much of modern economic analysis, markets are supreme as the institutions that allow individuals to best assert their interests in proportion to their wealth, as corporations best assert theirs.

We discussed in Chapter 22 the inherent value perpetuated by markets. It is one mechanism to allocate resources and sources of value among individuals and corporations. While most all of us are products of market economies, and the value of maximization of our self-interest has become ingrained in the liberal societies commended even by Mill himself, this emphasis on the potency of the individual to command resources to satisfy our self-interests is a relatively new concept that paralleled the departure from an environmental steady state toward a burgeoning Industrial Revolution.

In this classical economy, wealth is power. Such a monetary economy translates myriad human values into a one-dimensional and universal measure of value, namely money. Added to income and its power to purchase was utility and its capacity to satisfy. With that classical economic revolution, the political in political economy was dropped and the sole pursuit was ever-increasing efficiency and growth to further the satisfaction of human wants and the ability to procreate the human species at the expense of other ecosystem values not incorporated into the values of the individual.

This transformation of a political economy into our modern economy occurred over a half-century period of 1870 to about 1920 when transitions in our understanding of science were also occurring. While it is at times too compelling to divine analogies between a field as normative as politics or power and as positive as the physical sciences, a significant and meaningful analogy nonetheless exists. Rather than broad acceptance of politics and economics as a duality, they are instead treated separately as a dichotomy.

We discussed the example of Einstein's greater insights into physics that rewrote and replaced classical Newtonian mechanics. Einstein's fantastic Annus Mirabilis in 1905 is regarded as revolutionary, but it was in many ways evolutionary. In 1887, Albert A. Michelson and Edward W. Morley published the results of an experiment that contradicted the prevailing theory that the Earth was moving through an invisible aether wind. A few decades earlier, in 1860, Gustav Kirchhoff formulated the concept of black bodies that absorb energy and must thus reradiate it, in the form of heat or a temperature, to maintain a steady state. By 1900, *Lord Kelvin* noted that both these concepts defied physics as understood to that point. Einstein's special theory of relativity and his *photoelectric effect*, both published in 1905, solved these gaps in scientific understanding.

In the photoelectric effect, Einstein asserted the rays of light James Clerk Maxwell had described in 1861 and 1862 as a propagating wave was also a particle. Maxwell argued that frequency of light is a measure of the number of times a wave oscillates each second, while the amplitude of the wave gives the intensity of light. Maxwell's insights are some of the most profound in physics and remain accepted and applied today.

Einstein argued that such a method does not explain certain phenomena, such as the observation that light seems to act as discrete quanta at times rather than continuous waves. Einstein proposal be considered as simultaneously both a wave and a particle. To resolve such quanta as also traveling as a wave, Einstein and others argued that the position of such a quantum particle must radiate according to a probability function that appears as a wave.

Einstein was standing on the shoulders of giants among physicists, but he nonetheless produced a dilemma we still debate today. This concept creates what we now know as *duality*. Scientists often cannot at all times pin an observation down to be one thing or another. Indeed, with regard to light or other quantum phenomena, the closer one tries to pin down its location, the less reliably we can know other important aspects such as its velocity. The very nature of close scrutiny disturbs the system we are attempting to measure.

The central tenet of much of this text, and of classical physics and economics, is that the world appears simpler than it may actually be. Both economics and physics developed in the 1800s as a positive science, with little doubt that greater scrutiny will produce more exacting understanding.

Einstein created the notion that even the hardest of sciences has both positive and normative elements. Even the Theory of Relativity is based on the notion that reality is best formulated based on the observations relative to an observer. This insight allowed him to make sense out of Michelson and Morley's observation that the speed of light seems to be a constant that is immune to the drift of the aether wind. If our position as an observer or a participant matters, then our values affect our observations and realizations.

If physics is not so cut-and-dried and precise as our human experiences aspire for it to be, nor is economics. A second realization of Einstein was that history matters. For instance, the path light takes is not linear and absolute but depends on what it has passed in its journey to us. Even if the speed of light is constant to all observers, its frequency and orientation depend on the path it takes and what it has left in its wake. While we may know its frequency and orientation, these two metrics describe only a small part of its significance. Similarly, the temperature of a gas in a flask may describe the entropy of a closed system but not the myriad adjustments that brought the collection of molecules to that point.

Economics also reduces complex interactions into overly simplistic measures, most typically denominated in money, income, and wealth. Reduction of complexity to simple measures loses the richness and compromises that brought the system to a given point. For instance, reliance on the benefits in a benefit–cost analysis, or the Gross Domestic Product of an economy tells us little about the redistributions of income that brought the economic equilibrium to that point.

Nor can we glean how to reverse any injustices that led to an economic equilibrium just as we cannot reverse the path of light by observing its frequency and orientation. Neils Bohr, one of the founders of quantum physics, and with whom, among others, Einstein argued "God does not play dice with the Universe ," extended the notion that physical phenomena are not additive. A snapshot of a system cannot reflect the many ways that its components have added up to the current state. Bohr noted that physical systems and nature are not simply additive. Elements in the formation of an equilibrium complement each other in complicated ways.

Bohr's extension of duality and the complementarity principle was perhaps inspired partly by his father. Christian Bohr was a well-known physiologist who was a leading researcher in the scientific process of respiration. Yet, he also subscribed to the importance of vital forces, those unknowable elements others may call Gaia, God, or Mother Nature that reduce the determinism of positive science to a component rather than the essence.

This ***complementarity principle***, Bohr argued, applies to the dichotomy of the biological and the physio-chemical world. A combination of physics and chemistry are at the basis of molecular interactions that create the rich biology around us. But, from there, organisms evolve and interact in myriad ways, many of which are more random than determinative. Darwin's theory of evolution has myriad accidental and circumstantial components that make the sum of positive science random and more normative.

Similarly, the state of an economy, which may describe the sum of human exchanges at a given time, tells us little about the judgments and values, minor thefts and major externalities left uninternalized that brought the economy to that point. Yet, much of traditional economics puts these normative values aside and focuses solely on the equilibrium at a given point, and perhaps into the immediate future.

Strict adherents to the positive sciences even may minimize the dichotomy between the rational world and the irrational world that defies rational understanding to a simpler world of the rational and the not-yet-rational. Such a strong attachment to the belief that all things are ultimately rationally determined creates a bias toward scientific conclusions to the exclusion of irrational influences. Among economists there is a joke that describes a person searching underneath a streetlight for the keys to his car. When a bystander asks him where he lost his keys, he points to the dark park behind them. When the bystander then further asks why he is looking here rather than there, he responds "This is where the light is."

Classical physics and economics both shine a bright light on rationality on the physical world and perspectives on the human condition. But both must acknowledge there are large gaps. In physics, the greatest question is the possibility of dark matter and dark energy. In economics, it is the values and the politics that contribute according to the complementarity principle to the current economic state that defines us.

Western cultures have been especially influenced by the certitude of physics and economics. Our engineered technologies have created income and wealth unimaginable in Mill's era, and all the Western World subscribes to one degree or another to the prosperity and immediacy of modern economies, with its positive discount rates, Fisher's impatience, and profound effect on future generations precluded from incorporating their values into today's economic equilibrium. These other normative forces are essentially denied because of our faith in free markets.

These unincorporated or ignored values, perhaps of the injustices imposed on the environment, on past peoples, or on future generations, are not omitted by all peoples. Carlos Castaneda described the *Tonal* and the *Nagual* as the duality of the reality of people today and the everything beyond our reality. Those of faith may be governed by the decisions that brought them to a point today, but may also be influenced by their interpretation of personal faith and spirit. Among North American First Nations peoples, some believe in *Haudenosaunee Life* that acts as a bond across clans and generations. Rick Hill Sr., a member of the Tuscarora First Nations band of the Iroquoian family in New York, U.S.A., and Ontario, Canada, observed[2]:

> If you ask me what is the most important thing that I have learned about being a Haudenosaunee, it's the idea that we are connected to a community, but a community that transcends time.
>
> We're connected to the first Indians who walked on this earth, the very first ones, however long ago that was. But we're also connected to those Indians who aren't even born yet, who are going to walk this earth. And our job in the middle is to bridge that gap. You take the inheritance from the past, you add to it, your ideas and your thinking, and you bundle it up and shoot it to the future. And there is a different kind of responsibility. That is not just about me, my pride and my ego, it's about all that other stuff. We inherit a duty, we inherit a responsibility. And that's pretty well drummed into our heads. Don't just come here expecting to benefit. You come here to work hard so that the future can enjoy that benefit.

This connectedness of the present to the past and future, our complementarity across humankind, is embodied in the *Seventh Generation Philosophy*, states that there is a bond between our conditions now on the practices seven generations ago, and in our actions today on those that follow us seven generations hence. The faithkeeper G. Peter Jemison of the *Seneca Nation* of New York adds:

> We really do see ourselves as part of a community, the immediate community, the Native American community, but part of your nation and the Confederacy. And if you have been given responsibilities within that structure, you must really attend to those responsibilities. You start to think in terms of the people who come after me. Those faces that are coming from beneath the earth that are yet unborn, is the way we refer to that. They are going to need the same things that we have found here, they would like the earth to be as it is now, or a little better.
>
> Everything that we have now is the result of our ancestors who handed forth to us our language, the preservation of the land, our way of life and the songs and dances. So now we will maintain those and carry those on for future generations.

There are many such examples of environmental and intergenerational interconnectivity, in Hindu, Christian, Muslim, and Jewish faiths, and in the beliefs of many indigenous species. While a large share of humankind may aspire to the unidimensional pursuit of the affluence that is the common denominator measure of modern market economies, between a large minority and a majority of peoples also subscribe to normative values that may forever defy rational modeling.

Toward an Ecological Economics

The market economy allocates resources within and across human generations. We have described how it performs the allocation based on a combination of economic power and needs. Yet, even the goods and services consumed in a free-market economy in an MDC are typically provided in markets and by government in equal measure. If we include the thousands of non-market economic transactions we may each make in a given year, or of consumptions that affect those of others outside of markets, then markets likely represent a surprisingly small share of our economic exchanges.

If money then represents only a minority of the exchanges humans enjoy, it defies logic that so much of our economic science is devoted to the determination and documentation of monetary exchanges. Most of what we consume, and some of what we produce, are determined through democracies and power structures. Our understanding of these relatively darker quarters of the human condition cannot always or perhaps even often be described by the bright light of economic rationality.

To better describe the complementarity of human decisions, we must instead employ broader techniques that have been described as *positional analysis*. Only by augmenting traditional economic power, as measured by income and wealth, with positional indicators of power and of the effectiveness of democracy can we better understand our collective economic decisions and guide our intergenerational policies.

This broader awareness of the context of individuals and groups within and across peoples and generations and within our broader ecosystem is collectively described by *ecological economics*. Such an approach recognizes that even the internalization of externalities that lies at the root of correcting deficiencies in market economies still does not create efficient economic decision-making. We must not only correct market deficiencies but also incorporate non-market values and allocation mechanisms.

Likewise, how democracies function to allocate non-market goods and services, within and across generations must also be addressed. Yet, we can employ techniques drawn from the political science of domestic politics, perhaps based on the veil of ignorance of John Rawls, or his theories of interactions of peoples and of international law, and the ethos that we discover with regard to human interaction with the broader environment, to create a more holistic ecological economics that embraces these normative and the traditional positive elements.

It is this more holistic approach that some peoples still maintain but the Renaissance of the Western World has learned to neglect. Our market economies are instead driven by the rational model of benefit–cost analysis, from the largest scale in decisions of corporations that may forever define our economic and environmental future, to the smallest scale of our myriad decisions in which we individually weigh what we have to gain from a transaction with the opportunity cost of what we must sacrifice.

We have been conditioned by market economies to use the unidimensional monetary value as the common yardstick for these decisions. Even environmental economics preserves reliance on our neoclassical economic techniques by converting environmental values into a willingness to pay, measured in monetary terms. Those values, especially of future generations, may then be undervalued or depreciated, or, if measurement is intrinsically difficult, ignored completely.

We rely on democracies and governments to better make decisions over issues that defy market measurement or correct the inherent economic injustices that may arise in a market economy. They cannot fully fill the gaps left by classical economics that are analogous to the dark matter in cosmology which we know must be important but we do not yet know how to observe, measure, or incorporate into our modeling.

Economic theory tells us that the omission of important components or micro markets from a larger economy typically distorts the market equilibrium. Our inability to incorporate complementarity within a generation, the path dependence through past generations that brought us to today, or our disregard for future generations all affect our intergenerational equilibrium. Often, these distortions create irreversible results, perhaps even over the expected duration of human existence, such as the effects of decisions since the Industrial Revolution on global warming. Such *irreversibility* and *path dependency* are disregarded and unextractable in the collection of unidimensional values such as price, income, and wealth distribution in a snapshot of a market economy.

There is growing awareness of the need to broaden economics into a more holistic, inclusive, and normative study of ecological economics. It shall require greater recognition of values not evident in market measures. Such values are embodied in the seventeen Sustainable Development Goals (SDGs) as espoused in 2015 by the United Nations General Assembly as an elaboration of their Millennium Development Goals. Realization of these goals requires political will at both the domestic and global levels, and, while they depend on the actions of corporations, they shall require an expansion of our application of economics.

Rather than attempting to augment classical market-oriented economics to internalize non-market externalities, ecological economics is instead an embracement of both the positive aspects of traditional economics and the normative aspects of justice, intergenerational equity, and the challenges to reverse past economic and societal decisions. It more closely resembles the political economy debates of the nineteenth century than the cost–benefit and market-oriented analyses since then.

Concerns about normative aspects of our interaction with the broader environment were described by Smith but reached a strident tone through Malthus' dismal population prophecy. By 1919, Otto Neurath argued that the market economy did not properly distinguish between human-made physical capital and naturally endowed natural capital, and failed to properly incorporate the needs of future generations into economic decisions today. His analysis was decidedly against the grain of an increasingly traditional and positive approach championed by free marketeers such as Friedrich Hayek. This divergence in economics was evolving as nations such as the Soviet Union were adopting central planning techniques that eschewed reliance on markets in favor of allocation of a broader set of resources based on the value of physical and natural capital rather than on their market prices.

In *The Soviet Economy in Danger*, Leon Trotsky argued that[3]:

> If a universal mind existed, of the kind that projected itself into the scientific fancy of Laplace—a mind that could register simultaneously all the processes of nature and society, that could measure the dynamics of their motion, that could forecast the results of their inter-reactions—such a mind, of course, could a priori draw up a faultless and exhaustive economic plan, beginning with the number of acres of wheat down to the last button for a vest. The bureaucracy often imagines that just such a mind is at its disposal; that is why it so easily frees itself from the control of the market and Soviet democracy. But, in reality, the bureaucracy errs frightfully in its estimate of its spiritual resources.

However, Trotsky recognized the significant informational challenges of such a holistic approach, challenges that modern positional analysis attempts to resolve through more inclusive democratic participation in decisions relating to our interaction with the environment. Trotsky noted that[4]:

> The innumerable living participants in the economy, state and private, collective and individual, must serve notice of their needs and of their relative strength not only through the statistical determinations of plan commissions but by the direct pressure of supply and demand.

In the same era of debate following the formation of the first large-scale centrally planned economy, the scientist Frederick Soddy added in 1926 in his book *Wealth, Virtual Wealth, and Debt*[5] that a consumptive economy could not continue to use finite natural capital in a steady state manner. Soddy, who had won the Nobel Prize in Chemistry in 1921 for his discovery of isotopes that represented forms of a single element differentiated by their number of neutrons in the nucleus, argued that energy was necessary to create natural resources, and hence the natural economy is ever-constrained by the laws of entropy and thermodynamics, unlike the financial capital of money and debt that knows no such limitations.

This realization that our human-made and natural capital differ in that natural capital typically follows the one-way dictates of entropy became the central tenet of Nicolas Georgescu-Roegen and his mentee, Herman Daly half a century later, beginning with Georgescu-Roegen's book *The Entropy Law and the Economic Process.*[6] This more inclusive approach to humankind's relationship to our environment also spawned studies such as *Limits to Growth* in 1972 and *Small is Beautiful* in 1973.

Observations from these approaches and the realization of the uniqueness of natural capital have motivated much of the discussion in this text. Ecological economics differentiates itself from classical economics in that it is premised on stocks of natural capital and our human footprint upon nature rather than the flow approach of traditional economic models. In the face of the overwhelming and compelling influence of market economics even on such non-market challenges as global warming, the broader values incorporated in ecological economics are valuable, but also vexing when juxtaposed with conventional thinking (Fig. 27.1).

This is because ecological economics explicitly includes governance and political processes, and the normative values inherent in democratic decision-making, as elements of appropriate analyses. Given their acknowledgment of the unique role natural capital plays in sustainability, ecological economics rejects the notion of weak sustainability and its argument that enhanced

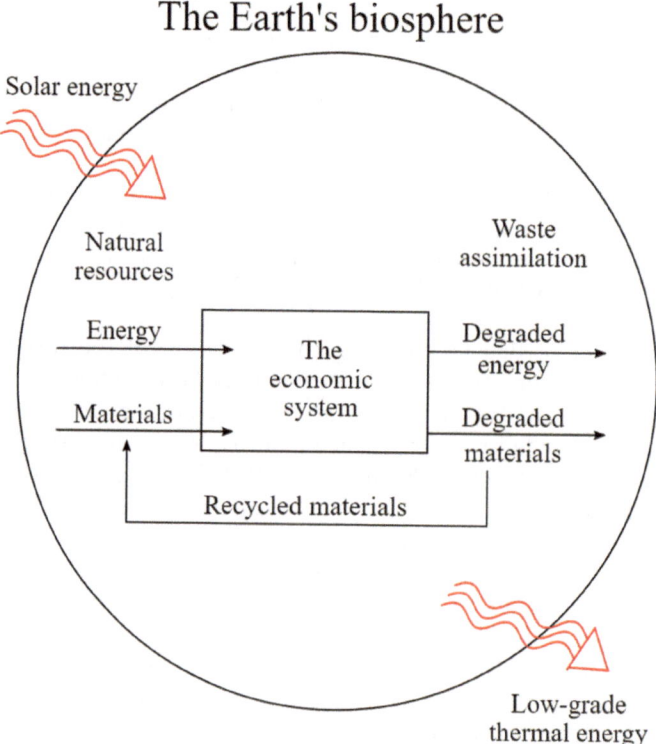

Fig. 27.1 The earth's biosphere from an ecological economics perspective (Courtesy: Д.Ильин: vectorization, translation, CC0, via Wikimedia Commons)

human-made capital can compensate for losses in natural capital. Some approaches to ecological economics provide a pathway to bridge the gap between weak and strong sustainability through conservation principles rather than the preservation inherent in strong sustainability.

This broader approach creates significant challenges to market economies oriented around classical economics. The modern corporation that matured in parallel with neoclassical economics has made many of the decisions that have depleted natural capital. From our study of Coase's *Nature of the Firm*, we know that corporations act in ways that reallocate property and avoid the internalization of externalities, all within the black box of the corporate veil. By producing in ways that impose costs on others within a generation or across generations, the corporation generates profits at times by imposing costs on others.

Ecological economists argue that to simply internalize externalities going forward does not remedy the losses and redistributions of income from past failures to internalize these costs. This criticism also recognizes the significant informational and scientific burden imposed to successfully internalize externalities. Such internalization cannot redress the past acts that have shifted powers within democracies as described by positional analyses. Nor can losses of species be remedied ex-post, especially given the complex relationships in the natural world and our very limited understanding of the long-term effects of a natural equilibrium perturbed by human decision-making.

The other concern is in the reliance of market mechanisms to solve these externalities. Ecological economists point out that the mapping of complex multidimensional interactions and decisions at the interface of the natural world and humankind into the linear and unidimensional measure of money is problematic. Hence, using money and markets to internalize or reverse these decisions is equally illogical. Even policies designed to remedy past actions will inevitably have normative consequences. Adaption and remedy to a problem arising from global warming will often create winners and losers. These entities may well be different from the losers and winners of the past environmental transgressions.

For instance, humankind can substitute new technologies for the loss of natural capital, but those who benefit from such new technologies are likely not those who were harmed. If weak sustainability could work in the aggregate, it still warps the balance of power and prosperity. For these reasons, voluntary or market measures, as well-intentioned as they may be, may ultimately fail to redress the displacements caused by humankind's broad and deep footprint on the natural world.

While elements of the market economy can be effective in discouraging future risky acts, they are not tools that are sufficient to remedy damage to the ecosystem. They generate additional problems in the resulting distribution of profit and power. Instead, as the United Nations efforts acknowledge, democratic processes that ultimately control the greatest share of human decision-making must be an integral part of sustainability solutions.

From this perspective of the central role of democracy, the education of humankind is the essential component. From education flows the environmental ethics and morals that drives our democracies and recognizes our mutual interdependencies within humankind and the broader ecosystem that contains and supports us. Only by acknowledging the path that brought us to this point and by recognizing an interconnection with our ecosystem that is likely more profound than we can imagine, will we successfully navigate our collective future which we can ill-afford to get wrong. It begins with the recognition of broader ecological values. When conventional institutions fail future generations, we are wise to consider unconventional approaches.

Notes

1. Mill, John Stuart (1848). *Principles of Political Economy*, John W. Parker, Publisher.
2. https://www.pbs.org/warrior/content/timeline/opendoor/roleOfChief.html, accessed March 8, 2023.
3. Trotsky, Leon (October 1932). "The Soviet Economy in Danger." Marxists Internet Archive, https://www.marxists.org/archive/trotsky/1932/10/sov econ.htm, retrieved March 9, 2023.
4. Trotsky, Leon (October 1932). "The Soviet Economy in Danger." Marxists Internet Archive, https://www.marxists.org/archive/trotsky/1932/10/sov econ.htm, retrieved March 9, 2023.
5. Soddy, Frederick (1933). *Wealth, Virtual Wealth, and Debt*, Britons Publishing Company, https://www.fadedpage.com/books/20140873/html.php, accessed March 9, 2023.
6. Georgescu-Roegen, Nicholas (1971). *The Entropy Law and the Economic Process* (Full book accessible at Scribd), Harvard University Press, Cambridge, Massachusetts, https://archive.org/details/entropylawe00nich/page/n3/mod e/1up, accessed March 9, 2023.

CHAPTER 28

A Critique of Sustainability Theories and Paths Forward

Dramatic growth in economies and populations since the onset of the Industrial Revolution has created unprecedented challenges to our natural capital. Before the onset of the Industrial Revolution and the Anthropocene Epoch, humans lived as the other species of our ecosystem, in relative harmony with nature. The world's human ecosystem has grown more than sixfold since then, and now represents more than 90% of the biomass of all mammals on Earth. The effects of rapid human system growth and the resulting extensive damage to the natural capital on which all living things depend have now outstripped the ability of our environment to function in a steady state.

At the essence of human and ecosystem survival is the preservation of our natural capital. The approach in this book has been to consider natural capital as a stock of an asset upon which all life on the planet depends. Such natural capital may be used sustainably if the ecosystem evolves naturally, or will be depleted over time if humans consume with little regard for the fixed supply of depletable natural capital and the needs of future generations.

However, people are not united in their regard for the future nor in our self-interest to compete for resources today. Were the peoples of the globe to unite in our recognition that we all share a planet and we each affect others in many of our most significant human decisions, we could avoid the accelerating depreciation of our natural capital. Such a recognition of our mutual dependence overarches our national concerns over domestic economies and our sustainability. In doing so, we may navigate a path that is both sustainable and equitable.

The increasing immediacy of climate peril arising from our failure to recognize a creeping and accelerating degradation of our atmosphere. Our faith in

C. Read, *Understanding Sustainability Principles and ESG Policies*, https://doi.org/10.1007/978-3-031-34483-1_28

and marvel of science has been an essential element of humankind, especially since the Renaissance Era, the Enlightenment, and the Industrial Revolution. In our Anthropocene Epoch, and more than at any time in the past, scientists have come forward to provide essential analyses to our policymakers and corporations with regard not only to the science of global warming but also to its social science and policy.

THE TRIUMPH OF SCIENCE

The United Nations Intergovernmental Panel on Climate Change is now on its sixth major iteration of its climate change assessment reports. These reports represent the deepest and broadest collaboration among scientists and social scientists worldwide. While the scientific method is based on a healthy dose of constant skepticism and collegial challenges, what has emerged has been an unusually cohesive and increasingly sophisticated consensus about one of the most critical aspects of our natural capital, the state of our atmosphere.

Their conclusion is that humankind will begin to see natural processes take over to accelerate global warming if human processes cannot limit our contribution to the content of carbon dioxide equivalents in our atmosphere. This concentration will soon double the natural background concentration evident before 1850. Should the carbon dioxide equivalent be held to less than 500 parts per million in our atmosphere, from its June 2023 level of 425 ppm and a level of 280 ppm before the Industrial Revolution, we may be able to hold average global warming to perhaps 1.5° Celsius and avoid the dire consequences should global warming exceed 2° Celsius. The summer of 2023 does not bode well, though. All but four days in a ninety day period were the hottest dates in the global average temperature record. The other four dates ranked second hottest globally.

Scientists and social scientists recognize that the political process is far more fraught with uncertainty than their scientific conclusions, though. Accordingly, the IPCC recommendations include analyses of what humankind and the ecosystem can expect based on various assumptions of how nations of peoples will respond to scientific conclusions. These reactions range from disregard for the finiteness of our resources and instead the consumption of finite stocks of natural capital at an accelerating rate as previously underdeveloped nations share in global prosperity, to adjustments arising from the most sobering absorption of scientific conclusions.

These necessary adjustments require global emissions of greenhouse gases to return to 2005 levels by the year 2035, if not earlier. If so, scientists believe that global warming can remain at or below the 1.5 degree threshold scientists believe necessary to avoid the worst consequences of global warming.

Between these two extremes are intermediate levels of acceptance and collective human behavior modification. The likelihood of which scenario emerges as most closely aligned to our collective efforts to avert the worst consequences of global warming ultimately depend on whether MDCs can assure their LDC cousins do not follow their greenhouse gas-emitting paths to prosperity.

The Brundtland Commission realized that success on behalf of our planet does not depend solely on our technical ability to construct a sustainable path. Rather, success depends critically on how we collectively ensure that all people can share in the aspirations and human rights enjoyed by some of our people. This redistribution of opportunity requires the most developed countries to divert some of their financial capital earned through a disregard for our shared environment so that less developed countries can develop with a new regard for our natural capital.

We have treated in this text the challenges of sustainability from a number of dimensions. While the science is complex, society is increasingly accepting the inconvenient truths that groups such as the Intergovernmental Panel on Climate Change have realized for some time. We have seen how science dictates the finiteness of the natural capital available to humankind, and we documented the role of abundant sustainable energy in moderating climate change and recovering our natural capital so it may also be enjoyed by future generations. Uncertainties around these scientific prognoses are narrowing with each iteration of IPCC reports.

However, the social science and politics of sustainability remain far less certain. If we accept the sustainability ethic that each generation has a responsibility to the next and those who follow, then we are left with two remaining issues to resolve.

A Stranglehold of Traditional Economics

The first issue is whether economic analysis is up to the task.

The neoclassical evolution of economics at the end of the nineteenth century was a turning point for the discipline. Social commentators such as Thomas Malthus, Adam Smith, and Thorstein Veblen were replaced by technicians who explored all aspects of economic decisions made by rational and self-interested actors at a given point in time. Neoclassical models strived for greater efficiency in these static transactions, but they were ill-adapted to deal with the eventual loss of natural resources, the lack of well-established property rights, or the inability to ensure properly priced and internalized negative externalities.

These problems are generally in the category of what economists call non-convexities. Monopolies, incomplete or ineffective markets or property rights, externalities, and public goods owned and managed by peoples rather than allocated by markets are all examples of such non-convexities that defy the assumptions of mainstream neoclassical free-market economics.

While we have seen that economists can offer guidance to policymakers on appropriate economic decisions from an intergenerational perspective, many of our institutions are designed to optimize the subset of human values most easy to quantify, such as income in our bank accounts rather than natural capital in

Nature's bank. Too often, humans focus on those elements we can measure and disregard what is more difficult to quantify. Discourse between diverse individuals and groups helps to fill the gaps in our foresight. We are then better able to see the world from multiple perspectives and foster a better understanding of how to proceed.

This more complete picture must include the valuation and even potentially the creation of markets for many things of value that have now gone neglected in our market-based economies. Our values must certainly include construction of markets for the myriad externalities we often ignore within each generation and the intergenerational externalities we impose upon those who follow us. They also include more explicit appreciation of the environment and its myriad amenities and on the diversity of the ecosystem. Market economies are far from fully embracing such an expansive view, but many more now see that the incompleteness of markets induces humans that rely too much on markets to wreak havoc within a more extensive economic model.

Traditional economics also does not delve too deeply into the black box of production. Instead, it assumes that our factors of production can be continuously substituted for one another to compensate for the scarcity of one with greater employment of another. While human-made capital such as labor and machines can be treated that way, natural capital is typically inherently scarce. We cannot make more minerals if we use the elements in the ground, and it takes centuries to repair an atmosphere that is pushed into a temperature regime and sea level that is seen only once every hundred millennia.

For these reasons, it is a far safer proposition to attempt to not tempt nature with anthropogenic forces to the point that we cannot remedy nature's wrath. Scientists understand this, and economics can accommodate it, but humans tend to make more immediate and thoughtful choices, despite our mortality.

These failures in the assumptions of traditional economics can be remedied. For instance, we can establish property rights when they don't otherwise exist, and we can calculate and properly internalize externalities. Failures of monopolists or competitors to properly extract natural capital in a sustainable manner can be corrected in the direction of an omnipotent central planner.

The greatest challenge remains the internalization of the intergenerational externality. The free-market functions when those who rely on the goods and services markets may offer have the agency to fully realize their economic aspirations. However, the most challenging aspect of sustainability is the creation of a facility to represent the preferences of future generations who have an equally strong, or perhaps even a stronger stake in the outcome of economic decisions today Yet, future generations do not have the agency to demonstrate their willingness to pay for a well-functioning atmosphere absent of accelerating global warming, or resources that have long since been consumed.

It is not that the tools of economics cannot offer insights into the nature of the intergenerational problem. Indeed, we have discovered that natural capital will be better employed if the economy adopted a social discount rate close to zero or, if we accept weak sustainability, the royalties of resource extraction are set aside as a permanent fund for all generations to follow. We can also benefit by moving away from the income approach of economic well-being and toward a social and natural balance sheet that manages assets and liabilities rather than income and expenses.

We have developed tools to show how resources must be managed for maximum benefit both now and well into the future. In a static world, the competitive model can be shown to both maximize efficiency and maximize the sum of consumers' surplus, the net of our willingness to pay for our goods and services over the competitive price we must pay, and producers' surplus, the net of what producers receive net of their marginal costs of production.

But, when markets are distorted by incompleteness and non-convexities and when those with interests in market outcomes without mechanisms to express their interests, transactions are confined to generations and nations in the right place at the right time to benefit from such important transactions as natural capital extraction.

For these additional reasons, an omnipotent planner must determine the optimal level of such transactions across generations to avoid these pitfalls arising from market failures. The tools of intertemporal optimization can inform these decisions of an omnipotent planner and can correct market failures. The mere imposition of a socially appropriate discount rate is inadequate. Economists well-versed in intertemporal economics know what must be done, but the free markets of mortals are unable to follow the correct path of economic efficiency using the invisible hand that so well functions in static markets.

This reality leaves us with a prescription that may be unpalatable for some. Long-term sustainability necessarily requires intervention in free markets so that the agency of long-term stakeholders is evident. It is worth exploring to determine whether these missing stakeholders can somehow be empowered in ways such as Ronald Coase may propose. If natural capital goes to the highest bidder today because stakeholders representing future generations are unable to bid, then it is possible that a public surrogate could fill that gap. Certainly, no mortal and contemporaneous agent could do so because they have no ability to capitalize on the value that may be gleaned at some point well into the future.

Given that such free-market participation will likely remain incomplete, the solution that Ronald Coase may prefer to avoid may be necessary. If markets fail, some entity must instead create quotas and command-and-control those aspects of markets that involve natural capital.

If so, the balance between the capitalist and the centrally planned components of the mixed-free market economy must shift. Dynamic optimization techniques and such insights as the Genuine Savings model and the Hartwick Rule can guide these decisions that must be made by an omnipotent central planner, however uncomfortable that shift may be.

Such a solution shifts the emphasis from the traditional measure of well-being, the Gross Domestic Product and the income a nation produces, to a balance sheet approach that instead sums the assets humankind has assembled to sustain economic well-being and ecosystem sustainability. Such an assets approach does not preclude the valuation of the environment and the ecosystem, even if incentives may be required to instill broader ecosystem values. However, the convenient emphasis on income generated at the national level, or profits at the corporate level, is misplaced. Such a measure has the advantage that it is simple to calculate, but it is no proxy for the goal of humankind to sustain happiness.

Hopefully, we can take some comfort in the thought that our social institutions and private corporations are tools to ultimately fulfill the needs of humankind and, in turn, the needs of the entire ecosystem.

A Failure of Leadership

There remains the second challenge to resolve to ensure intergenerational efficiency. It requires the political will to follow the policy prescriptions of scientists and social scientists. Because science and social science analyses are positive rather than normative, their intentions and accuracy should not be challenged as mere opinion, even if society may engage in vigorous discussion over assumptions and policy prescriptions.

Adoption of the accommodations sustainability requires will require much of capital markets, permanent and sovereign funds, government coffers, and the functioning of free markets. We live in mixed free-market economies that balance the needs and efficiencies of free markets and the abilities and domains of the public sector and politics.

Sufficient political will may be hard to muster, though. Critical to any well-functioning society is the need to allocate property, capital, wealth, and resources. Within market-based societies, we take for granted the process of allocations. People are accustomed to their ability to purchase goods and services of their choice, so long as they are willing to pay the price and have the wealth to obtain the property.

For most goods, the title is mere possession, and for other items, such as vehicles and homes, the government may register the title and will arbitrate its exchange. Citizens elect the government to protect these titles and property rights and pursue the values of our generation. We also expect our government to provide myriad goods and services that for which the free market fails to provide in the way we consider fair. The roads we drive on, the airwaves we enjoy, and even the schools that educate our children or the police and fire departments that keep us safe, and the armies that protect our borders, are all provided to us without regard for our income or ability to pay.

This combination of the provision of private and public goods constitutes a mixed economy. A large share of our free-market system in the United States is actually government administered, and paid for through taxes and fees to favor current voters. This is a precarious balance at the best of times. It is unlikely that a generation would elect leaders who must tell the electorate that their economic aspirations must be moderated or put aside to protect the interests of generations to come. Even those who may just be born today and may well be adversely affected by economic decisions today are unable to exercise their franchise until they reach the age of majority. The same reasons that we mortals maintain discount rates well above the socially ethical rate also cause us to endorse political leaders who will pursue our short-term interests, even if to the detriment of our children's children.

It is possible that an enlightened society can reflect these intergenerational values and accordingly elect officials who will promote intergenerational values. In the ideal, we know that our utility functions in principle reflect an appreciation for the environment and future generations to some extent. However, in democracies, the majority must maintain values that reflect the needs of future generations in ways that are institutionalized so they may prevail through human cycles of abundance and shortages. We have seen that our values toward the environment can best be described as regard for a luxury good we can appreciate when our wealth and economic security are high. This universality of values is unlikely over all economic times and for a consistent and sufficient share of the population.

Finally, we have witnessed the inability of nations to assemble to make the economic sacrifices necessary to mitigate global warming or sufficiently fund the costs climate change already induced based on the loss and damage principle identified in COP 27. Political leaders of the Most Developed Countries will likely continue to balk at meaningful solutions to global warming, given the scale and costs they must incur. Our leaders typically feel a responsibility to their electorate, not the demands of other nations or the entire planet. This creates a global free-rider problem in that we may all recognize what must be done, but no country is willing to pay its fair share unless all others do so as well, if at all.

One might look with some optimism at the success of the Montreal Protocol to deal with the problem and costs of ozone depletion or the resolution of members of the European Union to adopt the Polluter Pays Principle. However, these agreements were easier to address than the existential challenges to our natural capital and sustainability such as global warming. The agreement is easier when the stakes are lower.

The stakes are now exceedingly high. Political leaders of the world's most powerful economies that garnered their wealth because of unfettered and artificially inexpensive fossil fuels are unlikely to endorse such high-stakes transfers valued in the tens or hundreds of trillions of dollars.

Of course, our generation will never muster the political will to solve the problem of intergenerational sustainability unless we are aware of and well-versed in the issues. The brightest light of all is the degree of concern and activism among young people who sense that their futures have been squandered by the decisions of those before them.

GLOBAL CHALLENGES

At the global level, this recognition of the need to transfer knowledge, technology, and financial capital from the world's wealthiest to the poorest nations is one that introduces a number of political challenges, both at the domestic level among the people of wealthy nations and at the global level with regard to international relationships and negotiations among peoples. These challenges must succeed if MDCs expect LDCs to participate in our collective decarbonization and sustainability efforts.

Clearly, the globe's poorest nations cannot develop in sustainable and environmentally benign ways if they are not afforded the financial means to avoid consumption of resources such as fossil fuels that fail to incorporate the costs they impose on sustainability. Population growth is most spectacular among the nations that have yet to contribute to the pending climate change peril, but only because these nations have yet to be afforded rapid economic growth and the dramatic depletion of natural capital that has historically complemented growth.

Sustainable development must then function within the juxtaposition of science, social science, and politics, all at a time when the Earth is in potential peril. The sciences and social sciences recognize this challenge and have proposed viable solutions through the combination of new and sustainable technologies, especially for energy production. However, these technologies remain expensive, not in the levelized cost of these energy sources over the long run, but instead in the need to divert very large investments to bring online innovations and permit humankind to benefit earlier from their long-term benefits arising from low-cost sustainable energy. The hidden cost of delay impedes this transition, as does the necessary obsolescence of natural capital that must remain in the ground, the fossil fuels that have generated the vast majority of global warming.

Any change, regardless of its necessity, creates winners, losers, uncertainty, and economic displacements. An entity's ability to cope with uncertainty and displacements is a measure of resiliency. Invariably the poorest among us are typically also the most vulnerable and least resilient. If efforts to move to a more sustainable path are able to better protect our natural ecosystem as they lift billions of people out of poverty in the coming generations, they will continue to tax the resiliency of those who can least afford to develop sustainably. It is this dichotomy that represents our greatest challenge in the path toward greater sustainability.

Our scientists and social scientists understand this dichotomy but they also realize that the solution is not as simple as a tidy scientific equation. We can model the cost of carbon–neutral policies in the less developed world as we simultaneously combat poverty and starvation and preserve the ecosystem. However, we lack the political will to make the sacrifices necessary. For the first time, the recent 27th Annual Conference of Parties (COP27) ended with a statement that recognized atmospheric sustainability requires transfers between MDCs and LDCs, not to support the poor but to create a viable path for LDCs to develop sustainably and without repeating past mistakes of MDCs. The dialog has begun, even if sums casually mentioned remain just a tenth of one percent of what may be necessary to successfully navigate the necessary path.

Human circumstances often intervene to delay the best of intentions. A global pandemic that destabilized all peoples regardless of national wealth, and a war in Europe that ratchetted up the cost of fossil fuels and food upon which the least wealthy depend, has changed the focus of global peoples, hopefully just temporarily. Yet, despite these challenges, and others that will also surely redirect our attention, sustainability, energy security, and global resiliency remain the prevailing existential challenge that only worsens with inadequate attention.

With human challenges of one form or another diverting our human and economic attention, science continues to march on with innovations in insights and technology. We have seen that while much of our natural capital is fixed, with abundant sustainable energy, this natural capital need not be consumed by one generation at the expense of all subsequent generations. We can better shepherd our natural resources through recycling and through greater efficiencies, and we can still enjoy the products of the fossil fuel left in the ground by using carbon dioxide removed from the atmosphere to synthesize the hydrocarbons used to produce our plastics, medicines, and myriad other products. Access to sustainable energy permits us to also produce sustainable fuels that can allow us to continue to enjoy such human conveniences as long-distance air travel or even space exploration that rely on hydrocarbons but can be equally powered with synthetic fuels made with abundant sustainable energy. We know how to solve these challenges, if only we muster the collective will.

CORPORATE OPPORTUNITIES

These potential successes do not depend solely on innovations from the same scientists who offer us pathways toward a sustainable future. Private sector corporations perform an essential role in innovation. While these corporations are typically the extractors of resources and the producers of products we consumers purchase that embody our past ambivalence for sustainability, our collective success depends on both the vision of people and the collaboration with corporations. Increasingly, more companies are recognizing that:

1. Sustainable practices can be profitable, especially if we consider social profits as well as traditional financial profitability. Corporations are increasingly becoming aware of the need to operate sustainably while they remain sensitive to the need to respond to their stakeholders who wish to purchase, supply, work, and invest with firms that share emerging sustainability values. Some corporations wholeheartedly embrace the Environment, Social, Governance (ESG) paradigm as the responsible

way for corporations to operate within the expectations of civil society. Other corporations simply observe that communicating their ESG values is good for business. Increasingly, corporations recognize that they work within a corporate ecosystem of mutually dependent suppliers, demanders, employees, and investors and must converge upon a sense of shared values within this corporate ecosystem as part of a broad human system.

2. Corporations' stakeholders demand more sustainable practices and their regulators are demanding corporations act with greater transparency in their ESG policies and practices. For instance, customers, suppliers, and investors are increasingly held responsible for a new set of sustainability values and must ensure that the corporations with which they align account for shared values with transparency and sincerity.

3. Greater sustainability requires additional public education. Public policy depends not only on shared values but also on a shared basis to comprehend and effectively communicate these values. The degree to which societal sophistication in sustainability has increased over the past generation is reassuring. Corporations have perhaps lagged in this social discussion that first found its roots in groups of people, then our communities and nonprofit entities, government, and its regulating agencies. ESG values are now embraced by a growing and substantial number of our largest corporations. These corporations have assembled associations that educate their members and stakeholders and instruct them on pathways toward sustainability. By the end of the 2020s, a majority of major corporations and most regulators will have pledged to better communicate sustainability values. Given that sustainability is such a shared value, education is an essential component.

4. The corporate sector must view itself as an essential element of a private/public collaboration. Scientists and social scientists have been mobilized to provide blueprints for a trajectory toward sustainability. Governments and corporations are critical entities that allow us to converge to and remain on such a trajectory. While some corporations cling to the Friedman Doctrine as a rationale for the separation of their profit-making goals and the broader goals of society, the sustainability challenge that has accompanied too rapid economic growth has forced a reconsideration of the corporate sector as a partner in society. This realization creates opportunities for collaboration that have not before been explored.

5. Corporations and institutions are the engines of innovation. While government can promote and satisfy the desire for its people to be educated and creative, our most successful economic engines are those that bring innovations to market. This great power as innovation gatekeeper also imposes an equal responsibility. Corporations can take risks to adopt new and sustainable technologies, but society may need to provide incentives to reduce costs and indemnifications to reduce risks.

6. As economies become increasingly complex, so do our supply chains and our dependence on a wide variety of factors of production. The resulting circular flow of income ensures that corporations produce the income that flows to others and provides the means to purchase their products. As all elements of such a circular economy recognize their mutual interdependence, our values must converge. This notion of intradependence in a domestic economy can be extended through global trade to create greater interdependence among economies along with the shared value of sustainability.

These challenges are not insignificant, but humans have demonstrated in the past that overwhelming and multigenerational challenges can be met with incredible human ingenuity. The science of sustainability demonstrates that there exists both a tremendous challenge but also incredible opportunities. Education of the nuances of what is admittedly a very complicated challenge will be just as important as the need for a fully engaged public sector that collaborates effectively with the private sector. I believe it can be done. I trust you do too.

GLOSSARY

MATHEMATICAL GLOSSARY

B aggregate consumption of an environmental amenity.

β a measure of corporate return relative to market returns and the risk-free return to compensate for corporate risk.

b pollution abatement.

C aggregate consumption of a human-made good or service.

c fixed marginal cost.

CS consumer surplus.

D accumulated pollution damage.

δ rate of capital depreciation.

e pollution emissions.

f_H marginal increase in production arising from an additional unit of harvest of a renewable resource.

f_K marginal increase in production arising from an additional unit of human-made capital.

f_S marginal increase in production arising from an additional unit of a fixed resource.

f_x marginal increase in production arising from an additional unit of extraction of a fixed resource.

G renewable resource growth rate.

GS Genuine Savings.

H aggregate harvest value of a renewable resource.

h_i individual harvest rate of a fixed resource.

I income or budget.

K amount of human-made physical capital.

L amount of available human capital.

$l(M)$ aggregate human capital as a product of education investment.

λ shadow price of Lagrange Multiplier.

m absolute value of demand curve slope.

M education investment.
NB net benefits.
p price.
p_c choke price.
PS producer surplus.
q_i amount of consumption of a good i.
r market discount or interest rate.
r_f risk-free rate of return.
r_m market rate of return.
R stock of a renewable resource.
ρ rate of time preference.
S stock of a fixed resource.
U aggregate or present value of utility.
u individual or per period utility.
W aggregate or intertemporal wealth.
X aggregate extraction rate of a fixed resource.
x_i individual extraction rate of a fixed resource.

Glossary

3Ps See People, Planet, Prosperity.
Abiotic a fixed natural resource that cannot be replenished through biological means.
Abiotic natural resource a form of natural capital that is not biologically based and cannot reproduce to replenish itself.
abiotic resource A reproducible biological natural resource.
Absorptive capacity the ability of the Earth to repair damage such as caused by pollution.
Actinides radioactive byproducts of nuclear fission that can have half-lives of radioactive persistence from decades to thousands of years.
Activation energy an energy hurdle necessary to initiate an exothermic reaction.
Adaptation remediation of damages caused by global warming.
Adaptation the capacity to adjust to damage caused by natural events by changing patterns of property use.
Albedo effect the warming of the Arctic region that occurs as decreasing ice coverage allows the darker ocean to absorb more insolation.
Alkane carbon and hydrogen groups.
Allocation a quantity of a good, service, income, or property provided as part of an economic action.
Anthropocene Epoch a geological period from the time in which humans began to significantly impact the Earth's ecosystems.
Anthropogenic a result originating from human activity.
Argon the third most abundant chemical element in the atmosphere, with the symbol Ar and atomic weight 18.

Aromatics hydrocarbons constituted by lighter distillates.

Asphaltics petroleum products constituted by heavier distillates.

Atlantic Meridional Overturning Circulation a prevailing ocean current that distributes warmer tropical ocean water toward the North Atlantic.

Atmosphere layers of gas that envelope the planet through gravity and provide necessary elements for biological life.

Backstop technology a new or as-yet uneconomical technology that can replace another technology or resource as it increases in price.

Balance sheet an account of assets and liabilities.

Bang for the buck a measure of the value created for a given expense.

bargaining asymmetries differences in negotiating power or effectiveness.

Basel Committee on Banking Supervision (BCBS) an international bank regulatory agency based in Basel, Switzerland.

Battery storage a capacity to store energy in chemical bonds that can be released in the form of electricity.

Best available technologies a regulatory requirement to adopt the best economically affordable technology.

Big Bang a short and intense period 13.8 billion years ago that constituted the formation of mass that constitutes our universe.

Biomass organic material that can be converted to a fuel to provide energy.

Biotic resources natural capital that is organic and can replenish itself through biological processes.

Black swan event a very low probability event with high costs if it occurs.

Blue hydrogen the creation of hydrogen molecules by stripping methane molecules of their hydrogen atoms, with carbon as a byproduct.

Blue Marble the iconic image of the Earth from the vantage of Apollo 17 in space in 1972.

BOE barrel of oil equivalent as a measure to compare the energy content of various hydrocarbons relative to a barrel of oil.

Boreal forest the world's largest forest spanning much of the area above the 60th parallel.

Breeder Reactor a type of nuclear reactor that converts one form of nuclear fuel into a larger amount of another form of nuclear fuel.

Bromofluorocarbon an ozone-depleting molecule related to CFCs that acts to neutralize ozone in the atmosphere.

Brundtland Commission the United Nations World Commission on Environment and Development headed by Norwegian politician Gro Brundtland that released a report in 1987 called Our Common Future which articulates a goal of sustainable economic development.

Budget constraint an income constraint on the amount of goods and services that can be purchased in a period.

Business-as-usual a benchmark measure of a scenario that considers modifications to current human decisions.

Capacity factor the equivalent share of time a device can generate full power.

Cap-and-trade a market-oriented mechanism that restricts polluters to emission rights purchased and subject to a quota that permits trading among polluters.

Capital Asset Pricing Model a technique to predict the value of a security based on its past price variability relative to the overall variability of the market basket of securities.

Carbon an element with the symbol C and atomic number 6 that is the basis for most life on Earth.

Carbon curse poor economic performance despite rich fossil fuel endowments.

Carbon dioxide a molecule that combines one atom of carbon with two atoms of oxygen, with the formula CO_2. This molecule has strong internal bonds and reacts little with other atoms but when suspended in the atmosphere acts to reflect back heat radiating from the Earth.

Carbon dioxide equivalent a measure that compares the relative heat-reflecting potency of various greenhouse gases to their carbon dioxide reflectivity equivalent.

Carbon dioxide sink natural or engineered processes that absorb and sequester atmospheric carbon dioxide.

Carbon market a financial market that exchanges the right to emit carbon dioxide into the atmosphere.

Carbon neutral a physical or economic process that absorbs the same amount of carbon dioxide as it emits.

Carbon sequestration diversion of atmospheric carbon dioxide into permanently storable and stable forms or compounds.

Carbon sink a natural or humanmade process that has the ability to absorb carbon dioxide.

Carbon tax a surcharge price imposed on the emission of carbon dioxide.

Carrying capacity capacity of Earth systems to naturally recover from emissions.

CFC see chlorofluorocarbon.

Chlorofluorocarbon a nonflammable and nontoxic molecule that combines carbon, chlorine, and fluorine and acts as a solvent or refrigerant gas to move heat from a source to a sink.

Choke price the maximal willingness to pay for a resource based on the availability of a substitute resource.

Civil suit a lawsuit between two parties with an economic issue in conflict.

Clean Air Act a piece of legislation first passed in the United States in 1970 that provided the first holistic regulation of emissions into the atmosphere from mobile and stationary sources.

Clean Water Act a piece of legislation first passed in the United States in 1972 that provided the first holistic regulation of emissions into the water in ways that damage the chemical, biological, or physical qualities of bodies of water.

Climate the characteristics of temperature, air pressure, humidity, cloud cover, and wind movement across a region or the planet over an extended period of time.

Climate change changes in the climate induced by natural or human-induced phenomena.

Chlorofluorocarbon an ozone-depleting molecule that acts to neutralize ozone in the atmosphere.

Club of Rome an assemblage of thought leaders who seek solutions to complex global issues. The Club of Rome sponsored the Limits to Growth analysis on the Earth's sustainability in 1972.

Coase Theorem an assertion that, under certain ideal circumstances, conflicts and exchanges of property rights can be resolved through negotiation and bargaining rather than through markets.

Cobalt an element found in small quantities in the Earth's crust with the symbol Co and atomic number 27. It is an important component of many industrial processes and in the manufacture of some types of high-energy-density batteries.

Cobb–Douglas production function a specific pattern of combinations of factors of production to produce goods or services named after its developers Charles Cobb and Paul Douglas between 1927 and 1947. It is often employed in economic models because of its useful mathematical properties.

Coefficient of expansion the rate of density change of a molecule as its temperature increases.

Coefficient of Performance (COP) the amount of heat recovered relative to energy used.

Cogeneration the use of excess heat for other human purposes that is given off from the combustion of fuels in the generation of electricity.

Common heritage of all mankind principle by the UNCLOS to share commonly held resources.

Complementarity principle a concept that asserts all properties of an item or event cannot be observed simultaneously.

Conference of Parties a governing body formed to negotiate international treaties or agreements and is used annually by countries that are party to the United Nations climate agreements.

Conference on the Human Environment a United Nations group charged with assessment of the global environment.

Conservation a sustainability concept that advocates for no net depletion of the natural capital stock.

Convention of the Law of the Sea A United Nations sea treaty that is administered by a body appointed by the UN.

COP see Conference of Parties.

Coriolis Force an apparent force that results from the perspective of a frame of reference as an object rotates. It drives the observed swirling motion of winds in the atmosphere.

Corporate Social Responsibility a set of principles that guide business decisions in response to the values of its stakeholders and society.

Corporation a legal entity permitted by a state to conduct commerce based on certain specified rights, privileges, and obligations.

Cost of capital a calculation of the necessary rate of return for which a company can raise capital for investment purposes.

Costate variable the implicit price imposed on the element optimized as a constraint is adjusted.

Coulombic Efficiency (CE) the ratio of recoverable energy relative to the energy absorbed in an energy storage device.

Credit rating a ranking of the ability of an entity to pay a debt as specified in its conditions.

Credit risk a factor that may hinder the ability of an entity to pay a debt obligation.

CSR an abbreviation for Corporate Social Responsibility that assigns to a modern corporation the responsibility for various social values beyond profit maximization.

Deforestation the removal of forest tracts and their potential ability to remove carbon dioxide from the atmosphere through photosynthesis.

DEI see Diversity, Equity, and Inclusion.

Depletable a resource for which the stock declines steadily with extraction.

Differential equation an equation that relates variables to rates of change of variables, often over time.

Diminishing marginal returns the economic observation that the additional employment of an economic variable will result in smaller increases in another economic variable.

Diminishing marginal utility a concept that states the additional enjoyment declines as more of a good or service is consumed.

Discount factor the sum $(1 + r)$ that is used to convert a future value to a present value.

Discount rate the rate by which humans and businesses reduce the future value of an economic variable in present terms.

Discounting the process of devaluing a future economic variable in present terms.

Dismal Prophecy a label attached to Thomas Malthus' prediction that the geometric growth of a population will eventually outstrip the arithmetic growth of resources upon which it depends.

Dismal science a label assigned to the discipline of economics arising from Malthus' feast and famine prophecy.

Dispatchable Power the ability to either store and release or modulate power output to fill gaps between electrical energy supply and demand.

Distributional effect the result of the distribution of income or wealth as economic relationships are modified.

Diversification a reduction in risk by distributing economic activity across a number of uncorrelated instruments.

Diversity, Equity, and Inclusion the consideration of human values in support of diversity, equity or fairness, and inclusiveness in human relations and economic decisions.

Dobsonmeter a surface-based device that is part of a network to track the level of ultraviolet rays reaching the Earth's surface.

Drought an extended period of little or no participation across a region.

Duality a quality that an event can be described using two different paradigms.

Dutch disease a phenomenon that observes rich resource endowments may result in a decline in national economic well-being.

Dynamics the observation of physical or economic relationships over time.

Earnings at risk the share of earnings of a corporation that can vary due to risk and uncertainty.

Earth capital a form of factors that permit a capacity to produce and are created by nature.

Earth Day an annual recognition held on April 22 of every year since 1970 to increase awareness of the need to protect the Earth and its ecosystems.

Ecoclimatology the interdisciplinary study of climatology from an ecosystem perspective.

Ecohydrology the interdisciplinary study of hydrology from an ecosystem perspective.

Ecological economics the study of the Earth's ecosystem using the tools of economics.

Ecology the scientific study of the environment.

Economic impatience the tendency to discount the value of future economic variables.

Economic justice the degree to which economic institutions ensure the fair and equitable distribution of wealth and resources.

Economics the collection of techniques to manage the world around us.

Economies of Scale the realization of greater efficiencies as output is expanded.

Economies of Scope the realization of greater efficiencies as a wider area of products are produced.

Ecosystem a holistic systemic view of the natural and human environment.

Efficiency the process that measures and optimizes the amount of output for the least amount of inputs.

Endogenous growth theory a theory that argues economic growth is derived internally through innovation rather than externally through the employment of additional resources.

Endothermic a chemical process that absorbs heat energy.

Energy the capacity to perform work to enhance or produce goods, services, or mass using Albert Einstein's equation $E = mc^2$.

Enthalpy a thermodynamic variable that measures the combination of total heat in a system as a product of the energy derived from the combination of pressure and volume.

Entropy a thermodynamic variable that measures the unavailability of the capacity of a system's thermal energy to produce work because of its state of disorder.

Environment the natural and human-made surroundings that promote life and maintain the ecosystem.

Environmental justice the observation that the state of the environment affects regions and groups of people unequally, and degradation of the environment often imposes the greatest costs on those least able to afford them.

Environmental, Social, and Governance a paradigm that asserts economic and financial value is dependent on factors that affect the environment, society, and the ways in which humans govern our actions.

Equilibrium the tendency toward a balanced steady state in a process or reaction.

Equimarginal principle an economic result that shows the discounted present value of the surplus earned by the extraction of a resource is equal across all extraction periods.

Equity a measure of fairness, or the residual value attributed to shareholders in the difference between a corporate asset value and its associated liabilities.

ESG see Environmental, Social, and Governance.

Ethical hedonism a philosophical argument that human decisions should be made primarily based on ethical concerns.

Eunice Newton Foote a nineteenth-century scientist from New York State who discovered that gases held in a sealed tube rise in temperature when exposed to insolation.

Eutectic mixture a homogenous combination of substances in specified quantities that creates a change to a liquid phase.

Evolutionary economics an economic theory coined by Thorstein Veblen that asserts economic behavior is determined by the combination of individuals and society.

Exothermic a chemical process that gives off heat energy.

Externality the consequence of an economic decision or transaction on others not included in the transaction that benefits (positive externality) or is harmed (negative externality) by the action.

Feast and famine a prediction of Malthus' Dismal Prophecy that population will oscillate between growth during times of abundance followed by famine and population decline as the Earth's carrying capacity is exceeded by population growth.

Federal Reserve the central bank of the United States charged with ensuring a well-functioning commercial banking system and the maintenance of favorable economic conditions arising from low and predictable inflation and unemployment.

Feedback the process by which an action creates reactions that tend to reinforce and amplify the action (positive feedback) or reduce and attenuate the action (negative feedback).

Financial Stability Board an international body enabled by a consortium of nations and entities and designed to monitor financial systems and recommend ways to enhance their stability.

First price sealed bid auction an auction in which bidders must provide a sealed bid without knowledge of the bids of other auction participants.

Fiscal risk the financial and corporate risk that arises from governmental fiscal policy.

Fishery a fishing ground or region where fish are harvested.

Fission the process of a splitting of an atom bombarded by energetic particles that result in two or more smaller atoms or particles of combined mass less than the original atom. The small difference in mass is converted to a large amount of energy according to Einstein's $E = mc^2$ equation.

Fixed resource a natural resource in fixed supply in nature that cannot be quickly replaced, if at all.

Flow battery an energy storage and production unit that converts chemical to electrical energy by permitting the flow of an electrolyte between an anode and a cathode. To recharge, the electrolyte is replenished in the reverse process and stored until needed.

Flow pollutants pollutants for which the Earth can cleanse within a relatively short time frame.

Flywheel a rapidly spinning mass that stores rotational kinetic energy to be later converted to electrical or mechanical energy.

For all humankind a reference to a criterion that is judged based on its effect on all humans.

force a capacity to induce acceleration upon an object.

FRB see Federal Reserve Board.

Free-for-all catch-as-catch-can the degeneration of an economic process to non-cooperation in which participants do not acknowledge or regard the effect of their choices on other participants and instead operate to maximize individual gain.

Free-rider effect the result when participants in an economic process conclude they can benefit from a collective action even if they refuse to pay a shared cost of the activity.

Friedman Doctrine an assertion by economist Milton Friedman that corporations should pursue the sole goal of profit generation and leave to shareholders decisions of how to use these profits to support their societal values.

Fusion the process of forcing two smaller atoms or particles together to form a larger atom of a mass less than the sum of masses of the smaller atoms or particles. The small difference in mass is converted to a large amount of energy according to Einstein's $E = mc^2$ equation.

Game-theoretic the use of game theory to analyze socioeconomic phenomena.

Gas turbine a piece of physical capital that converts the chemical energy of a fuel through combustion to rotational mechanical energy through pressure on rotating blades.

GDP see Gross Domestic Product.

Gen I the first iteration of prototype commercial nuclear fission reactors constructed in the 1950s and 1960s.

Gen II the second iteration of commercial nuclear fission reactors constructed in the 1960s and 2010s with enhancements in safety and efficiency over previous generation reactors.

Gen III the third iteration of commercial advanced light-water nuclear fission reactors commercialized beginning in the 2020s with enhancements in safety and efficiency over previous generation reactors.

Gen IV the fourth iteration of commercial nuclear fission reactors based on advanced fission reactions that are more efficient and safer than previous generation reactors. While mostly still in development, they also provide the possibility to extract energy and render safer the nuclear waste left by previous reactors generations.

Generally Accepted Accounting Principles a series of standardized and broadly accepted accounting rules.

Generation IV International Forum a forum to enhance research and development of new fission reaction systems.

Genetically Modified Organism commercial seeds and organisms that have been modified to enhance productivity and profitability.

Genuine savings a measure of economic savings closely aligned to the effective capacity of resources and assets to produce on a sustainable basis.

Geoengineering a set of technologies designed to offset global warming.

Geothermal the use of heat generated below the Earth's surface to provide heat to humans and commerce.

GHG see Greenhouse Gases.

GHG Protocol an entity that generates and communicates greenhouse gas emission accounting standards.

Gibbs Free Energy a measure of energy that can be extracted or may be required for a closed system chemical reaction.

Gilded Age a period in the United States between its Civil War and World War I that was characterized by a high rate of economic growth and wealth concentration.

Global Compact a United Nations initiative among Chief Executive Officers concerned about sustainability.

Global Reporting Initiative (GRI) an independent association of organizations responsible for enhancing sustainability.

Global warming the process by which the global average temperature rises due to changes in the atmosphere arising from natural or human processes.

Gospel of Wealth a policy recommended by Andrew Carnegie in which a share of the wealth generated by industry magnates is returned to society in the form of philanthropy.

Governance Risk corporate risk arising from corporate leader actions.

Great oxygenation event a period of relatively rapid increase in atmospheric oxygen as a result of oxygen-emitting cyanobacteria in an early stage of the Earth's development.

Great rusting also called the Great Oxidation Event that oxidized exposed metals as a result of oxygen produced through photosynthesis by cyanobacteria from three billion to one billion years ago.

Green bond a bond issued to support projects that yield demonstrable benefits to the environment.

Green finance financial markets and tools designed to promote sustainability.

Green hydrogen molecular hydrogen that is created typically through electrolysis of water using sustainable energy.

Green New Deal a set of public policy proposals designed to address climate change.

Greenhouse Effect the warming of the Earth arising from the insulating effect of certain atmospheric gases.

Greenhouse gas a gas suspended in the atmosphere that reflects radiated heat back to the Earth.

Greenpeace an international environmental group formed in Vancouver in 1971.

Greenwashing techniques designed to deceptively communicate and exaggerate the effectiveness of actions by corporations to stem global warming or enhance the environment.

Gross Domestic Product a measure of annual income and spending in an economy as a proxy for the creation of economic happiness.

Group of 20 Financial Stability Board (G20 FSB) an association that promotes financial stability among G20 nations.

Gulf Stream a flow of warm Atlantic Ocean water from the Caribbean to Europe that has a significant effect on regional climates.

GWh Gigawatt-hour, a measure of electrical energy and capacity to do work.

Hamilton's Method a technique to optimize the path of a trajectory of values that can be used to determine the most efficiency path of resource extraction over time.

Hamiltonian the extension of Lagrange's Method of optimization subject to a constraint to dynamic systems that evolve over time.

Hartwick Rule a rule of thumb that states consumption can be optimized over time if various balances of the sum of natural and human-made resources can be maintained over time.

Haudenosaunee Life a spirituality that respects all generations, past, present, and future.

Heat Balance of the Earth's Surface the distribution of heat energy sources and sinks across a system represented by the Earth's surface.

Heat of combustion (exothermic) heat that can be given off when a molecule is oxidized.

Heat pump a method to transfer heat from a colder to a warmer body through the mechanism of using mechanical energy to pump a heat transfer fluid. Such a movement of heat can be more efficient than the production of the same amount of heat.

Helium a scarce and nonreactive element with the symbol He and atomic weight of 2.

Hotelling's Rule an equation that defines the optimal path of prices of a fixed non-renewable resource over time.

Human capital the capacity of humans to produce goods and services.

Human enlightenment a state of knowledge that appreciates the full richness of human knowledge and interactions.

Human resources the treatment of human capacity as a resource in production.

humankind the consideration of values as they affect all humans.

human-made capital reproducible capacities produced through human rather than natural factors.

Hurricane a cyclonic weather phenomenon that arises when winds rotate rapidly counterclockwise in the Northern Hemisphere and clockwise in the Southern Hemisphere around a very low pressure area.

Hydroelectric the conversion of water pressure into electrical energy through turbines.

Hydrogen the most abundant and simplest atom in the universe, with a symbol H and an atomic weight of 1. It is usually held in molecular form H_2 as a gas or in combination with other elements in molecules.

Hydrogen Economy a subsector of the energy economy that relies on the production, transportation, and use of hydrogen as a fuel.

IAM see Integrated Assessment Model.

income statement an analysis of the flows of revenues and expenses.

Indifference Curve a combination of goods and services that can be consumed in a way that yields an equal level of human satisfaction.

Individual Transferable Quota a property right conferred to a natural abiotic resource that can be traded in a market but which constrains total usage of the resource based on a limit that is managed for resource sustainability.

Industrial Revolution a period of accelerating capital investment and output that begins with innovations in the nineteenth century.

Insolation the level of solar power impinging on a geographical location, expressed in watts per square meter.

Institutional Investors Group on Climate Change a group of large investment managers devoted to sustainability.

Integrated assessment model a model that combines physical and social science variables to formulate optimal environmental policy.

Interest rate a financial return sufficient to meet the requirements for the supply and demand of financial capital.

Intergenerational factors that affect multiple generations of the human population.

Intergenerational efficiency the partitioning of economic values across time so that benefits accrue equally to every generation.

Intergovernmental Panel on Climate Change a body of scientists and social scientists appointed by the United Nations to assess scientific and social factors related to climate change.

Intergovernmental Science-Policy Platform on Biodiversity and Ecosystem Services (IPBES) an international organization committed to the combination of science and public policy to address ecosystem management.

International Seabed Authority a committee assigned to regulate the use of shared sea beds in international waters.

Intertemporal variables that evolve over time.

intertemporal equilibrium an equilibrium sustained over time.

intertemporal rate of time preference the relative value assigned to various points in time.

Intertemporal rate of time preference the rate at which an individual or organization compares an economic value in one period relative to another period.

Intragenerational variables that span generations.

intragenerational efficiency a principle that generates efficiency at one point in time.

IPCC see Intergovernmental Panel on Climate Change.

Isoquant a combination of factors that yield a single level of output.

ITQ see Individual Transferable Quota.

Joule the unit of energy in the meter-kilogram-second (MKS) metric system.

Kilowatt-hour a unit of energy equal to 3.6 million joules.

Kyoto Protocol an agreement among nations committed to the United Nations Framework Convention on Climate Change in 2005.

Lagrange Method a technique developed by Joseph-Louis Lagrange to optimize the value of a function that is subject to constraints.

Lagrangian multiplier a costate variable in a maximization that measures the sensitivity of a function's value to changes in the value of a constraint.

Latent heat the heat absorbed or given off when a compound changes phases between solid, liquid, and vapor or gaseous state.

Latent heat of combustion the amount of heat given off in the combustion of a given mass of a fuel.

Legal Risk risks arising from legal liabilities.

Less Developed Countries (LDCs) the group of countries that have not sustained economic development equal to More Developed Countries.

Levelized Cost of Energy (LCOE) a comparison of all costs of energy production on an average cost per kilowatt-hour.

Lifecycle the view of a resource, product, or consumption pathway over the life of an entity or person.

Lifetime wealth a calculation of the value of the sum of future income discounted to the present.

Limits to Growth an influential body of research first published in 1972 sponsored by the Club of Rome that demonstrated the finiteness of fixed resources to meet human needs.

Liquidity risk financial or institutional risk arising if an otherwise healthy entity cannot temporarily balance its revenue with expenses.

Lithium-Ion a type of reaction that can store and release electrical energy with a relatively high energy density per unit of mass in a battery.

Logistics curve a particular S-shaped mathematical function that occurs when populations grow subject to a constraint.

Long-wave cycle theory an evolutionary economic theory that argues innovations which enhance human productivity comes in waves.

loss and damage a compensation principle advocated by LDCs to compensate for the costs they suffer from climate change.

Malthusian Trap the result that increased population tends to outstrip the rate of increase in agricultural production.

Marginal rate of substitution a measure of the rate one would sacrifice one good to obtain another good while satisfaction is held constant.

Marginal user cost an implicit cost that represents sacrifices imposed on later periods for consumption of resource decisions made in an earlier period.

Marginal utility a measurement of the increased satisfaction arising from the consumption of one additional unit of a good or service.

Market risk a measure of potential losses that arise because of adverse movements in the price of a financial asset or product.

mass a fundamental characteristic of matter that is affected by gravity and maintains momentum.

Materiality issues that have a substantial effect on the financial, reputational, legal, or economic performance of a company.

Maximum Sustainable Yield the maximum amount that can sustainably be extracted from a well-managed abiotic resource.

Mesozoic Era a period from 252 to 66 million years ago, meaning "middle life." It includes the era of the dinosaurs.

Methane the most common hydrocarbon fuel made up of carbon and hydrogen in the formula CH_4.

Millenium Development Goals a set of environmental and social goals to be attained by the year 2000.

minimax solution a solution to a non-cooperative game that minimizes the gain to winning parties and maximizes the outcome to losing parties.

Mitigation reduction in the impact of global warming by reducing factors that cause it.

MKS units a unit of physical measures based on the meter, kilogram, and second.

Modern Portfolio Theory a technique to aggregate and order a set of security investments to minimize risk for a given desired return.

Monopolist a sole seller of a good or service that can use its exclusivity to adjust output and hence manipulate its price and revenue generated.

Montreal Protocol a conference as part of the United Nations Environment Programme that fomented an agreement in 1987 to phase out the production and consumption of ozone-depleting gases.

More Developed Countries (MDCs) a group of countries that have enjoyed a high level of sustained economic development.

Mother Nature an affectionate and respectful nickname for the Earth and its natural processes.

Maximum Sustainable Yield (MSY) the attainment of maximum yield from a renewable resource through careful management of the resource stock.

Multiplier an economic concept that measures the ratio of the sum of direct, indirect, and induced consequences across a market, markets which serve the market, and others who benefit from these transactions relative to the direct consequences themselves.

Naphthalene an aromatic hydrocarbon containing benzene rings.

Natural Capital the capacity of natural resources and factors to produce goods, services, and natural amenities.

Natural Capital Protocol a decision-making protocol to guide organizations in sustainability assessments.

Natural Resource Curse the observation that nations with rich endowments of natural resources are unable to manage their natural wealth in a sustainable way.

Nature of the Firm a theory by Ronald Coase that describes the advantages a corporation or firm has in their ability to operate internally without the constraints placed on market transactions.

Nature's bank a metaphor that considers natural resource assets in a way analogous to the assets on a balance sheet or in a bank account.

Negative feedback a feedback process that tends to be self-attenuating.

Net National Product (NNP) Gross Domestic Product net of capital asset depreciation.

Net-Zero Tracker an entity that monitors actions of publicly listed corporations that allow global warming to be restricted to no more than 2° Celsius.

Nicomachean Ethic a philosophical theory advocated by Aristotle that advocates for individuals to use their personal property to advance their interests and society's interests.

Non-cooperative equilibrium an equilibrium established between parties motivated by their self-interest.

Normative a value that is person-dependent and cannot be established from scientific principles.

Nuclear power the conversion of the energy that holds together atoms into other useful forms of energy.

Obsolescence risk the risk that an obsolete technology or product will penalize corporate profits.

Oikos a Greek language term that literally means house but metaphorically refers to the environment around us.

Operational risk the risk to the systems and operations of a corporation arising from uncertainties and internal organizational flaws.

Opportunity cost the amount that must be given up to earn a benefit in an economic transaction.

Optimization a technique to derive maximal efficiency out of a process of a mathematical model.

Options Pricing Theory a method that prices options on securities based on their pattern of past volatility.

Organization for Economic Cooperation and Development a multinational organization of.

Organization of Petroleum Exporting Countries nations that assembled to exercise their collective market power and act as a cartel to influence the price of oil.

Original position the human tendency to consider one's position in society or an economic arrangement when an outcome is influenced or determined.

Oxidizing atmosphere an atmosphere containing sufficient oxygen to react to oxidize minerals and organic materials.

Ozone an oxygen ion capable of absorbing ultraviolet energy.

Ozone hole an area usually concentrated near the Earth's poles in which the atmosphere's quantity of ultraviolet ray-protecting ozone ion $O3 +$ is depleted.

Ozone layer a thin upper layer of ionized oxygen gas molecules that are able to absorb ultraviolet-C rays, convert them into heat by splitting an oxygen atom, followed by recombination into an ozone molecule.

Paleozoic Era the period of initial animal life 541 to 252 million years ago.

Paradox of Plenty see the Natural Resource Curse.

Paraffin waxy solid distillates of petroleum.

Pareto Optimality an economic principle that justifies a transaction if the value it creates is sufficient to indemnify the losses incurred by those damaged economically from the transaction.

Paris Agreement a legally binding treaty on climate change agreed to by the Conference of Parties 21 in Paris on December 12, 2015. Sometimes called the Paris Climate Accord, it specified an upper limit for global warming of 2.0° Celsius, with a preferred limit of 1.5 degrees, which would require a 50% reduction in anthropogenic greenhouse gas emissions from 2005 levels by the year 2030.

People, Planet, and Prosperity (3Ps) a United Nations initiative to document human actions based on their implications on humans, the ecosystem, and economic development.

Permafrost a layer of surface or subsurface ground that remains permanently frozen and hence acts to prevent the decay of adjoining biomass.

Permanent Fund a financial fund that accumulates as natural capital is depleted to ensure future generations are able to enjoy the same level of consumptive enjoyment as past generations.

Personal property items or rights owned by an individual.

Photovoltaics materials capable of producing electricity from photons.

Physical capital humanmade machinery that provides a capacity to produce goods and services.

Physical risk the risk to entities of damage arising from physical phenomena such as fire, wind, flooding, storms, and coastal erosion.

Pink hydrogen hydrogen produced in the disassociation of elements in water at high temperatures encountered in a nuclear power plant.

Planet, People, and Prosperity a paradigm advocated by the United Nations that defines a series of principles and acts that can simultaneously improve the Earth, its people, and our mutual prosperity.

Political economy the analysis of the meeting of individual and collective human needs using markets and political processes.

Polluter Pays Principle (PPP) a principle that those who damage the environment are responsible for environmental remedy regardless of their intentions.

Pollution the release of effluents that damage the environment, people, the atmosphere, and/or the ecosystem.

Population the size of a biological species, often a measure of the number of people coexisting on Earth.

Population crash a sudden and significant decline in an ecosystem component, usually due to an external shock or from non-linear responses to interactions in an interrelated ecosystem.

Positive feedback a feedback process that tends to be self-amplifying.

Potential energy stored energy that has the potential to be released to perform work.

Present value the conversion of a value or cost that occurs in the future into an equivalent value today.

Principle of microscopic reversibility a principle in chemical reactions, also known as the principle of detailed balancing, in which the rate of individual reactions in the forward and reverse direction are equalized to hold the overall reaction in equilibrium.

Prisoner's dilemma a game-theoretic concept of non-cooperative equilibrium that produces total value inferior to an equilibrium attainable through collusion or cooperation.

Private Cost the cost of an action absorbed by a private entity. This cost may not represent the entire social cost if externalities exist.

Private Marginal Benefits benefits accruing to an individual or entity without regard to benefits accruing to others.

Private Marginal Costs costs incurred by an individual or entity without regard to external costs to others.

Production function a mathematical description of how various factors of production can be combined in different proportions to yield a useful output of a good or service.

Property rights the legal rights and privileges associated with ownership of an item or resource.

Public policy principles that govern social and economic actions in the best interest of society.

Pumped Hydro an energy storage method that uses energy to pump fluid to a higher elevation and subsequently releases the resulting potential energy when needed.

Qualitative a factor that can be described but not directly measured.

Quantitative a factor that lends itself to direct measurement and hence management.

Race to Zero a campaign sponsored by the United Nations Framework Convention on Climate Change to encourage civic and corporate leaders to adopt policies to help ensure global carbon neutrality by the year 2050.

Rate of growth the mathematical measurement of the pace an economic or a natural system may increase over time.

Reactor in the context of nuclear fission, a piece of physical capital that enables energy production from the splitting of atoms.

Redox-flow see Flow Battery.

Redox-Flow battery a storage battery that charges and discharges by oxidizing and reducing a material that can flow and be stored.

Reducing atmosphere an atmosphere with insufficient oxygen content to support oxidation.

Regulator an entity charged as a matter of public policy to monitor individuals or corporations to ensure they meet policy objectives.

Regulatory Risk risk resulting from changes in or imposition of regulations on an organization.

Renewable resource a resource that can be harvested sustainably with good resource management.

Representative Concentration Pathway a greenhouse gas concentration trajectory defined by the IPCC to meet specified atmospheric temperature endpoints.

Reputational risk the damage to an entity that can arise if its values are not aligned with those of its stakeholders.

Resiliency a measure of the ability of an entity to withstand uncertainty, risk, and economic shocks by maintaining a sufficient distance between their operating point and a point beyond which viability is threatened.

Risk appetite the ability or willingness of an entity to tolerate risk to increase their returns.

Risk culture the set of internal policies or understandings that affect how an entity manages risk.

Risk the unpredictability and variability of value or profits facing an entity arising from forces beyond their direct control.

Risk adjusted return on capital—the necessary rate of return on capital invested to compensate for the amount of risk exposure.

Safe Drinking Water Act U.S. legislation in 1974 designed to protect the public's drinking water from harmful effluents.

Schumpeterian Growth Paradigm the theory advanced by Joseph Schumpeter that is based on the premises that innovation resulting from entrepreneurial investments designed to expand market power produces long-run growth and displaces legacy technologies.

Schumpeterian Hypothesis the assertion by Joseph Schumpeter that large firms have a comparative innovation advantage over small firms. See also Schumpeterian Growth Paradigm.

Second Law of Thermodynamics the thermodynamic principle that heat moves from a warmer to a colder body and that energy is necessary to reverse the direction of this flow.

Second price sealed bid auction an auction technique in which the highest bid in a sealed auction must only pay the price bid by the second highest bidder. The process is more likely to induce bidders to bid according to their willingness to pay rather than on the strategy they project on other bidders.

Securities and Exchange Commission a U.S. regulatory body with the mission to ensure that publicly traded companies follow securities laws and that material information about a corporation is released to the investing public so security prices properly reflect corporate value. See materiality.

Sensitivity analysis the exploration of changes in the assumptions or inputs of a model and its outputs and implications.

Sequestration a technique to remove and isolate a substance such as carbon dioxide to prevent harm to a physical system.

Seventh Generation Philosophy a principle derived from Haudenosaunee (Iroquois) principles that require decisions to be assessed based on their consequences to seven generations into the future.

Severance tax a tax or royalty imposed on a fixed stock resource presumably to compensate future generations from the loss of access to abundance of the natural capital.

Shadow price an internal valuation of relative actions or constraints.

Shadow price the estimated or internal valuation of a good or service beyond a direct market valuation.

Shared Socioeconomic Pathway a possible scenario for subsequent exploration that embodies a set of assumptions with regard to possible social policies that may be adopted that may have implications on climate change.

Shareholder an owner of shares of a corporation.

Side payment a payment to entice a participant in an economic transaction to behave in a way consistent with the preferences of the payer.

Silent Spring an influential book published in 1962 by ecologist Rachel Carson that documented the harmful effects of increased pesticide use in agriculture on local ecosystems.

Small is Beautiful an influential book published in 1973 by E.F. Schumacher that calls for the curtailment of excessive consumption. His ecological economics thesis is similar to the critiques of Thorstein Veblen.

Small Modular Reactor a new class of small self-contained nuclear reactors under exploration that are proposed to be cheaper, safer, and more efficient.

Social bonds bonds issued to raise funds for investments of social value.

Social Cost a measure of all the costs, both explicit and intangible, that arise from an economic decision. The difference between a market and a social cost is the externalities a process may generate.

Social discount rate the optimal rate by which an immortal society should discount the future. Under the argument that one generation is no less valuable than another, economists assert that this measure should be zero.

Social Marginal Benefits benefits arising from an economic activity and conferred upon all social members of an economy.

Social Marginal Costs cost arising from an economic activity and conferred upon all social members of an economy.

Social Responsibility is to Increase its Profits the Friedman Doctrine that the sole objective is to increase profits for shareholders so they alone may decide how to use their profits to generate personal or social welfare.

Soil organic matter soil that arises from the decay of organic matter and has the capacity to sequester carbon.

Solar energy the power of the sun's energy, or insolation, on the Earth. Solar panels are able to extract upward of 25% for commercial solar panels and upwards of 35% for solar concentrators.

Solow Growth Model an economic model that describes the evolution of capital and consumption over time.

Sovereign the authority to act unilaterally with respect to the domestic policies of a nation.

Sovereign funds permanent funds administered by agents of a sovereign nation.

Specific heat the amount of heat absorbed or given off when the temperature of a mass is changed by a given amount (usually one degree).

Specific Heat Capacity quantity of heat that can be absorbed to raise a kilogram of a material one degree Celsius.

SSP see Shared Socioeconomic Pathway.

Stakeholder an entity that derives value from the actions of a corporation or public entity.

State variable a variable that describes the mathematical state of a dynamic system.

Stated Policies Scenario benchmarks used by the International Energy Agency to describe sustainability actions over time.

Steady state an equilibrium point in an economic or natural process that maintains constant values of critical variables over time.

Stock pollutants pollutants that continue to accumulate over time for which the Earth can cleanse only very slowly, if at all.

Stranded Asset an asset that is prematurely devalued or converted to a liability because of changes in social policy or preferences. For instance, a complete movement to sustainable energy may convert a natural asset such as a fossil fuel reservoir to a valueless stranded asset or even to a liability that requires constant monitoring.

Strategic risk external or internal events that jeopardize the realization of an entity's corporate strategy.

Stress test the imposition of detrimental factors or assumptions on a model of a system to determine the system's resiliency to shocks.

Strong reparation the requirement that losses due to physical, economic, or property damage are indemnified by property replacement.

Strong sustainability the requirement that extraction of fixed stocks of natural capital is maintained for the enjoyment of all generations.

Substitutability the ability to ensure a constant flow of production or enjoyment by rebalancing in different proportions the factors that create the flow.

Supply chain the network of upstream and downstream providers and transformers of factors of production that result in a final good. Such networks are typically carefully managed for maximum efficiency but can be vulnerable to uncertainties.

Sustainability the ability to employ available resources to ensure a consistent enjoyment of the products they generate.

Sustainability Bonds investment funds that finance various green initiatives.

Sustainable Development Goals a set of goals originally formulated to ensure viable and sustainable economic development by developing and least developed nations to protect the planet, end poverty and economic injustice, and ensure mutual health and prosperity.

Sustainable Development Scenario scenarios developed by the International Energy Association to advance sustainable energy, manage global warming, and enhance air quality in ways consistent with the Paris Agreement.

Symbiosis a mutually interdependent biological system in which separate organisms benefit from and provide benefits to each other.

Taskforce on Climate-Related Financial Disclosures an entity created by the Financial Stability Board to enhance reporting of climate-related financial information.

TCFD see Task Force on Climate-Related Financial Disclosures.

Technology a term used to describe the ability to combine factors in particular proportions to produce goods or services.

Terminal condition an endpoint to be met in a dynamic process.

Thermal energy storage the use of physical masses with high specific heat to store heat energy and fill the gap between heat supply and demand.

Thermo-chemical storage a thermal energy storage system that can be compact and efficient because it relies on chemical conversions or phase transitions to store and release energy.

Three Lines of Defense three tiers of protection of a commercial entity from risks that may threaten financial viability. These include management processes, risk management and compliance, and internal audit functions.

Thünen's Rings concentric rings of varying economic intensity used to support an economy in the center of an isolated state, as defined by Johann Heinrich von Thünen in 1826.

Tidal power the use of the kinetic energy of a rising and falling tied to generate electric energy.

Time preference the human regard of mortals for the present relative to the future.

Total maximum daily load a regulatory condition imposed on emitters that specify the maximum allowable emission of various pollutants.

Tragedy of the Commons the result of destruction of a shared natural resource if its enjoyment and use are not regulated.

Tragedy of the Horizons a speech by the former Bank of England and Bank of Canada chief executive Mark Carney that links global warming and climate change to financial stability.

Transactions cost the fixed and lump sum costs related to the facilitation of a transaction beyond the value of the exchange itself.

Transition Risk the various risks that arise as an organization transitions from one sustainability or other regimes to another.

Triple Bottom Line an accounting framework developed in 1994 by John Elkington that adds social and environmental results to the traditional financial results reported by businesses, nonprofits, and government agencies. In the following year, he described TBL as the 3Ps, people, planet, and profits or prosperity.

Two-period model a simplified model of resource usage over time that demonstrates principles in the interaction of two periods.

UNCLOS see United Nations Conference of the Law of the Sea.

United Nations Biodiversity Conference a conference of interested parties that convenes every year to discuss biodiversity under the auspices of the United Nations.

United Nations Conference of the Law of the Sea an international agreement under the auspices of the United Nations that establishes the principle that the oceans beyond the territorial waters of sovereign nations must be managed for the common good.

United Nations Conference on the Human Environment a group of interested parties convened by the United Nations to discuss sustainability of the human environment.

United Nations Environment Programme Financial Initiative a network of financial institutions such as banks, insurers, and investment funds that help guide participants in the global financial system toward sustainable economics.

United Nations Framework Convention on Climate Change a United Nations framework formed in 1992 to prevent further human damage to the climate system.

United Nations Principles for Responsible Investment a United Nations supported network of investors working to promote sustainability.

Value-at-Risk a statistical quantification of potential financial losses that could occur within a portfolio or position with a specified probability over a specific time frame.

Vanadium an element with the symbol V and atomic number 23 that is a promising component of an inexpensive and efficient battery storage technology. See Flow battery.

Vapor pressure the equilibrium pressure created by the tendency of a given vapor to escape confinement at a given temperature.

VAR see value at risk.

Veil of Ignorance a technique John Rawls used to neutralize one's original position in the determination of public policy that best serves all of society.

Venture capital funding speculative capital raised for high-risk and high-reward projects.

Vulnerability the degree to which a given shock can impose significant harm on an individual or entity.

WACC see the weighted average cost of capital.

Water cycle the process by which water is shuttled between the ocean, atmosphere, and land to support life and dictate the climate and weather.

Water Pollution Control Act U.S. legislation passed first in 1972 as an extension of the Federal Water Pollution Control Act of 1948 that sought to address growing public concern over effluent discharge in waterways.

Water Vapor water in a gaseous state.

Watt a unit of power in the MKS system that is equivalent to the energy in joules per second.

Wave power techniques used to convert the kinetic energy of waves into electrical energy to perform work.

WBCSD see World Business Council for Sustainable Development.

Weak reparation the requirement that losses due to physical, economic, or property damage are indemnified by compensation that may not be a replacement but may instead keep the whole entity damaged.

Weak sustainability the requirement that extraction of fixed stocks of natural capital is replaced with other forms of capital, such as human, physical, or technological capital for the enjoyment of future generations.

Weighted Average Cost of Capital the determination of the cost of capital available to a corporation for the purposes of investment based on their traditional capital sources and risk culture.

WHO see World Health Organization.

Wind turbine a series of blades that rotate from the force of the wind and are able to generate electricity or do mechanical work.

Work the application of force over a distance, measured in joules in the meter-kilogram-second (MKS) units system.

World Bank an international financial institution that was established as part of the Bretton Woods Conference to unite allies in 1944 by integrating economies. The World Bank was formed to provide grants and loans to low and middle-income countries so they may pursue economic development.

World Business Council for Sustainable Development an assembly of non-governmental organizations and businesses with the goal of enhancing sustainability in accordance with the United Nations Global Compact.

World Economic Forum an assembly of global thought leaders, elected officials, and corporate leaders who pool their interests and perspectives to foster concepts and projects that create social value.

World Health Organization a United Nations agency with the goal of the promotion of health to protect the vulnerable and enhance world safety.

World Meteorological Organization a specialized agency of the United Nations that was formed in 1947 to succeed the International Meteorological Organization by 1951 to establish an integrated network of Earth system observation facilities to monitor weather, climate, and water resources.

World Resources Institute a global nonprofit organization that assembles civic, business, and societal leaders to research and implement solutions that protect people and the ecosystem.

World Trade Organization (WTO) an international organization that enforces trade treaties between the world's trading nations.

Worldwatch Institute an environmental research group.

Index